AF322892

Poisson Point Processes

Roy L. Streit

Poisson Point Processes

Imaging, Tracking, and Sensing

 Springer

Roy L. Streit
Metron Inc.
1818 Library St
Reston Virginia 20190-6281, USA
r.streit@ieee.org
www.roystreit.com

ISBN 978-1-4419-6922-4   e-ISBN 978-1-4419-6923-1
DOI 10.1007/978-1-4419-6923-1
Springer New York Dordrecht Heidelberg London

Library of Congress Control Number: 2010932610

Printed on acid-free paper

Springer is part of Springer Science+Business Media (www.springer.com)

To my wife, Nancy

Preface

A constructive approach to Poisson point processes (PPPs) is well suited to those who seek to learn quickly what they are, how they are used, and—above all—whether or not they can help the reader in whatever application they are working on. To help readers reach these goals, PPPs are defined via a two-step realization, or simulation, procedure. This procedure is subsequently used to obtain many wonderful properties of PPPs.

The approach is pedagogically efficient because it uses a constructive definition of a PPP in place of the traditional PPP axioms that include, for example, independent scattering. The two approaches are in fact mathematically equivalent. The question of elegance is a matter of taste. Many readers will find the constructive definition easier to understand than an axiomatic one, especially on first encounter. In any event, the simulation approach is taken here because it provides an efficient entrée to PPPs and point processes generally.

There are now several books on PPPs and related topics. Most seem to be written by mathematicians striving for mathematical rigor, and these are valuable books for that very reason. Many practitioners, however, do not have the background to read them without a serious commitment of time and effort. What I felt was lacking was a book that made PPPs accessible to many mathematically capable readers and that provided insight and technique without recourse to an axiomatic measure-theoretic approach. Therefore, while I was writing the book, the audience foremost in my mind was graduate students, especially beginning students in electrical engineering, computer science, and mathematics. The real audience for this book is, of course, anyone who wants to read it.

PPPs came only slowly to my attention and from many sources, most of which relate in one way or another to specific passive sonar problems. One was two dimensional power spectral estimation, especially k-ω beamforming for towed linear arrays. Another was the relationship between sonar beamforming and computed tomography (especially the projection-slice theorem). Other problems were single target tracking in clutter and the even harder problem of multiple target tracking. A final problem was target detection by a distributed field of sonobuoys. The pattern of the book is clearly visible in this progression of topics.

The book relates what I have learned about PPPs in a way I hope others will find worthwhile. It is a calibrated exposition of PPP ideas, methods, and applications that would have facilitated my own understanding, had I but known the material at the outset.

I am indebted to many individuals and institutions for their help in writing this book. I thank Prof. Don Tufts (University of Rhode Island) for the witty, but appropriate, phrase "the alternative tradition in signal processing." This phrase captures to my mind the novelty I feel about the subject of PPPs. I thank Dr. Wolfgang Koch (Fraunhofer-FKIE/University of Bonn) for cluing me in to the splendid aphorism that begins this book. It is a great moral support against the ubiquity of the advocates of simulation. I thank Dr. Dale Blair (Georgia Tech Research Institute, GTRI) for suggesting a tutorial as a way to socialize PPPs. That the project eventually became a book is my own fault, not his.

I thank Dr. Keith Davidson (Office of Naval Research) for supporting the research that became the basis Chapter 6 on multitarget intensity tracking. I thank Metron, Inc., for providing a nurturing mathematical environment that encouraged me to explore seriously the various applications of PPPs. Such working environments are the result of sustaining leadership and management over many years.

I thank Dr. Lawrence Stone, one of the founders of Metron, for many helpful comments on early drafts of several chapters. These resulted in improvements of content and clarity. I thank Dr. James Ferry (Metron) for helpful discussions over many months. I have learned much from him. I thank Dr. Grant Boquet (Metron) for his insight into wedge products, and for helping me to learn and use LaTeX. His patience is remarkable. I also thank Dr. Lance Kaplan (US Army Research Laboratory), Dr. Marcus Graham (US Naval Undersea Warfare Center), and Dr. Frank Ehlers (NATO Undersea Research Centre, NURC) for their encouragement and helpful comments on early drafts of the tutorial that started it all.

I thank my wife Nancy, our family Adam, Kristen, Andrew, and Katherine, and our ever-hopeful four-legged companions, Sam and Eddie, for their steadfast love and support. They are first to me, now and always.

Reston, Virginia   Roy L. Streit
February 14, 2010

Contents

Part III   Beyond the Poisson Point Process

Chapter 1
Introduction

Nothing is as practical as a good theory.[1]

James Clerk Maxwell

Abstract The purpose of the book is to provide an accessible discussion of multidimensional nonhomogeneous Poisson point processes. While often overlooked in the literature, new applications are bringing them to greater prominence. One chapter is devoted to developing their basic properties in a constructive manner. The approach is the reverse of the usual abstract approach, and it greatly facilitates understanding of PPPs. Two chapters discuss intensity estimation algorithms, with special attention to Gaussian sums, and the Cramér-Rao bound on unbiased estimation error. Three chapters are devoted to applications from medical imaging, multitarget tracking, and distributed network sensing. A final chapter discusses non-Poisson point processes for modeling spatial correlation between the points of a process.

Keywords Poisson point processes (PPPs) · Tomography · Positron emission tomography (PET) · Single photon emission computed tomography (SPECT) · Multitarget tracking · Distributed sensor detection · Communication diversity · Maximum likelihood estimation · Cramer-Rao Bound · Marked PPP · Germ-grain models · Binomial point process

Poisson point processes (PPPs) are very useful theoretical models for diverse applications involving the geometrical distribution of random occurrences of points in a multidimensional space. Both the number of points and their locations are modeled as random variables. Nonhomogeneous PPPs are designed specifically for applications in which spatial and/or temporal nonuniformity is important. Homogeneous PPPs are idealized models useful only for applications involving spatial or temporal uniformity.

The exposition shows that nonhomogeneous PPPs require little or no additional conceptual and mathematical burden above that required by homogeneous PPPs.

---

[1] Physics community folklore attributes this aphorism to Maxwell, but the available reference [68] is to Kurt Lewin (1890-1947), a pioneer of social, organizational, and applied psychology.

R.L. Streit, *Poisson Point Processes*, DOI 10.1007/978-1-4419-6923-1 1,
© Springer Science+Business Media, LLC 2010

The discussion is readily accessible to a broad audience—constructive mathematical tools are used throughout. Abstractions that contribute to mathematical rigor but not to insight and understanding are left to the references.

PPPs are highly flexible models with a growing number of new applications. Interesting modern applications typically involve both nonhomogeneous and multidimensional PPPs. In some applications, PPPs are extremely well matched to the physics and the engineering system. Tomography, especially positron emission tomography, is an excellent example of this kind. In other applications, PPP approximations capture the most important aspects of a problem and manage to avoid or circumvent crippling computational bottlenecks. Recent multitarget tracking applications are wonderful examples of this kind. In still other applications, PPPs are used primarily to gain insight into the behavior of very complex systems. Examples of this kind include detection coverage and inter-sensor communication in distributed sensor networks.

By presenting diverse applications in a common PPP framework, unanticipated connections between them are exposed, and directions for new research are revealed. Some of these connections are presented here for the first time. As engineers and physicists know well, mathematics and its applications are deeply satisfying when they work together.

1.1 Chapter Parade

The book divides into three distinct parts, together with several appendices. The first part comprises the first four chapters. These chapters discuss several aspects of PPPs, ranging from fundamental properties to inference and estimation. The PPPs themselves, not their applications, are the focus of the discussion.

Applications come into their own in the second part. This part deals with applications in imaging, tracking, and sensing. Imaging means almost exclusively tomography; tracking means multitarget and multisensor tracking; sensing means largely detection coverage and communication diversity with distributed sensor networks. The third part is an effort to balance the overall perspective by discussing point processes that find use in applications but are not PPPs. Only a few of the many interesting classes of point processes are mentioned.

1.1.1 Part I: Fundamentals

Chapter 1 introduces PPPs and the justification for the emphasis on multidimensional processes. It discusses several ways to restrict the too general concept of a random set so that it is useful in theory and application. This puts the PPP into proper perspective relative to other point processes.

Chapter 2 discusses several of the useful and important properties of nonhomogeneous PPPs. The general PPP is defined via a two-step simulation procedure. The important properties of PPPs follow from the simulation. This approach enables those new to the subject to understand quickly what PPPs are about; however, it is the reverse of the usual textbook approach in which the simulation procedure is derived almost as an afterthought from a few "idealized" assumptions. The style is informal throughout. The main properties of PPPs that are deemed most useful to practitioners are discussed first. Many basic operations are applied to PPPs to produce new point processes that are also PPPs. Several of the most important of these operations are superposition, independent thinning, nonlinear mappings, and stochastic transformations such as transition and measurement processes. Examples are presented to assist understanding.

Chapter 3 discusses estimation problems for PPPs. The defining parameter of a PPP is its intensity function, or intensity for short. When the intensity is known, the PPP is fully characterized. In many applications the intensity is unknown and is estimated from data. This chapter discusses the case when the form of intensity function is specified in terms of a finite number of parameters. The estimation problem is to determine appropriate values for these parameters from given measured data. Maximum likelihood (ML) and maximum a posteriori (MAP) methods are the primary estimation methods explored in this book. ML algorithms for estimating an intensity that is specified as a Gaussian sum are obtained by the method of Expectation-Maximization (EM). Gaussian sums, when normalized to integrate to one, are called Gaussian mixtures, and are widely used in applications to model probability density functions (pdfs). Two different kinds of data are considered—PPP sample data that comprise the points of a realization of a PPP, and histogram data that comprise only the numbers of points that fall into a specified grid of histogram cells.

Chapter 4 explores the quality of estimators of intensity, where quality is quantified in terms of the Cramér-Rao Bound (CRB). The CRB is a lower bound on the variance of any unbiased estimator, and it is determined directly from the mathematical form of the likelihood function of the data. It is a remarkable fact that the CRB for general PPP intensity estimation takes a simple form. The CRB of the Gaussian sum intensity model in Chapter 3 is given in this chapter. The CRB is presented for PPP sample data, sometimes called "count record" data as well as histogram data. Sample data constitute a realization of the PPP, so the points are i.i.d. (independent and identically distributed).

1.1.2 Part II: Applications

Chapter 5 discusses both emission and transmission tomography. Positron emission tomography (PET) is presented first because of the excellent match between PET and the PPP model. The maximum likelihood algorithm for PET is derived. The original algorithm [110] dates to 1982, and there are now many variations of it, but all are called Shepp-Vardi algorithms. It is justly famous, not least because it is (said to be) the basis of all currently fielded PET medical imaging systems. It was independently discovered in image deconvolution some years earlier, first by

Richardson in 1972 [102] and then again by Lucy in 1974 [72]. In these applications it is known as the Richardson-Lucy algorithm.

SPECT (single photon emission computed tomography) is much more commonly used diagnostically than PET. The reconstructed image is estimated from multiple snapshots made by a movable gamma camera. A reconstruction algorithm for SPECT based on EM was derived by Miller, Snyder, and Miller [82] in 1985. Loosely speaking, the algorithm averages several Shepp-Vardi algorithms, one for each gamma camera snapshot; that is, it is a multi-snapshot average. It is presented in Section 5.4.

Transmission tomography (called computed tomography (CT)) is discussed in Section 5.5. A reconstruction algorithm based on EM was derived by Lange and Carson [65] in 1984. While based on EM, its detailed structure differs significantly from that of Shepp-Vardi algorithms. CRBs for PET and CT are presented in Section 5.6.

Chapter 6 presents multitarget tracking applications of PPPs. The multitarget state is modeled as a PPP. The Bayesian posterior point process is not a PPP, but it is approximated by a PPP. It is called an intensity filter because it recursively updates the intensity of the PPP approximation. An augmented state space enables "on line" estimates of the clutter and target birth PPPs to be produced as intrinsic parts of the filter. The information update of the intensity filter is seen to be identical to the first step of the Shepp-Vardi algorithm for PET discussed in Chapter 5.

The PHD (Probability Hypothesis Density) filter is obtained from the intensity filter by modifying the posterior PPP intensity with a priori knowledge of the clutter and target birth PPPs. The PHD filter was first derived by other methods in the multitarget application by Mahler in a series of papers beginning about 1994. For details, see [76] and the papers referenced therein. The relationship between Mahler's method and the approach taken here is discussed.

The multisensor intensity filter is also presented in Chapter 6. The PPP multiple target model is the same as in the single sensor case. The sensors are assumed conditionally independent. The resulting intensity filter is essentially the same as the first step of the Miller-Snyder-Miller algorithm for SPECT given in Chapter 5. Sensor data are, by analogy, equivalent to the gamma camera snapshots.

Chapter 7 discusses distributed networked sensors using an approach based on ideas from stochastic geometry, one of the classic roots of PPPs. These results typically provide ensemble system performance estimates rather than a performance estimate for a given sensor configuration. The contrast between point-to-event and event-to-event properties is highlighted, and Slivnyak's Theorem is discussed as a method that relates these two concepts. Threshold effects are dramatic and provide significant insights. For example, recent results from geometric random graph theory show that—with very high probability—randomly distributed networks very abruptly achieve communication diversity as the sensor communication range increases. This result guarantees that the overwhelming majority of random sensor distributions achieve (do not achieve) communication diversity if the sensor communication range R is larger (smaller) than some threshold, say R_{thresh}.

1.1.3 Part III: Beyond the Poisson Point Process

Chapter 8 reviews several interesting point processes that are not PPPs. Non-Poisson point processes are needed for high fidelity, realistic models in many different applications, including signal processing. Perhaps the most widely used are the marked PPPs in which a "mark" is associated with each point of a PPP realization. Other models provide various kinds of spatial correlation between the points of a process. They are needed for problems in which points are not distributed independently. An example is the set of points that are the centers of nonoverlapping hard spheres. Many useful processes obtained from PPPs by various devices are described in this chapter. Other point processes are at best only distantly related to PPPs. A prominent example of this kind are Gibbs processes. They model spatial correlation via a specified "energy" function defined on the points of the process.

Chapter 9 provides a brief sketch about directions for further work in applications and methods. Imaging technologies that overcome the Rayleigh resolution limit will probably provide a wealth of new applications of nonhomogeneous and multidimensional PPPs. New methods in statistical thinking, for example, Markov Chain Monte Carlo (MCMC) [24], will probably make non-Poisson processes even more widely used in applications than they currently are.

1.1.4 Appendices

Several appendices contain material that would interfere with the flow of the exposition in one way or another. Others contain material that supplements the text.

- Appendix A briefly presents the Expectation-Maximization algorithm. The iterative majorization method is used to provide a strictly numerical, non-probabilistic insight into EM.
- Appendix B discusses the numerical solution of the conditional mean equations. These equations arise in a variety of maximum likelihood estimation problems.
- Appendix C gives an overview of general Bayesian filtering. The linear-Gaussian Kalman filter is given as a special case.
- Appendix D gives an alternative Bayesian derivation of the intensity filter that stresses symmetry, marginalization, and the "mean field" approximation. It may also be helpful to readers who seek to read the original literature.
- Appendix E derives the conditional cluster intensity filter for multitarget tracking via the EM method. It is a parametric filter based on superposition of conditional Gaussian single target PPP intensities.
- Appendix F presents a novel marked PPP model of the power spectrum of stationary time series. The model lends itself to EM estimation methods.

1.2 The Real Line Is Not Enough

Applications of interest in imaging, target tracking, and distributed sensing are typically of dimension greater than one. Consequently, with few exceptions, only properties of PPPs that hold in any finite dimension are discussed. Excluding PPPs on the real line is a loss since their applications are widespread and highly successful in telephony, beginning with Erlang (c. 1900) and Palm (1943), renewal theory, and queuing networks. Many of their properties depend crucially on the linear ordering of points on the real line and, therefore, do not extend to higher dimensional spaces. One dimensional PPPs are well and widely discussed in the literature.

The multidimensional perspective provides a purely technical reason to neglect one dimensional PPPs: all nonhomogeneous PPPs on the real line can be transformed into a homogeneous PPP (see the discussion of Section 2.10 below). The required transformation is the inverse of a nonlinear mapping and is of independent interest; however, the inverse function does not exist in dimensions larger than one. Thus, much of the innate distinctiveness of nonhomogeneous PPPs is lost on the real line. They come into their own in higher dimensional spaces.

1.3 General Point Processes

The notion of a random set is too general to be particularly useful, and it is typically restricted in some fashion. There are many ways to do this. In stochastic geometry, a random set is often a "randomly" chosen member of a class of sets deemed interesting. The class of all triangles in the plane is an example of such a class. A traditional question in this case is "What is the probability that a random triangle is acute?" A harder question is "What is the probability that a random quadrilateral is convex?"

A more elaborate example is the class of unions of finite numbers of discs in the plane. In this case a random set is the union of a finite number of closed discs where the number of discs, as well as their centers and radii, are selected according to specified random procedures. This kind of model is often called a Boolean model, or sometimes a germ-grain model [16, Section 9.1.3], [123, Chapter 3]. The *germs* are the centers of the discs and the *grains* (also called primary grains) are the discs themselves.

Germs and grains can take very general forms—the grains need not even be connected sets. The germs are typically chosen to be the points of a special kind of finite point process called a PPP, which is described below. A classic question for Boolean models is "What is the probability that a given point is covered by (contained in) at least one disc in the random set?" A closely related question is "What is the probability that a given point is covered by exactly $k \geq 1$ discs in the random set?" Sets generated by a Boolean model are a special kind of much more general class of sets called RACS (*random closed sets*) [79].

Random sets that belong to the class of sets containing only points of the state space $\mathcal{S}$ gives rise to point processes. A point process is thus a random variable whose realizations are sets in this class. For most applications point processes are defined on $\mathbb{R}^m$, $m \geq 1$.

A *finite* point process is a random variable whose realizations in any bounded subset $\mathcal{R}$ of $\mathcal{S}$ are sets containing only a *finite* number of points of $\mathcal{R}$. The number of points and their locations can be chosen in many ways. An important subclass comprises finite point processes with independently and identically distributed (i.i.d.) points.

There are many members of the class i.i.d. finite point processes, at least two of which are named processes. One is the binomial point process (BPP). The BPP is a finite point process in which the number of points n is binomially distributed on the integers $\{0, 1, \ldots, K\}$, where $K \geq 0$ is an integer parameter. The points of the BPP are located according to a spatial random variable X on $\mathcal{S}$ with probability density function (pdf) $p_X(x)$. Explicitly, for any bounded subset $\mathcal{R} \subset \mathcal{S}$, the probability of n points occurring in $\mathcal{R}$ is

$$\Pr[n] = \binom{K}{n} (p_{\mathcal{R}})^n (1 - p_{\mathcal{R}})^{K-n} , \tag{1.1}$$

where the probability $p_{\mathcal{R}}$ is given by

$$p_{\mathcal{R}} = \int_{\mathcal{R}} p_X(x)\,\mathrm{d}x . \tag{1.2}$$

The binomial coefficient is $\binom{K}{n} = \frac{K!}{n!(K-n)!}$. The points $\{x_1, \ldots, x_n\}$ of the BPP are i.i.d. samples from the pdf $p_X(x)$. The number of points and the pdf of the point locations are closely coupled via the parameter $p_{\mathcal{R}}$. Using well known facts about the binomial distribution shows that the mean, or expected, number of points in $\mathcal{R}$ is $p_{\mathcal{R}} K$ and the variance is $p_{\mathcal{R}} (1 - p_{\mathcal{R}}) K$. The number of points in any set $\mathcal{R}$ cannot exceed K; therefore, even though the points are i.i.d., there is a negative correlation in the distribution of the *numbers* of points in disjoint sets. Loosely, for $\mathcal{R}_1 \cap \mathcal{R}_2 = \varnothing$, if more points than expected are in $\mathcal{R}_1$, the fewer there are in $\mathcal{R}_2$.

The other named i.i.d. finite point process is the Poisson point process (PPP). As discussed at length in the next chapter, the number of points is Poisson distributed on the nonnegative integers $\{0, 1, 2, \ldots\}$. The defining parameter of the Poisson distribution on a subset $\mathcal{R}$ of $\mathcal{S}$ is closely coupled to a spatial random variable $X \mid \mathcal{R}$ restricted to (conditioned on) $\mathcal{R}$. In this instance, however, the pdf of $X \mid \mathcal{R}$ is proportional to a nonnegative function called the intensity function, and the proportionality constant is the expected number of points in a realization. PPPs are a narrowly specialized class of i.i.d. finite point processes. The property that makes them uniquely important is that PPPs on disjoint subsets of $\mathcal{S}$ are statistically independent. This property is called independent scattering (although it is also sometimes called independent increments). In this terminology, BPPs are not independent scattering processes.

The points in realizations of PPPs with an intensity function defined on a continuous space are distinct with probability one. The points of a PPP realization are in this case often referred to as random finite sets. However, PPPs are also sometimes defined on discrete spaces and discrete-continuous spaces (see Section 2.12). In such cases, the discrete points of the PPP can be repeated with nonzero probability. Sets do not, by definition, have repeated elements, so it is more accurate to speak of the points in a PPP realization as a random finite list, or multiset, as such lists are sometimes called. To avoid making too much of these subtleties, random finite sets and lists are both referred to simply as PPP realizations.

1.4 An Alternative Tradition

PPPs constitute an alternative tradition to that of the more widely known stochastic processes, with which they are sometimes confused. A stochastic process is a family $X(t)$ of random variables indexed by a parameter t that is usually, but not always, thought of as time. The mathematics of stochastic processes, especially Gaussian stochastic processes, is widely known and understood. They are pervasive, with applications ranging from physics and engineering to finance and biology.

Poisson stochastic processes (again, not to be confused with PPPs) provided one of the first models of white Gaussian noise. Specifically, in vacuum tubes, electrons are emitted by a heated cathode and travel to the anode, where they contribute to the anode current. The anode current is modeled as shot noise, and the fluctuations of the current about the average are approximately a white Gaussian noise process when the electron arrival rate is high [11, 101]. The emission times of the electrons constitute a one dimensional PPP. Said another way, the occurrence times of the jump discontinuities in the anode current are a realization of a PPP.

Wiener processes are also known as Brownian motion processes. The sample paths are continuous, but nowhere differentiable. The process is sometimes thought of more intuitively as integrated white Gaussian noise. Armed with this intuition it may not be too surprising to the reader that nonhomogeneous PPPs approximate the level crossings of the correlation function of Gaussian processes [48] and sequential probability ratio tests.

The concept of independent increments is important for both point processes and stochastic processes; however, the concept is not exactly the same for both. It is therefore more appropriate to speak of independent scattering in point processes and of independent increments for stochastic processes. As is seen in Chapter 2, every PPP is an independent scattering process. In contrast, every independent increments stochastic process is a linear combination of a Wiener process and a Poisson process [119]. Further discussion is given in Section 2.9.1. The mathematics of PPPs is not as widely known as that of stochastic processes.

PPPs have many properties such as linear superposition and invariance under nonlinear transformation that are useful in various ways in many applications. These fundamental properties are presented in the next chapter.

Part I
Fundamentals

Chapter 2
The Poisson Point Process

Make things as simple as possible, but not simpler.[1]
Albert Einstein *(paraphrased)*,
On the Method of Theoretical Physics, *1934*

Abstract Properties of multidimensional Poisson point processes (PPPs) are discussed using a constructive approach readily accessible to a broad audience. The processes are defined in terms of a two-step simulation procedure, and their fundamental properties are derived from the simulation. This reverses the traditional exposition, but it enables those new to the subject to understand quickly what PPPs are about, and to see that general nonhomogeneous processes are little more conceptually difficult than homogeneous processes. After reviewing the basic concepts on continuous spaces, several important and useful operations that map PPPs into other PPPs are discussed—these include superposition, thinning, nonlinear transformation, and stochastic transformation. Following these topics is an amusingly provocative demonstration that PPPs are "inevitable." The chapter closes with a discussion of PPPs whose points lie in discrete spaces and in discrete-continuous spaces. In contrast to PPPs on continuous spaces, realizations of PPPs in these spaces often sample the discrete points repeatedly. This is important in applications such as multitarget tracking.

Keywords Event space · Intensity function · Orderly PPP · Realizations · Likelihood functions · Expectations · Random sums · Campbell's Theorem · Characteristic functions · Superposition · Independent thinning · Independent scattering · Poisson gambit · Nonlinear transformations · Stochastic transformations · PPPs on discrete spaces · PPPs on discrete-continuous spaces

Readers new to PPPs are urged to read the first four subsections below in order. After that, they are free to move about the chapter as their fancy dictates. There is a lot of information here. It cannot be otherwise for there are many wonderful and useful properties of PPPs.

---

[1] What he really said [27]: "It can scarcely be denied that the supreme goal of all theory is to make the irreducible basic elements as simple and as few as possible without having to surrender the adequate representation of a single datum of experience."

R.L. Streit, *Poisson Point Processes*, DOI 10.1007/978-1-4419-6923-1_2,
© Springer Science+Business Media, LLC 2010

The emphasis throughout the chapter is on the PPP itself, although applications are alluded to in several places. The event space of PPPs and other finite point processes is described in Section 2.1. The concept of intensity is discussed in Section 2.2. The important concept of orderliness is also defined. PPPs that are orderly are discussed in Sections 2.3 through 2.11. PPPs that are not orderly are discussed in the last section, which is largely devoted to PPPs on discrete and discrete-continuous spaces.

2.1 The Event Space

The points of a PPP occur in the state space $\mathcal{S}$. This space is usually the Euclidean space, $\mathcal{S} = \mathbb{R}^m$, $m \geq 1$, or some subset thereof. Discrete and discrete-continuous spaces $\mathcal{S}$ are discussed in Section 2.12. PPPs can be defined on even more abstract spaces, but this kind of generality is not needed for the applications discussed in this book.

Realizations of PPPs on a subset $\mathcal{R}$ of $\mathcal{S}$ comprises the number $n \geq 0$ and the locations $x_1, \ldots, x_n$ of the points in $\mathcal{R}$. The realization is denoted by the ordered pair

$$\xi = (n, \{x_1, \ldots, x_n\}).$$

The set notation signifies only that the ordering of the points x_j is irrelevant, but *not* that the points are necessarily distinct. It is better to think of $\{x_1, \ldots, x_n\}$ as an unordered list. Such lists are sometimes called multisets. Context will always make clear the intended usage, so for simplicity of language, the term *set* is used here and throughout the book.

It is standard notation to include n explicitly in ξ even though n is determined by the size of the set $\{x_1, \ldots, x_n\}$. There are many technical reasons to do so; for instance, including n makes expectations easier to define and manipulate.

If $n = 0$, then ξ is the trivial event $(0, \varnothing)$, where $\varnothing$ denotes the empty set. The event space is the collection of all possible finite subsets of $\mathcal{R}$:

$$\mathcal{E}(\mathcal{R}) = \{(0, \varnothing)\} \cup_{n=1}^{\infty} \left\{ (n, \{x_1, \ldots, x_n\}) : \quad x_j \in \mathcal{R}, \quad j = 1, \ldots, n \right\}.$$
$$(2.1)$$

The event space is clearly very much larger in some sense than the space $\mathcal{S}$ in which the individual points reside.

2.2 Intensity

Every PPP is parameterized by a quantity called the intensity. Intensity is an intuitive concept, but it takes different mathematical forms depending largely on whether the state space $\mathcal{S}$ is continuous, discrete, or discrete-continuous. The continuous case is discussed in this section. Discussion of PPPs on discrete and discrete-continuous spaces $\mathcal{S}$ is postponed to the last section of the chapter.

A PPP on a continuous space $\mathcal{S} \subset \mathbb{R}^n$ is *orderly* if the intensity is a nonnegative function $\lambda(s) \geq 0$ for all $s \in \mathcal{S}$. If $\lambda(s) \equiv \alpha$ for some constant $\alpha \geq 0$, the PPP is said to be homogeneous; otherwise, it is nonhomogeneous. It is assumed that

$$0 \leq \int_{\mathcal{R}} \lambda(s)\,ds < \infty \tag{2.2}$$

for *all* bounded subsets $\mathcal{R}$ of $\mathcal{S}$, i.e., subsets contained in some finite radius m-dimensional sphere. The sets $\mathcal{R}$ include—provided they are bounded—convex sets, sets with "holes" and internal voids, disconnected sets such as the union of disjoint spheres, and sets that are interwoven like chain links.

The intensity function $\lambda(s)$ need not be continuous, e.g., it can have step discontinuities. The only requirement on $\lambda(s)$ is the finiteness of the integral (2.2). The special case of homogeneous PPPs on $\mathcal{S} = \mathbb{R}^m$ with $\mathcal{R} = \mathcal{S}$ shows that the inequality (2.2) does not imply that $\int_{\mathcal{S}} \lambda(s)\,ds < \infty$. Finally, in physical problems, the integral (2.2) is a dimensionless number, so $\lambda(s)$ has units of number per unit volume of $\mathbb{R}^m$.

The intensity for general PPPs on the continuous space $\mathcal{S}$ takes the form

$$\lambda_D(s) = \lambda(s) + \sum_{j=1}^{\infty} w_j\,\delta(s - a_j), \qquad s \in \mathcal{S}, \tag{2.3}$$

where $\delta(\,\cdot\,)$ is the Dirac delta function and, for all j, the weights w_j are nonnegative and the points $a_j \in \mathcal{S}$ are distinct: $a_i \neq a_j$ for $i \neq j$. The intensity $\lambda_D(s)$ is not a function in the strict meaning of the term, but a "generalized" function. It is seen in the next section that the PPP corresponding to the intensity $\lambda_D(s)$ is orderly if and only if $w_j = 0$ for all j; equivalently, a PPP is orderly if and only if the intensity $\lambda_D(s)$ is a function, not a generalized function.

The concept of orderliness can be generalized so that finite point processes other than PPPs can also be described as orderly. There are several nonequivalent definitions of the general concept, as discussed in [118]; however, these variations are not used here.

2.3 Realizations

The discussion in this section and through to Section 2.11 is implicitly restricted to orderly PPPs, that is, to PPPs with a well defined intensity function on a continuous space $\mathcal{S} \subset \mathbb{R}^m$. Realizations and other properties of PPPs on discrete and discrete-continuous spaces are discussed in Section 2.12.

Realizations are conceptually straightforward to simulate for bounded subsets of continuous spaces $\mathcal{S} \subset \mathbb{R}^m$. Bounded subsets are "windows" in which PPP realizations are observed. Stipulating a window avoids issues with infinite sets; for example, realizations of homogeneous PPPs on $\mathcal{S} = \mathbb{R}^m$ have an infinite number of points but only a finite number in any bounded window.

Every realization of a PPP on a bounded set $\mathcal{R}$ is an element of the event space $\mathcal{E}(\mathcal{R})$. The realization ξ therefore comprises the number $n \geq 0$ and the locations $\{x_1, \ldots, x_n\}$ of the points in $\mathcal{R}$.

A two-step procedure, one step discrete and the other continuous, generates (or, simulates) one realization $\xi \in \mathcal{E}(\mathcal{R})$ of a nonhomogeneous PPP with intensity $\lambda(s)$ on a bounded subset $\mathcal{R}$ of $\mathcal{S}$. The procedure also fully reveals the basic statistical structure of the PPP. If $\int_{\mathcal{R}} \lambda(s)\, ds = 0$, ξ is the trivial event. If $\int_{\mathcal{R}} \lambda(s)\, ds > 0$, the realization is obtained as follows:

Step 1. The number $n \geq 0$ of points is determined by sampling the discrete Poisson random variable, denoted by N, with probability mass function given by

$$p_N(n) \equiv \frac{\left(\int_{\mathcal{R}} \lambda(s)\, ds\right)^n}{n!}\, e^{-\int_{\mathcal{R}} \lambda(s)\, ds}. \tag{2.4}$$

If $n = 0$, the realization is $\xi = (0, \varnothing)$, and Step 2 is not performed.

Step 2. The n points $x_j \in \mathcal{R}$, $j = 1, \ldots, n$, are obtained as independent and identically distributed (i.i.d.) samples of a random variable X on $\mathcal{R}$ with probability density function (pdf) given by

$$p_X(s) = \frac{\lambda(s)}{\int_{\mathcal{R}} \lambda(s)\, ds} \quad \text{for } s \in \mathcal{R}. \tag{2.5}$$

The output is the ordered pair $\xi_o = (n, (x_1, \ldots, x_n))$. Replacing the ordered n-tuple $(x_1, \ldots, x_n)$ with the set $\{x_1, \ldots, x_n\}$ gives the PPP realization $\xi = (n, \{x_1, \ldots, x_n\})$.

The careful distinction between ξ_o and ξ is made to avoid annoying, and sometimes confusing, problems later when order is important. For example, it is seen in Section 2.4 that the pdfs (probability density functions) of ξ_o and ξ differ by a factor of $n!$. Also, the points $\{x_1, \ldots, x_n\}$ are i.i.d. when conditioned on the number n of points. The conditioning on n is implicit in the statement of Step 2.

For continuous spaces $\mathcal{S} \subset \mathbb{R}^m$, an immediate consequence of Step 2 is that the points $\{x_1, \ldots, x_n\}$ are distinct with probability one: repeated elements are allowed in theory, but in practice they never occur (with probability one). Another way to say this is that the list, or multiset, $\{x_1, \ldots, x_n\}$ is a set with probability one. The statement fails to hold when the PPP is not orderly, that is, when the intensity (2.3) has one or more Dirac delta function components. It also does not hold when the state space $\mathcal{S}$ is discrete or discrete-continuous (see Section 2.12).

An acceptance-rejection procedure (see, e.g., [56]) is used to generate the i.i.d. samples of (2.5). Let

$$\alpha = \max_{s \in \mathcal{R}} \frac{p_X(s)}{g(s)}, \tag{2.6}$$

where $g(s) > 0$ is any bounded pdf on $\mathcal{R}$ from which i.i.d. samples of $\mathcal{R}$ can be generated via a known procedure. The function $g(\cdot)$ is called the importance function. For each point x with pdf g, compute $t = p_X(x)/(\alpha\, g(x))$. Next, generate a uniform variate u on $[0, 1]$ and compare u and t: if $u > t$, reject x; if $u \le t$, accept it. The accepted samples are distributed as $p_X(x)$.

The acceptance-rejection procedure is inefficient for some problems, that is, large numbers of i.i.d. samples from the pdf (2.5) may be drawn before finally accepting n samples. As is well known, efficiency depends heavily on the choice of the importance function $g(\cdot)$. Table 2.1 outlines the overall procedure and indicates how the inefficiency can occur. If inefficiency is a concern, other numerical procedures may be preferred in practice. Also, evaluating $\int_{\mathcal{R}} \lambda(s)\,\mathrm{d}s$ may require care in some problems.

Table 2.1 Realization of a PPP with intensity $\lambda(s)$ on bounded set $\mathcal{R}$

- Preliminaries
  - Select importance function $g(s) > 0,\ s \in \mathcal{R}$
  - Set efficiency scale $c_{eff} = 1000$ (corresponds to a 0.1% acceptance rate)
- *Step 1.*
  - Compute $\mu = \int_{\mathcal{R}} \lambda(s)\,\mathrm{d}s$
  - Compute

$$\alpha = \max_{s \in \mathcal{R}} \frac{\lambda(s)}{\mu\, g(s)}$$

  - Draw random integer $n \in \{0, 1, 2, \dots\}$ from Poisson distribution with parameter μ

$$\Pr[n] = e^{-\mu}\frac{\mu^n}{n!}$$

  - IF $n = 0$, STOP
- *Step 2.*
  - FOR $j = 1 : c_{eff}\, n$
    - Draw random sample x with pdf g
    - Compute

$$t = \frac{\lambda(s)}{\alpha\, \mu\, g(s)}$$

    - Draw random sample u with pdf Uniform$[0, 1]$
    - ACCEPT x, if $u \le t$
    - REJECT x, otherwise
    - Stop when n points are accepted
  - END FOR
- If number of accepted samples is smaller than n after computing $c_{eff}\, n$ draws from g, then find a better importance function or make c_{eff} larger and accept the inefficiency.

Example 2.1 The two step procedure is used to generate i.i.d. samples from a PPP whose intensity function is nontrivially structured. These samples also show the difficulty of observing this structure in small sample sets. Denote the multivariate Gaussian pdf on $\mathbb{R}^m$ with mean μ and covariance matrix Σ by

$$\mathcal{N}(s\,;\,\mu,\,\Sigma) \;=\; \frac{1}{\sqrt{\det(2\pi\,\Sigma)}}\,\exp\left(-\frac{1}{2}(s-\mu)^T\,\Sigma^{-1}\,(s-\mu)\right). \qquad (2.7)$$

Let $\varXi$ be the bivariate PPP whose intensity function on the square $\mathcal{R} \equiv [-4\,\sigma,\,4\,\sigma]^2$ is, for $s \equiv (x,\,y)^T \in \mathcal{R}$,

$$\lambda(s) \;\equiv\; \lambda(x,\,y) \;=\; \frac{a}{64\,\sigma^2} + b\,f(x,\,y), \qquad (2.8)$$

where $a = 20$, $b = 80$, and

$$f(x,\,y) = \begin{cases} 0, & \text{if } -\frac{6}{5}\sigma \le y \le -\frac{4}{5}\sigma \\ \mathcal{N}\left(\begin{bmatrix} x \\ y \end{bmatrix};\,\begin{bmatrix} 0 \\ 0 \end{bmatrix},\,\sigma^2\begin{bmatrix} 1 & 0 \\ 0 & 1 \end{bmatrix}\right), & \text{otherwise.} \end{cases}$$

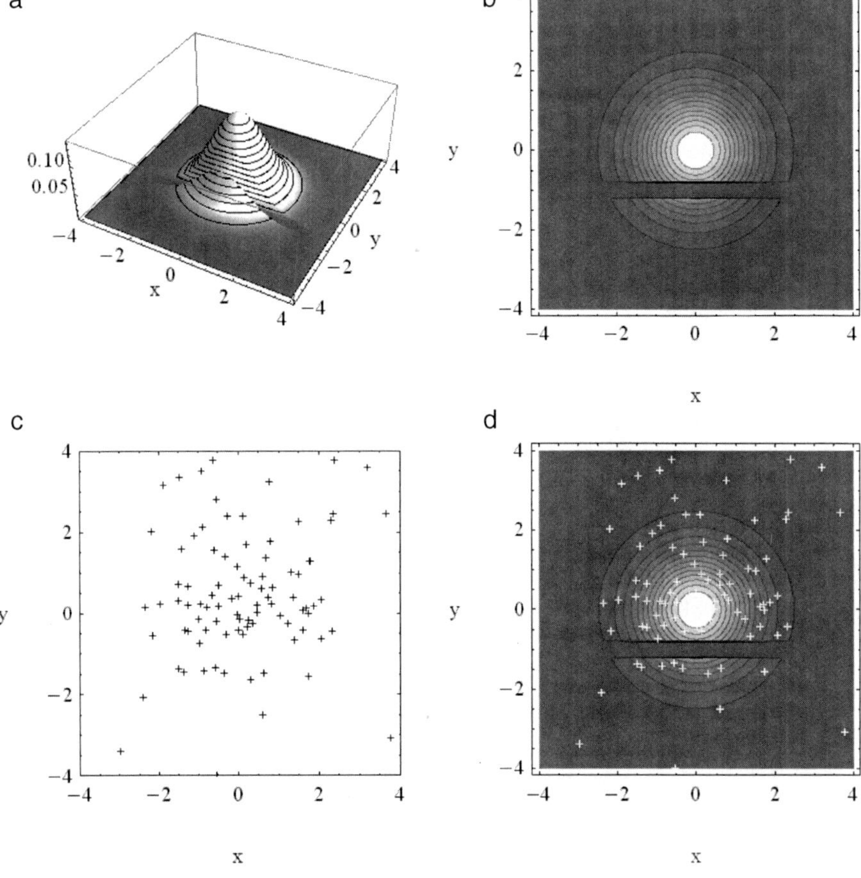

Fig. 2.1 Realizations of the PDF (2.9) of the intensity function (2.8) for $\sigma = 1$, $a = 20$, and $b = 80$. Samples are generated by the acceptance-rejection method. The prominent horizontal notch in the intensity is hard to see from the samples alone

For $\sigma = 1$, numerical integration gives the mean intensity $\mu = \int_{\mathcal{R}} \lambda(x, y)\, dx\, dy = 92.25$, approximately. A pseudo-random integer realization of the Poisson discrete variable (2.4) is $n = 90$, so 90 i.i.d. samples of the pdf (cf. (2.5))

$$p_X(s) \equiv p(x, y) = \lambda(x, y) / 92.25 \tag{2.9}$$

are drawn via the acceptance-rejection procedure with $g(x, y) = 1/(64\,\sigma^2)$. The pdf (2.9) is shown as the 3-D plot in Fig. 2.1a and as a set of equispaced contours in Fig. 2.1b, respectively Fig. 2.1c and 2.1d, show the 90 sample points with and without reference to the intensity contours.

The horizontal "notch" is easily missed using these 90 samples in Fig. 2.1c. The detailed structure of an intensity function can be estimated reliably only in special circumstances, e.g., when a large number of realizations is available, or when the PPP has a known parametric form (see Section 3.1).

2.4 Likelihood Function

The random variable Ξ with realizations in $\mathcal{E}(\mathcal{R})$ for *every* bounded subset $\mathcal{R}$ of S is a PPP if its realizations are generated via the two-step procedure. Let $p_\Xi(\xi)$ denote the pdf of Ξ evaluated at $\Xi = \xi$. Let $\Xi \equiv (N, \mathcal{X})$, where N is the number of points and $\mathcal{X} \equiv \{x_1, \ldots, x_N\}$ is the point set. Let the realization be $\xi = (n, \{x_1, \ldots, x_n\})$. From the definition of conditioning,

$$p_\Xi(\xi) = p_N(n)\, p_{\mathcal{X}|N}(\{x_1, \ldots, x_n\} \mid n), \tag{2.10}$$

where $p_N(n)$ is the unconditional probability mass function of N given by (2.4). The conditional pdf of $\mathcal{X} \mid N$ is

$$p_{\mathcal{X}|N}(\{x_1, \ldots, x_n\} \mid n) = n! \prod_{j=1}^{n} p_X(x_j), \tag{2.11}$$

where X is the random variable corresponding to a single sample point whose pdf is (2.5). The $n!$ in (2.11) arises from the fact that there are $n!$ equally likely ordered i.i.d. trials that generate the unordered set $\mathcal{X}$. Substituting (2.4) and (2.11) into (2.10) gives the pdf of Ξ evaluated at $\xi = (n, \{x_1, \ldots, x_n\}) \in \mathcal{E}(\mathcal{R})$:

$$
\begin{aligned}
p_\Xi(\xi) &= p_N(n)\, p_{\mathcal{X}|N}(\{x_1, \ldots, x_n\} \mid n) \\
&= \exp\left(-\int_{\mathcal{R}} \lambda(s)\, ds\right) \frac{\left(\int_{\mathcal{R}} \lambda(s)\, ds\right)^n}{n!}\, n! \prod_{j=1}^{n} \frac{\lambda(x_j)}{\int_{\mathcal{R}} \lambda(s)\, ds} \\
&= \exp\left(-\int_{\mathcal{R}} \lambda(s)\, ds\right) \prod_{j=1}^{n} \lambda(x_j), \qquad \text{for } n \geq 1.
\end{aligned}
\tag{2.12}
$$

The likelihood of the trivial event is the special case $n = 0$ of (2.4), so that $p_\Xi(\xi = (0, \varnothing)) = \exp(-\int_\mathcal{R} \lambda(s)\,ds)$. The pdf of Ξ is parameterized by the intensity function $\lambda(s)$. Any positive scalar multiple of the pdf is "the" likelihood function of Ξ.

The expression (2.12) is used in estimation problems involving measured data sets for which data order is irrelevant. For ordered data, from (2.11),

$$p_{\mathcal{X}|N}(x_1, \ldots, x_n \mid n) \equiv \frac{1}{n!} p_{\mathcal{X}|N}(\{x_1, \ldots, x_n\} \mid n) \tag{2.13}$$

$$= \prod_{j=1}^{n} p_X(x_j) \tag{2.14}$$

$$= \prod_{j=1}^{n} \frac{\lambda(x_j)}{\int_\mathcal{R} \lambda(s)\,ds}. \tag{2.15}$$

Let $\xi_o = (n, (x_1, \ldots, x_n))$. Using (2.15) and the definition of conditioning gives

$$p_\Xi(\xi_o) = p_N(n)\, p_{\mathcal{X}|N}(x_1, \ldots, x_n \mid n) \tag{2.16}$$

$$= \frac{1}{n!} \exp\left(-\int_\mathcal{R} \lambda(s)\,ds\right) \prod_{j=1}^{n} \lambda(x_j), \quad \text{for } n \geq 1. \tag{2.17}$$

This notation interprets arguments in the usual way, so it is easier to understand and manipulate than (2.12). For example, the discrete pdf $p_N(n)$ of (2.4) is merely the integral of (2.17) over $x_1, \ldots, x_n$, but taking the same integral of (2.12) requires an additional thought to restore the missing $n!$.

The argument ξ_o in (2.17) is written simply as ξ below. This usage may cause some confusion, since then the left hand side of (2.17) becomes $p_\Xi(\xi)$, which is the same as the first equation in (2.12), a quantity that differs from it by a factor of $n!$. A similar ambiguity arises from using the same subscript $\mathcal{X} \mid N$ on both sides of (2.13). Context makes the intended meaning clear, so these abuses of notation will not cause confusion.

In practice, when the number of points in a realization is very large, the points of a PPP realization are often replaced by a smaller data set. If the smaller data set also reduces the information content, the likelihood function obtained in this section no longer applies. An example of a smaller data set (called histogram count data) and its likelihood function is given in Section 2.9.1.

2.5 Expectations

Expectations are decidedly more interesting for point processes than for ordinary random variables. Expectations are taken of real valued functions $\mathcal{F}$ defined on the event space $\mathcal{E}(\mathcal{R})$, where $\mathcal{R}$ is a bounded subset of $\mathcal{S}$. Thus $\mathcal{F}(\xi)$ evaluates to a real number for all $\xi \in \mathcal{E}(\mathcal{R})$. The expectation of $\mathcal{F}(\xi)$ is written in the very general form

$$E[\mathcal{F}] \equiv E_\Xi[\mathcal{F}] = \sum_{\xi \in \mathcal{E}(\mathcal{R})} \mathcal{F}(\xi)\, p_\Xi(\xi), \tag{2.18}$$

where the sum, properly defined, is matched to the likelihood function of the point process. In the case of PPPs, the likelihood function is that of the two-step simulation procedure. The sum is often referred to as an "ensemble average" over all realizations of the point process.

The sum is daunting because of the huge size of the set $\mathcal{E}(\mathcal{R})$. Defining the expectation carefully is the first and foremost task of this section. The second is to show that for PPPs the expectation, though fearsome, can be evaluated explicitly for many functions of considerable application interest.

2.5.1 Definition

Let $\xi = (n, \{x_1, \ldots, x_n\})$. For analytical use, it is convenient to rewrite the function $\mathcal{F}(\xi) = \mathcal{F}(n, \{x_1, \ldots, x_n\})$ in terms of a function that uses an easily understood argument list, that is, let

$$\mathcal{F}(n, \{x_1, \ldots, x_n\}) \equiv F(n, x_1, \ldots, x_n). \tag{2.19}$$

The function F inherits an important symmetry property from $\mathcal{F}$. Let $\mathrm{Sym}(n)$ denote the set of all permutations of the first n positive integers. For all permutations $\sigma \in \mathrm{Sym}(n)$,

$$\begin{aligned}
F(n, x_{\sigma(1)}, \ldots, x_{\sigma(n)}) &= \mathcal{F}\left(n, \{x_{\sigma(1)}, \ldots, x_{\sigma(n)}\}\right) \\
&= \mathcal{F}(n, \{x_1, \ldots, x_n\}) \\
&= F(n, x_1, \ldots, x_n). \tag{2.20}
\end{aligned}$$

In words, $F(n, x_1, \ldots, x_n)$ is symmetric, or invariant, under permutations of its location arguments.

Using ordered argument forms in the expectation (2.18) gives

$$E[F] = \sum_{(n, x_1, \ldots, x_n)} F(n, x_1, \ldots, x_n)\, p_\Xi(n, x_1, \ldots, x_n). \tag{2.21}$$

The sum in (2.21) is an odd looking discrete-continuous sum that needs interpretation. The conditional factorization

$$p_\Xi(\xi) = p_N(n)\, p_{X|N}(x_1, \ldots, x_n \mid n)$$

of the ordered realization $\xi = (n, x_1, \ldots, x_n)$ provides the key—make the sum over n the outermost sum, and interpret "continuous sums" in a natural way as integrals over sets of the form $\mathcal{R} \times \cdots \times \mathcal{R}$. This gives the expectation of the function F as a nested pair of expectations. The first E_N is over N, and the second $E_{\mathcal{X}|N}$ is over $\mathcal{X}|N$. The expectation with respect to the point process Ξ is given by

$$E[F] \equiv E_N \left[E_{\mathcal{X}|N}[F] \right] \tag{2.22}$$

$$\equiv \sum_{n=0}^{\infty} p_N(n) \int_{\mathcal{R}} \cdots \int_{\mathcal{R}} F(n, x_1, \ldots, x_n) \, p_{\mathcal{X}|N}(x_1, \ldots, x_n | n) \, dx_1 \cdots dx_n.$$

$$\tag{2.23}$$

The expectation is formidable, but it is not as bad as it looks. Its inherently straightforward structure is revealed by verifying that $E[F] = 1$ for $F(n, x_1, \ldots, x_n) \equiv 1$. The details of this trivial exercise are omitted.

The expectation of non-symmetric functions is undefined. The definition is extended—formally—to general functions, say $G(n, x_1, \ldots, x_n)$, via its symmetrized version:

$$G_{\text{Sym}}(n, x_1, \ldots, x_n) = \frac{1}{n!} \sum_{\sigma \in \text{Sym}(n)} G(n, x_{\sigma(1)}, \ldots, x_{\sigma(n)}). \tag{2.24}$$

The expectation of G is defined by $E[G] = E\left[G_{\text{Sym}} \right]$. This definition works because G_{Sym} is a symmetric function of its arguments, a fact that is straightforward to verify. The definition is clearly compatible with the definition for symmetric functions since $G_{\text{Sym}}(n, x_1, \ldots, x_n) \equiv G(n, x_1, \ldots, x_n)$ if G is symmetric.

The expectation is defined by (2.23) for any finite point process with events in $\mathscr{E}(\mathcal{R})$, not just PPPs. For PPPs and other i.i.d. finite point processes (such as BPPs),

$$p_{\mathcal{X}|N}(x_1, \ldots, x_n | n) = \prod_{j=1}^{n} p_X(x_j), \tag{2.25}$$

so the expectation (2.23) is

$$E[F] \equiv \sum_{n=0}^{\infty} p_N(n) \int_{\mathcal{R}} \cdots \int_{\mathcal{R}} F(n, x_1, \ldots, x_n) \prod_{j=1}^{n} p_X(x_j) \, dx_1 \cdots dx_n. \tag{2.26}$$

PPPs are assumed throughout the remainder of this chapter, so the discrete probability distribution $p_N(n)$ and pdf $p_X(x)$ are given by (2.4) and (2.5).

The expected number of points in $\mathcal{R}$ is $E[N(\mathcal{R})]$. When the context clearly identifies the set $\mathcal{R}$, the expectation is written simply as $E[N]$. By substituting $F(n, x_1, \ldots, x_n) \equiv n$ into (2.26) and observing that the integrals all integrate to one, it follows immediately that

$$E[N] \equiv \sum_{n=0}^{\infty} n \, p_N(n)$$

$$= \int_{\mathcal{R}} \lambda(s) \, \mathrm{d}s . \tag{2.27}$$

Similarly, the variance is equal to the mean:

$$\mathrm{Var}[N] = \sum_{n=0}^{\infty} (n - E[N])^2 \, p_N(n)$$

$$= \int_{\mathcal{R}} \lambda(s) \, \mathrm{d}s. \tag{2.28}$$

The explicit sums in (2.27) and (2.28) are easily verified by direct calculation using (2.4).

2.5.2 Random Sums

Evaluating expectations presents significant difficulties for many choices of the function F. There are, fortunately, two important classes of functions whose expectations simplify dramatically. The first class comprises functions called random sums. They are especially useful in physics and signal processing. The expectations of random sums reduce to an ordinary integral over $\mathcal{R}$, a result that is surprising on first encounter.

Let $f(x)$ be a given real valued function. The random variable

$$F(\Xi) = \sum_{j=1}^{N} f(X_j) \tag{2.29}$$

is called a random sum. Given a realization $\Xi = \xi$, a realization of the random sum is given by

$$F(n, x_1, \ldots, x_n) = \sum_{j=1}^{n} f(x_j), \qquad \text{for } n \geq 1, \tag{2.30}$$

and, for $n = 0$, by $F(0, \varnothing) \equiv 0$. The special case of (2.30) for which $f(x) \equiv 1$ reduces to $F(n, x_1, \ldots, x_n) = n$, the number of points in $\mathcal{R}$. The mean of F is given by

$$E[F] = E\left[\sum_{j=1}^{N} f(X_j)\right] \tag{2.31}$$

$$= \int_{\mathcal{R}} f(x)\,\lambda(x)\,dx . \tag{2.32}$$

The expectation (2.32) is obtained by cranking through the algebra—substituting (2.30) into (2.26) and interchanging the sum over j and the integrals over $\mathcal{R}$ gives

$$E[F] = \sum_{n=0}^{\infty} p_N(n) \sum_{j=1}^{n} \int_{\mathcal{R}} \cdots \int_{\mathcal{R}} f(x_j) \prod_{j=1}^{n} p_X(x_j)\,dx_1 \cdots dx_n.$$

All but one of the integrals evaluates to 1, so

$$E[F] = \sum_{n=0}^{\infty} p_N(n)\, n \int_{\mathcal{R}} f(x_j)\, p_X(x_j)\,dx_j .$$

Substituting (2.4) and (2.5) and simplifying gives (2.32).

The result (2.32) also holds for vector valued functions f, i.e., functions such that $f(x) \in \mathbb{R}^m$. This is seen by separating f into components.

Let G be the same kind of function as F, namely,

$$G(\xi) = \sum_{j=1}^{n} g(x_j) , \tag{2.33}$$

where $g(x)$ is a real valued function. Then the expected value of the product is

$$E[F\,G] = \int_{\mathcal{R}} f(x)\,\lambda(x)\,dx \int_{\mathcal{R}} g(x)\,\lambda(x)\,dx + \int_{\mathcal{R}} f(x)\,g(x)\,\lambda(x)\,dx. \tag{2.34}$$

Before verifying this result in the next paragraph, note that since the means of F and G are determined as in (2.32), the result is equivalent to

$$\mathrm{cov}[F,\,G] \equiv E\left[(F - E[F])\,(G - E[G])\right]$$

$$= \int_{\mathcal{R}} f(x)\,g(x)\,\lambda(x)\,dx. \tag{2.35}$$

Setting $g(x) = f(x)$ in (2.34) gives the variance:

$$\mathrm{Var}[F] = E[F^2] - E^2[F]$$

$$= \int_{\mathcal{R}} f^2(x)\,\lambda(x)\,dx . \tag{2.36}$$

The special case $f(x) \equiv 1$, (2.36) reduces to the variance (2.28) of the number of points in $\mathcal{R}$.

The result (2.34) is verified by direct evaluation. Write

$$F(\xi)\,G(\xi) = \sum_{\substack{i,j=1 \\ i \neq j}}^{n} f(x_i)\,g(x_j) + \sum_{j=1}^{n} f(x_j)\,g(x_j). \tag{2.37}$$

The second term in (2.37) is (2.30) with $f(x_j)\,g(x_j)$ replacing $f(x_j)$, so its expectation is the second term of (2.34). The expectation of the first term is evaluated in much the same way as (2.32); details are omitted. The identity (2.34) is sometimes written

$$E\left[\sum_{\substack{i,j=1 \\ i \neq j}}^{N} f(X_i)\,g(X_j)\right] = \int_{\mathcal{R}} f(x)\,\lambda(x)\,\mathrm{d}x \int_{\mathcal{R}} g(x)\,\lambda(x)\,\mathrm{d}x. \tag{2.38}$$

The expression (2.38) holds for products of any number of functions.

For vector valued functions, f and g, the result (2.34) holds if g and G are replaced by g^T and G^T:

$$E[F\,G^T] = \int_{\mathcal{R}} f(x)\,\lambda(x)\,\mathrm{d}x \int_{\mathcal{R}} g^T(x)\,\lambda(x)\,\mathrm{d}x + \int_{\mathcal{R}} f(x)\,g^T(x)\,\lambda(x)\,\mathrm{d}x. \tag{2.39}$$

This is verified by breaking the product $F\,G^T$ into components.

2.6 Campbell's Theorem[2]

Campbell's Theorem is the classic keystone result for random sums that dates to 1909 [11]. It gives the characteristic function of random sums F of the form (2.30). The characteristic function is useful in many problems. For instance, it enables all the moments of F to be found by a straightforward calculation. The mean and variance of F given in the previous section are corollaries of Campbell's Theorem.

Slivnyak's Theorem is the keystone result for a random sum in which the argument of the function in the summand depends in a "holistic" way on the PPP realization. These sums are useful for applications involving spatially distributed sensor networks. The application context facilitates understanding the result, so discussion of Slivnyak's Theorem is postponed to Section 7.1.2.

---

[2] This section can be skipped entirely on a first reading of the chapter. The material presented is used only Chapter 8.

Under mild regularity conditions, Campbell's Theorem says that when θ is *purely imaginary*

$$E\left[e^{\theta F}\right] = \exp\left[\int_{\mathcal{R}} \left(e^{\theta f(x)} - 1\right) \lambda(x)\,dx\right], \qquad (2.40)$$

where $f(x)$ is a real valued function. The expectation exists for any complex θ for which the integral converges. It is obtained by algebraic manipulation. Substitute the explicit form (2.17) into the definition of expectation and churn:

$$E\left[e^{\theta F}\right] = \sum_{n=0}^{\infty} p_N(n) \int_{\mathcal{R}} \cdots \int_{\mathcal{R}} e^{\theta \sum_{j=1}^{n} f(x_j)}\, p_{X|N}(x_1, \ldots, x_n \mid n)\,dx_1 \cdots dx_n$$

$$= e^{-\int_{\mathcal{R}} \lambda(s)\,ds} \sum_{n=0}^{\infty} \frac{1}{n!} \int_{\mathcal{R}} \cdots \int_{\mathcal{R}} \left\{\prod_{j=1}^{n} e^{\theta f(x_j)} \lambda(x_j)\right\} dx_1 \cdots dx_n$$

$$= e^{-\int_{\mathcal{R}} \lambda(s)\,ds} \sum_{n=0}^{\infty} \frac{1}{n!} \left(\int_{\mathcal{R}} e^{\theta f(s)} \lambda(s)\,ds\right)^n$$

$$= e^{-\int_{\mathcal{R}} \lambda(s)\,ds} \exp\left[\int_{\mathcal{R}} e^{\theta f(s)}\,ds\right]. \qquad (2.41)$$

The last expression is obviously equivalent to (2.40). See [49, 57, 63] for further discussion.

The characteristic function of F is given by (2.40) with $\theta = i\omega$, where ω is real and $i = \sqrt{-1}$, and $\mathcal{R} = \mathbb{R}$. The convergence of the integral requires that the Fourier transform of f exist as an ordinary function, i.e., it cannot be a generalized function. As is well known, the moment generating function is closely related to the characteristic function [93, Section 7.3]. Expanding the exponential gives

$$E\left[e^{i\omega F}\right] = E\left[1 + i\omega F + (i)^2 \frac{(\omega)^2 F}{2!} + \cdots\right]$$

$$= 1 + i\omega E[F] + (i)^2 \frac{(\omega)^2}{2!} E[F^2] + \cdots,$$

assuming that integrating term by term is valid. Hence, by differentiation, the moment of order $n \geq 1$ is

$$E\left[F^n\right] = (-i)^n \frac{d^n}{d\omega^n} E\left[e^{i\omega F}\right]\Big|_{\omega=0}. \qquad (2.42)$$

The results (2.32) and (2.36) are corollaries of (2.42).

The joint characteristic function of the random sum F and the sum G defined via the function $g(x)$ as in (2.33) is

$$E\left[e^{i\,\omega_1 F + i\,\omega_2 G}\right] = \exp\left[\int_{\mathcal{R}}\left(e^{i\,\omega_1 f(x) + i\,\omega_2 g(x)} - 1\right)\lambda(x)\,dx\right]. \qquad (2.43)$$

To see this, simply use $\omega_1 f(x) + \omega_2 g(x)$ in place of $f(x)$ in (2.40). An immediate by-product of this result is an expression for the joint moments of F and G. Expanding (2.43) in a joint power series and assuming term by term integration is valid gives

$$E\left[e^{i\,\omega_1 F + i\,\omega_2 G}\right]$$

$$= E\left[1 + i\,\omega_1 F + i\,\omega_2 G + \frac{(i\,\omega_1)^2}{2!}F^2 + \frac{(i\,\omega_1)(i\,\omega_2)}{2!}FG + \frac{(i\,\omega_2)^2}{2!}G^2 + \cdots\right]$$

$$= 1 + i\,\omega_1 E[F] + i\,\omega_2 E[G]$$

$$+ \frac{(i\,\omega_1)^2}{2!}E[F^2] + \frac{(i\,\omega_1)(i\,\omega_2)}{2!}E[FG] + \frac{(i\,\omega_2)^2}{2!}E[G^2] + \cdots,$$

where terms of order larger than two are omitted. Taking partial derivatives gives the joint moment of order (r, s) as

$$E\left[F^r G^s\right] = (-i)^{r+s}\,\frac{\partial^r\,\partial^s}{\partial\omega_1^r\,\partial\omega_2^s}\,E\left[e^{i\,\omega_1 F + i\,\omega_2 G}\right]\Big|_{\omega_1 = \omega_2 = 0}. \qquad (2.44)$$

In particular, a direct calculation for the case $r = s = 1$ verifies the earlier result (2.34).

The form (2.40) of the characteristic function also characterizes the PPP; that is, a finite point process whose expectations of random sums satisfies (2.40) is necessarily a PPP. The details are given in the next subsection.

2.6.1 Characterization of PPPs

A finite point process is necessarily a PPP if its expectation of random sums matches the form given by Campbell's Theorem. Let $\varXi$ be a finite point process whose realizations $\xi = (n, \{x_1, \ldots, x_n\})$ are in the event space $\mathcal{E}(\mathcal{S})$. The pdf of $\varXi$ is $p_{\varXi}(\xi)$, and the expectation is defined as in (2.18). The expectation of the random sum

$$F(\varXi) = \sum_{j-1}^{N} f(X_j), \qquad n \geq 1, \qquad (2.45)$$

is assumed to satisfy Campbell's Theorem (with $\theta = -1$) for a sufficiently large class of functions f. This class of functions is defined shortly. Thus, for all f in this class, it is assumed that

$$E\left[e^{-F}\right] = \exp\left[\int_{\mathcal{R}}\left(e^{-f(x)} - 1\right)\lambda(x)\,dx\right] \tag{2.46}$$

for some nonnegative function $\lambda(x)$. The goal is to show that (2.46) implies that the finite point process $\varXi$ is necessarily a PPP with intensity function $\lambda(x)$. This is done by showing that $\varXi$ satisfies the independent scattering property for any finite number k of sets A_j such that $\mathcal{S} = \cup_{j=1}^{k} A_j$ and $A_i \cap A_j = \varnothing$ for $i \neq j$.

Consider a nonnegative function f with values $f_1,\ f_2,\ \ldots,\ f_k$ on the specified sets $A_1,\ A_2,\ \ldots,\ A_k$, respectively, so that

$$A_j = \{x : f(x) = f_j\}.$$

Let

$$m_j = \int_{A_j} \lambda(x)\,dx .$$

The right hand side of (2.46) is then

$$E\left[e^{-F}\right] = \exp\left[\sum_{j=1}^{k}\left(e^{-f_j} - 1\right)m_j\right]. \tag{2.47}$$

Observe that

$$\sum_{j=1}^{N} f(X_j) \equiv \sum_{j=1}^{k} f_j\, N(A_j), \tag{2.48}$$

where $N(A_j)$ is the number of points in A_j. For the given function f, the assumed identity (2.46) is equivalent to

$$E\left[e^{-\sum_{j=1}^{k} f_j N(A_j)}\right] = \exp\left[\sum_{j=1}^{k}\left(e^{-f_j} - 1\right)m_j\right]. \tag{2.49}$$

Let $z_j = e^{-f_j}$. The last result is

$$E\left[\prod_{j=1}^{k} z_j^{N(A_j)}\right] = \prod_{j=1}^{k} e^{m_j\,(z_j - 1)}. \tag{2.50}$$

By varying the choice of function values $f_j \geq 0$, the result (2.50) is seen to hold for all $z_j \in (0,\ 1)$.

The joint characteristic function of several random variables is the product of the individual characteristic functions if and only if the random variables are independent [93], and the characteristic function of the Poisson distribution with mean m_j is (in this notation) $e^{m_j(z_j-1)}$. Therefore, the counts $N(A_j)$ are independent and Poisson distributed with mean m_j. Since the sets A_j are arbitrary, the finite point process Ξ is a PPP.

The class of functions for which the identity (2.46) holds must include the class of *all* nonnegative functions that are piecewise constant, with arbitrarily specified values f_j, on an arbitrarily specified finite number of disjoint sets A_j. The discussion here is due to Kingman [63].

2.6.2 Probability Generating Functional

A *functional* is an operator that maps a function to a real number. With this language in mind, the expectation operator is a functional because $E[f] = \int_S f(x)\, dx \in \mathbb{R}$.

The Laplace functional evaluated for the function f is defined for finite point processes Ξ by

$$
\begin{aligned}
L_\Xi(f) &= E\left[e^{-F(\Xi)}\right] \\
&= E\left[e^{-\sum_{j=1}^N f(X_j)}\right].
\end{aligned}
\tag{2.51}
$$

The characteristic function of the random sum F is $L_\Xi(-i\omega f)$. As in Campbell's Theorem, f is a nonnegative function for which the expectation exists.

Mathematical discussions also commonly use the probability generating functional. For functions f such that $0 < f(x) \leq 1$, it is defined as the Laplace functional of $-\log f$:

$$
\begin{aligned}
G_\Xi(f) &= L_\Xi(-\log f) \\
&= E\left[\exp\left(\sum_{j=1}^N \log f(X_j)\right)\right] \\
&= E\left[\prod_{j=1}^N f(X_j)\right].
\end{aligned}
\tag{2.52}
$$

The probability generating functional is the analog for finite point processes of the probability generating function for random variables.

The Laplace and probability generating functionals are defined for general finite point processes Ξ, not just PPPs. If Ξ is a PPP with intensity function $\lambda(x)$, then

$$
G_\Xi(f) = \exp\left[\int_{\mathcal{R}} (f(x) - 1)\, \lambda(x)\, dx\right].
\tag{2.53}
$$

Probability generating functionals are used only in Chapter 8.

2.7 Superposition

A very useful property of independent PPPs is that their sum is a PPP. Two PPPs on S are superposed, or summed, if realizations of each are combined into one event. Let Ξ and Υ denote these PPPs, and let their intensities be $\lambda(s)$ and $v(s)$. If $(m, \{x_1, \ldots, x_m\})$ and $(n, \{y_1, \ldots, y_n\})$ are realizations of Ξ and Υ, then the combined event is $(m + n, \{x_1, \ldots, x_m, y_1, \ldots, y_n\})$. Knowledge of which points originated from which realization is assumed lost.

The combined event is probabilistically equivalent to a realization of a PPP whose intensity function is $\lambda(s) + v(s)$. To see this, let $\xi = (r, \{z_1, \ldots, z_r\}) \in \mathcal{E}(\mathcal{R})$ be an event constructed in the manner just described. The partition of this event into an m point realization of Ξ and an $r - m$ point realization of Υ is unknown. Let the sets P_m and its complement P_m^c be such a partition, where $P_m \cup P_m^c = \{z_1, \ldots, z_r\}$. Let $\mathbb{P}_m$ denote the collection of all partitions of size m. There are

$$\binom{r}{m} \equiv \frac{r!}{m!(r-m)!}$$

partitions in $\mathbb{P}_m$. The partitions in $\mathbb{P}_m$ are equally likely, so the likelihood of ξ is the sum over partitions:

$$p(\xi) = \sum_{m=0}^{r} \frac{1}{\binom{r}{m}} \sum_{P_m \in \mathbb{P}_m} p_{\Xi}(m, P_m)\, p_{\Upsilon}\left(r - m, P_m^c\right).$$

Substituting the pdfs using (2.12) and rearranging terms gives

$$p(\xi) = \frac{e^{-\mu}}{r!} \sum_{m=0}^{r} \sum_{P_m \in \mathbb{P}_m} \left(\prod_{z \in P_m} \lambda(z) \prod_{z \in P_m^c} v(z) \right),$$

where $\mu \equiv \int_R (\lambda(s) + v(s))\, ds$. The double sum in the last expression is recognized (after some thought) as an elaborate way to write an r-term product. Thus,

$$p(\xi) = \frac{e^{-\mu}}{r!} \prod_{i=1}^{r} (\lambda(z_i) + v(z_i)) . \tag{2.54}$$

Comparing (2.54) to (2.12) shows that $p(\xi)$ is the pdf of a PPP with intensity function given by $\lambda(s) + v(s)$.

More refined methods that do not rely on partitions show that superposition holds for a countable number of independent PPPs. The intensity of the superposed PPP is the sum of the intensities of the constituent PPPs, provided the sum converges. For details, see [63].

The Central Limit Theorem for sums of random variables has an analog for point processes called the Poisson Limit Theorem: the superposition of a large number of "uniformly sparse" independent point processes converges in distribution to a homogeneous PPP. These point processes need *not* be PPPs. The first statement and proof of this difficult result dates to the mid-twentieth century. For details on $\mathbb{R}^1$, see [62, 92]. The Poisson Limit Theorem also holds in the multidimensional case. For these details, see [15, 40].

Example 2.2 The sum of dispersed unimodal intensities is sometimes unimodal. Consider the intensity function

$$\lambda_c(x, y) = \sum_{i \in \{-1,0,1\}} \sum_{j \in \{-1,0,1\}} c \, \mathcal{N}\left(\begin{bmatrix} x \\ y \end{bmatrix}; \begin{bmatrix} i\,\Delta \\ j\,\Delta \end{bmatrix}, \sigma^2 \begin{bmatrix} 1 & 0 \\ 0 & 1 \end{bmatrix} \right), \qquad (2.55)$$

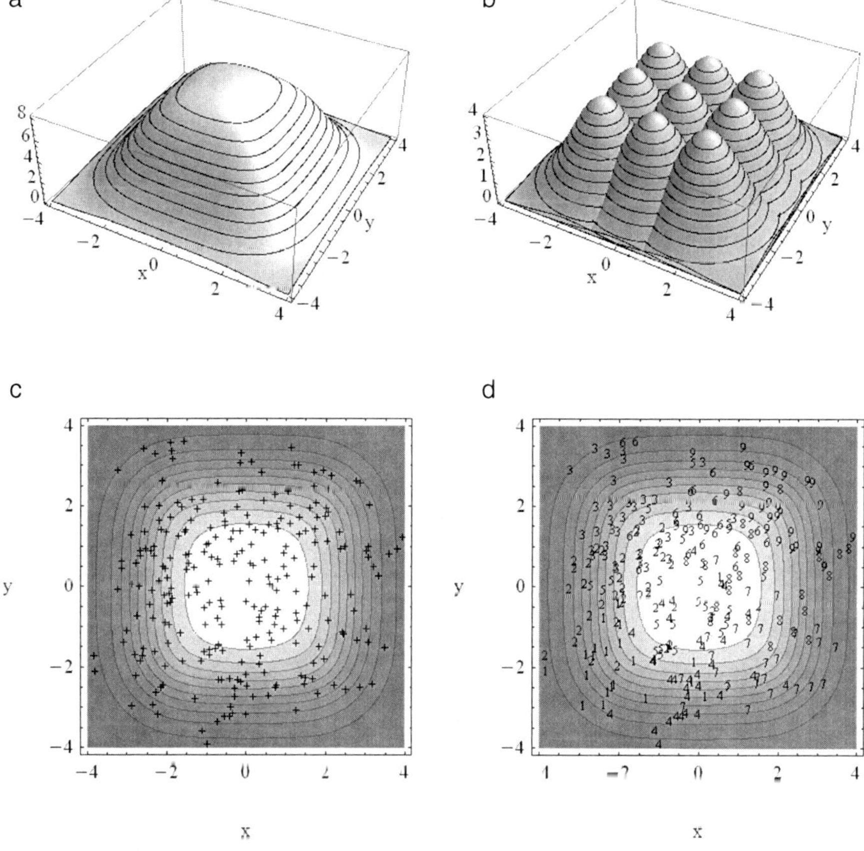

Fig. 2.2 Superposition of an equispaced grid of nine PPPs with circular Gaussian intensities (2.55) of equal weight and spread, $\sigma = 1$. Samples from the PPP components are generated independently and superposed to generate samples from the unimodal flat-topped intensity function

where $(x, y) \in \mathcal{R} \equiv [-4\sigma, 4\sigma]^2$, $\Delta = 1.75$, $\sigma = 1$, and $c = 25$. The nine term sum is unimodal, as is seen in Fig. 2.2a. The terms in the sum are proportional to truncated Gaussian pdfs; they are over-plotted (but not added!) in Fig. 2.2b. The means of the summands range from 24.3925 to 24.9968. The number of samples in the realizations of the nine discrete Poisson variates range from 21 to 39, with an average of 28.89 in this case.

The realizations of the nine component PPPs are shown in Fig. 2.2d using plotting symbols "1" to "9" to identify the originating component. All nine sample sets are combined and recolored green in Fig. 2.2c; the green samples are statistically equivalent to a realization of the intensity function shown in Fig. 2.2a.

2.8 Independent (Bernoulli) Thinning

Thinning is a powerful method for sculpting interesting and practical PPP intensities by reducing the number of points in the realizations. Let $\varXi$ be a PPP on $\mathcal{S}$. For every $x \in \mathcal{S}$, let $1 - \alpha(x)$, $0 \leq \alpha(x) \leq 1$, be the probability that a point located at x is removed, or culled, from any realization that contains it. For the realization $\xi = (n, \{x_1, \ldots, x_n\})$, the point x_j is retained with probability $\alpha(x_j)$ and culled with probability $1 - \alpha(x_j)$. The thinned realization is $\xi_\alpha = (m, \{x'_1, \ldots, x'_m\})$, where $m \leq n$ is the number of points $\{x'_1, \ldots, x'_m\} \subset \{x_1, \ldots, x_n\}$ that *pass* the Bernoulli test. The culled realization $\xi_{1-\alpha}$ is similarly defined. Knowledge of the number n of points in ξ is assumed lost. It is called Bernoulli, or independent, thinning because $\alpha(x)$ depends only on x.

The thinned process is a PPP with intensity function

$$\lambda_\alpha(x) = \alpha(x)\, \lambda(x). \tag{2.56}$$

To see this, consider first the special case that $\alpha(x)$ is constant on $\mathcal{R}$. Let

$$\mu = \int_\mathcal{R} \lambda(x)\, dx, \quad \mu_\alpha = \int_\mathcal{R} \lambda_\alpha(x)\, dx, \quad \beta = \frac{\mu_\alpha}{\mu}.$$

The probability that ξ_α has m points after thinning ξ is

$$\Pr[m \mid n] = \binom{n}{m} \beta^m\, (1 - \beta)^{n-m}, \quad m \leq n.$$

The number of points n in the realization of ξ is unknown, so

$$\Pr[m] = \sum_{n=m}^{\infty} \binom{n}{m} \beta^m (1 - \beta)^{n-m} \Pr[n] \tag{2.57}$$

$$= \sum_{n=m}^{\infty} \frac{n!}{m!(n-m)!} \beta^m (1 - \beta)^{n-m} \frac{\mu^n}{n!} e^{-\mu}$$

$$= \frac{(\beta \mu)^m}{m!} e^{-\mu} \sum_{n=m}^{\infty} \frac{((1 - \beta) \mu)^{n-m}}{(n - m)!}$$

$$= \frac{(\beta \mu)^m}{m!} e^{-\beta \mu} \equiv \frac{\mu_\alpha^m}{m!} e^{-\mu_\alpha}.$$

Thus, from (2.2), the number of points m is Poisson distributed with mean μ_α. The samples $\{x_1', \ldots, x_m'\}$ are clearly i.i.d., and a Bayesian posterior computation shows that their pdf is $\lambda_\alpha(x)/\mu_\alpha = \lambda(x)/\mu$.

The problem is harder if $\alpha(x)$ is not constant. A convincing demonstration in this case goes as follows. Break the set $\mathcal{R}$ into a large number of "small" nonoverlapping cells. Let $\mathcal{R}'$ be one such cell, and let

$$\mu = \int_{\mathcal{R}'} \lambda(x)\,dx, \quad \mu_\alpha = \int_{\mathcal{R}'} \lambda_\alpha(x)\,dx, \quad \beta = \frac{\mu_\alpha}{\mu}.$$

The probability that ξ_α has m points after thinning ξ is, by the preceding argument, Poisson distributed with mean μ_α. The samples $\{x_1', \ldots, x_m'\}$ are i.i.d., and their pdf on $\mathcal{R}'$ is $\lambda_\alpha(x)/\mu_\alpha$. Now extend the intensity function from $\mathcal{R}'$ to all $\mathcal{R}$ by setting it to zero outside the cell. Superposing these cell level PPPs and taking the limit as cell size goes to zero shows that $\lambda_\alpha(x)$ is the intensity function on the full set $\mathcal{R}$. Further details are omitted.

An alternative demonstration exploits the acceptance-rejection method. Generate a realization of the PPP with intensity function $\lambda(x)$ from the homogeneous PPP with intensity function $\Lambda = \max_{x \in \mathcal{R}} \lambda(x)$. Redefine $\mu_\alpha = \int_{\mathcal{R}} \lambda_\alpha(x)\,dx$, and let $|\mathcal{R}| = \int_{\mathcal{R}} dx$. The probability that no points remain in $\mathcal{R}$ after thinning by $\alpha(x)$ is

$$v(\mathcal{R}) = \sum_{n=0}^{\infty} \Pr[(n, \{x_1, \ldots, x_n\})\text{ and all points are thinned}]$$

$$= \sum_{n=0}^{\infty} e^{-\Lambda |\mathcal{R}|} \frac{\Lambda^n |\mathcal{R}|^n}{n!} \left[\int_{\mathcal{R}} \frac{1}{|\mathcal{R}|} \left(1 - \alpha(s) \frac{\lambda(s)}{\Lambda} \right) ds \right]^n$$

$$= e^{-\Lambda |\mathcal{R}|} \sum_{n=0}^{\infty} \frac{\Lambda^n |\mathcal{R}|^n}{n!} \left[1 - \frac{\mu_\alpha}{\Lambda |\mathcal{R}|} \right]^n$$

$$= e^{-\mu_\alpha}.$$

The void probabilities $v(\mathcal{R})$ for a sufficiently large class of "test" sets $\mathcal{R}$ characterize a PPP, a fact whose proof is unfortunately outside the scope of the present book.

(A clean, relatively accessible derivation is given in [136, Theorem 1.2].) Given the result, it is clear that the thinned process is a PPP with intensity function $\lambda_\alpha(x)$.

Example 2.3 Triple Thinning. The truncated and scaled zero-mean Gaussian intensity function on the rectangle $[-2\sigma,\, 2\sigma] \times [-2\sigma,\, 3\sigma]$,

$$\lambda_c(x,\, y) \,=\, c\,\mathcal{N}(x\,;\, 0,\, \sigma^2)\mathcal{N}(y\,;\, 0,\, \sigma^2),$$

is depicted in Fig. 2.3a for $c = 2000$ and $\sigma = 1$. Its mean intensity (i.e., the integral of λ_{2000} over the rectangle) is $\mu_0 = 1862.99$. Sampling the discrete Poisson variate with mean μ_0 gives, in this realization, 1892 points. Boundary conditions are imposed by the thinning functions

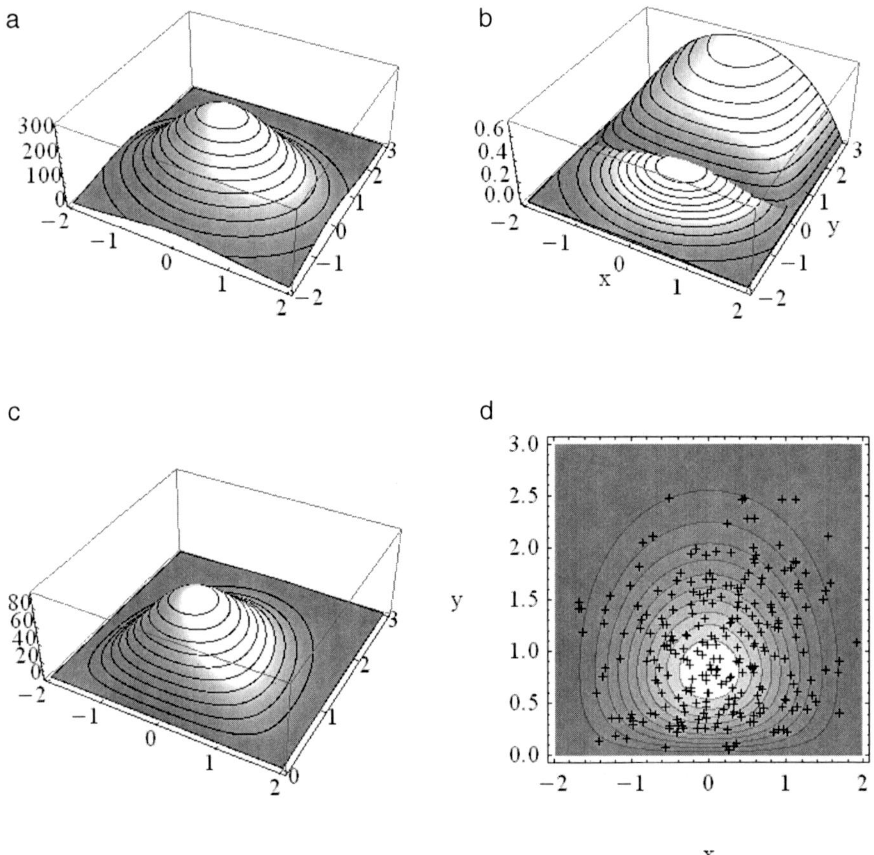

Fig. 2.3 Triply thinning the Gaussian intensity function by (2.58) for $\sigma = 1$ and $c = 2000$ yields samples of an intensity with hard boundaries on three sides

$$\alpha_1(x, y) = 1 - e^{-y} \qquad\qquad\qquad \text{if } y \geq 0$$
$$\alpha_2(x, y) = 1 - e^{x-2} \qquad\qquad\qquad \text{if } x \leq 2 \qquad\qquad (2.58)$$
$$\alpha_3(x, y) = 1 - e^{-x-2} \qquad\qquad\qquad \text{if } x \geq -2,$$

where $\alpha_j(x, y) = 0$ for conditions not specified in (2.58). The overall thinning function, $\alpha_1 \alpha_2 \alpha_3$, is depicted in Fig. 2.3b overlaid on the surface corresponding to λ_1. The intensity of the thinned PPP, namely $\alpha_1 \alpha_2 \alpha_3 \lambda_{2000}$, is nonzero only on the rectangle $[-2\sigma, 2\sigma] \times [0, 3\sigma]$. It is depicted in Fig. 2.3c. Thinning the 1892 points of the realization of λ_{2000} leaves the 264 points depicted in Fig. 2.3d. These 264 points are statistically equivalent to a sample generated directly from the thinned PPP. The mean thinned intensity is 283.19.

2.9 Declarations of Independence

Several properties of PPPs related to independence are surveyed in this section. Independent scattering is discussed first. It is most often used as one of the defining properties of PPPs. Since the two-step generation procedure defines PPPs, it is necessary to obtain it from the procedure. Thinning is the method used here. As mentioned earlier, PPPs are characterized by the form (see Campbell's Theorem) of the characteristic function of random sums. The easy way to see this relies on independent scattering, so this is the second topic.

The nest topic is Poisson's gambit. This is a hopefully not irreverent name for a surprising property of the Poisson distribution when it is used as a prior on the number of Bernoulli trials performed. The last topic speaks of the fact that a finite point process that satisfies independent scattering must have Poisson distributed numbers of points.

2.9.1 Independent Scattering

Independent scattering[3] is a fundamental property of point processes. It may seem somewhat obvious at first glance, but there is a small subtlety in it that deserves respect. The discussion hopefully exposes the subtlety and makes clear the importance of the result. In any event, the independent scattering property is very useful in applications, and often plays a crucial role in determining the mathematical structure of the likelihood function.

Let $\Xi \equiv \Xi(\mathcal{R})$ denote a point process on $\mathcal{R} \subset \mathbb{R}^m$, and let $\xi = (n, \{x_1, \ldots, x_n\})$ be a realization. It is not assumed that Ξ is a PPP. Let $A \subset \mathcal{R}$

---

[3] This name conveys genuine meaning in the point process context, but it seems of fairly recent vintage [84, Section 3.1.2] and [123, p. 33]. It is more commonly called independent increments, which can be confusing because the same name is used for a similar, but different, property of stochastic processes. See Section 2.9.1.

and $B \subset \mathcal{R}$ denote bounded subsets of $\mathcal{R}$. The point processes $\Xi(A)$ and $\Xi(B)$ are obtained by restricting realizations of Ξ to A and B, respectively. Simply put, the points in $\xi(A)$ are the points of ξ that are in $A \cap \mathcal{R}$, and the same for $\xi(B)$. This somewhat obscures the fact that the realizations ξ_A and ξ_B are obtained from the *same* realization ξ. Intuition may suggest that constructing ξ_A and ξ_B from the very same realization ξ will force the point processes $\Xi(A)$ and $\Xi(B)$ to be highly correlated in some sense. Such intuition is in need of refinement, for it is incorrect. This is the subtlety mentioned above.

Let ξ denote an arbitrary realization of a point process $\Xi(A \cup B)$ on the set $A \cup B$. The point process $\Xi(A \cup B)$ is an independent scattering process if

$$p_{\Xi(A \cup B)}(\xi) = p_{\Xi(A)}(\xi_A)\, p_{\Xi(B)}(\xi_B), \tag{2.59}$$

for all disjoint subsets A and B of $\mathcal{R}$, that is, for all subsets such that $A \cap B = \varnothing$. The pdfs in (2.59) are determined by the specific character of the point process, so they are not in general those of a PPP. The product in (2.59) is the reason the property is called *independent* scattering.

A nonhomogeneous multidimensional PPP is an independent scattering point process. To see this it is only necessary to verify that (2.59) holds. Define thinning probability functions, $\alpha(x)$ and $\beta(x)$, by

$$\alpha(x) = \begin{cases} 1, & \text{if } x \in A \\ 0, & \text{if } x \notin A \end{cases}$$

and

$$\beta(x) = \begin{cases} 1, & \text{if } x \in B \\ 0, & \text{if } x \notin B. \end{cases}$$

The point processes $\Xi(A)$ and $\Xi(B)$ are obtained by α-thinning and β-thinning realizations ξ of the PPP $\Xi(A \cup B)$, so they are PPPs. Let $\lambda(x)$ be the intensity function of the PPP $\Xi(A \cup B)$. Let $\xi = (n, \{x_1, \ldots, x_n\})$ be an arbitrary realization of $\Xi(A \cup B)$. The pdf of ξ is, from (2.12),

$$p_{\Xi(A \cup B)}(\xi) = e^{-\int_{A \cup B} \lambda(x)\, dx} \prod_{j=1}^{n} \lambda(x_j). \tag{2.60}$$

Because the points of the α-thinned and β-thinned realizations are on disjoint sets A and B, the realizations $\xi_A = (i, \{y_1, \ldots, y_i\})$ and $\xi_B = (n, \{z_1, \ldots, z_k\})$ are necessarily such that $i + k = n$ and

$$\{y_1, \ldots, y_i\} \cup \{z_1, \ldots, z_k\} = \{x_1, \ldots, x_n\}.$$

Because $\Xi(A)$ and $\Xi(B)$ are PPPs, the pdfs of ξ_A and ξ_B are

$$p_{\Xi(A)}(\xi_A) = e^{-\int_A \lambda(x)\,dx} \prod_{j=1}^{i} \lambda(y_j)$$

$$p_{\Xi(B)}(\xi_B) = e^{-\int_B \lambda(x)\,dx} \prod_{j=1}^{k} \lambda(z_j)\,.$$

The product of these two pdfs is clearly equal to that of (2.60). The key elements of the argument are that the thinned processes are PPPs, and that the thinned realizations are free of overlap when the sets are disjoint. The argument extends easily to any finite number of disjoint sets.

Example 2.4 Likelihood Function for Histogram Data. A fine illustration of the utility of independent scattering is the way it makes the pdf of histogram data easy to determine. Denote the cells of a histogram by $\mathcal{R}_1, \ldots, \mathcal{R}_K$, $K \geq 1$. The cells are assumed disjoint, so $\mathcal{R}_j \cap \mathcal{R}_j = \varnothing$ for $i \neq j$. Histogram data are nonnegative integers that count the number of points of a realization of a point process that fall within the various cells. No record is kept of the locations of the points within any cell. Histogram data are very useful for compressing large volumes of sample (point) data.

Denote the histogram data by $n_{1:K} \equiv \{n_1, \ldots, n_K\}$, where $n_j \geq 0$ is the number of points of the process that lie in $\mathcal{R}_j$. Let the point process Ξ be a PPP, and let $\Xi(\mathcal{R}_j)$ denote the PPP obtained by restricting Ξ to $\mathcal{R}_j$. The intensity function of $\Xi(\mathcal{R}_j)$ is $\int_{\mathcal{R}_j} \lambda(s)\,ds$. The histogram cells are disjoint. By independent scattering, the PPPs $\Xi(\mathcal{R}_1), \ldots, \Xi(\mathcal{R}_K)$ are independent and the pdf of the histogram data is

$$p(n_{1:K}) = \prod_{j=1}^{K} \exp\left(-\int_{\mathcal{R}_j} \lambda(s)\,ds\right) \frac{\left(\int_{\mathcal{R}_j} \lambda(s)\,ds\right)^{n_j}}{n_j!} \tag{2.61}$$

$$- \exp\left(\int_{\mathcal{R}} \lambda(s)\,ds\right) \prod_{j=1}^{K} \frac{\left(\int_{\mathcal{R}_j} \lambda(s)\,ds\right)^{n_j}}{n_j!}, \tag{2.62}$$

where $\mathcal{R} = \mathcal{R}_1 \cup \cdots \cup \mathcal{R}_K \subseteq \mathcal{S}$ is the coverage of the histogram. Estimation problems involving histogram PPP data start with expression (2.62).

Example 2.5 Poisson Distribution Without Independent Scattering. It is possible for a point process to have a Poisson distributed number of points in bounded subsets $\mathcal{R}$, but yet not satisfy the independent scattering property on disjoint sets, that is, it is not a PPP. An interesting example on the unit interval due to L. Shepp is given here (see [40, Appendix]).

Choose the number of points n in the interval [0, 1] with probability $e^{-\lambda} \lambda^n / n!$, where $\lambda > 0$ is the intensity function of a homogeneous PPP on [0, 1]. For $n \neq 3$, let the points be i.i.d., so their cumulative distribution function (CDF) is $F(c_1, \ldots, c_n) = c_1 \cdots c_n$, where $c_j \in [0, 1]$. For $n = 3$, the points are chosen according to the CDF

$$F(c_1, c_2, c_3) = c_1\, c_2\, c_3 + \varepsilon\, (c_1 - c_2)^2 (c_1 - c_3)^2 (c_2 - c_3)^2$$
$$\times\, c_1\, c_2\, c_3\, (1 - c_1)\, (1 - c_2)(1 - c_3). \tag{2.63}$$

The point process has realizations in the event space $\mathcal{E}([0,\ 1])$, but it is not a PPP because of the way the points are sampled for $n = 3$.

For any $c \in [0,\ 1]$, define the random variable

$$X_c(x) = \begin{cases} 1, & \text{if} \quad x < c \\ 0, & \text{if} \quad x \geq c. \end{cases} \tag{2.64}$$

The number of points in a realization of the point process in the interval $[a,\ b]$ conditioned on n points in $[0,\ 1]$ is

$$G_n(a,\ b,\ m) = \Pr\left[\text{exactly } m \text{ points of } \{x_1,\ \ldots,\ x_n\} \text{ are in } [a,\ b]\,\right]. \tag{2.65}$$

Using the functions (2.64),

$$G_n(a,\ b,\ m) = \binom{n}{m} \Pr\left[\{x_1,\ \ldots,\ x_m\} \in [a,\ b]\right]\, \Pr\left[\{x_{m+1},\ \ldots,\ x_n\} \notin [a,\ b]\right]$$
$$= \binom{n}{m} E\left[\prod_{j=1}^{m} \left(X_b(x_j) - X_a(x_j)\right) \prod_{j=m+1}^{n} \left(X_a(x_j) + X_1(x_j) - X_b(x_j)\right)\right].$$
$$\tag{2.66}$$

For $n \neq 3$, the points are i.i.d. conditioned on n, so for all $c_j \in [0,\ 1]$

$$E\left[X_{c_1}(x_1) \cdots X_{c_n}(x_n)\right] = F[c_1,\ \ldots,\ c_n] = c_1 \cdots c_n. \tag{2.67}$$

For $n = 3$, the product in $G_n(a,\ b,\ m)$ expands into a sum of expectations of products of the form (2.67) with c_j equal to one of the three values a, b, or 1. From the definition (2.63), it follows in this case that $F[c_1,\ c_2,\ c_3] = c_1\, c_2\, c_3$. Hence, (2.67) holds for all $n \geq 0$. Substituting this result into (2.66) and manipulating the result in the manner of (2.57) shows that the number of points in the interval $[a,\ b]$ is Poisson distributed with intensity $\lambda(b - a)$.

2.9.2 Poisson's Gambit

A Bernoulli trial is an idealized coin flip. It is any random variable with two outcomes: "success" and "failure." The outcomes are commonly called "heads" and "tails". Obviously, the names attached to the two outcomes are irrelevant here. The probability of heads is p and the probability of tails is $q = 1 - p$. Sequences of Bernoulli trials are typically independent unless stated otherwise.

Denote the numbers of heads and tails observed in a sequence of $n \geq 1$ independent Bernoulli trials by n_h and n_t, respectively. The sequence of Bernoulli trials is performed (conceptually) many times, so the observed numbers n_h and n_t are realizations of random variables, denoted by N_h and N_t, respectively. If exactly n trials are always performed, the random variables N_h and N_t are not independent because of the deterministic constraint

$$N_h + N_t = n.$$

However, if the sequence length n is a realization of a Poisson distributed random variable, denoted by N, then N_h and N_t are *independent* random variables! The randomized constraint

$$N_h + N_t = N$$

holds, but it is not enough to induce any dependence whatever between N_h and N_t.

This property is counterintuitive when first encountered, but it plays an important role in many applications. To give it a name, since one seems to be lacking in the literature, Poisson's gambit[4] is the assumption that the number of Bernoulli trials is Poisson distributed. Poisson's gambit is realistic in many applications, but in others it is only an approximation. The name is somewhat whimsical—it is not used elsewhere in the literature.

Invoking Poisson's gambit, the number N is an integer valued, Poisson distributed random variable with intensity $\lambda > 0$. Sampling N gives the length n of the sequence of Bernoulli trials performed. Then $n = n_h + n_t$, where n_h and n_t are the observed numbers of heads and tails. The random variables N_h and N_t are independent Poisson distributed with mean intensities $p\lambda$ and $(1-p)\lambda$, respectively. To see this, note that the probability of a Poisson distributed number of n Bernoulli trials with outcomes n_h and n_t is

$$\Pr[N = n,\ N_h = n_h,\ N_t = n_t] = \Pr[n]\ \Pr[n_h,\ n_t \mid n]$$

$$= e^{-\lambda}\ \frac{\lambda^n}{n!} \binom{n}{n_h} p^{n_h}\ (1 - p)^{n_t}$$

$$= \left\{ e^{-p\lambda}\ \frac{(p\,\lambda)^{n_h}}{n_h!} \right\} \left\{ e^{-(1-p)\lambda}\ \frac{((1-p)\lambda)^{n_t}}{n_t!} \right\}.$$

$$(2.68)$$

---

[4] A gambit in chess involves sacrifice or risk with hope of gain. The sacrifice here is loss of control over the number of Bernoulli trials, and the gain is independence of the numbers of different outcomes.

The final product is the statement that the number of heads and tails are independent Poisson distributions with the required parameters. For further comments, see, e.g., [52, Section 9.3] or [42, p. 48].

Example 2.6 Independence of Thinned and Culled PPPs. The points of a PPP that are retained and those that are culled during Bernoulli thinning are both PPPs. Their intensities are $p(x)\lambda(x)$ and $(1 - p(x))\lambda(x)$, respectively, where $p(x)$ is the probability that a point at $x \in S$ is retained. Poisson's gambit implies that the numbers of points in these two PPPs are independent. Step 2 of the realization procedure guarantees that the sample points are of the two processes are independent. The thinned and culled PPPs are therefore independent, and superposing them recovers the original PPP, since the intensity function of the superposition is the sum of the component intensities. In other words, splitting a PPP into two parts using Bernoulli thinning, and subsequently merging the parts via superposition recovers the original PPP.

Example 2.7 Coloring Theorem. Replace the Bernoulli trials in Example 2.6 by independent multinomial trials with $k \geq 2$ different outcomes, called "colors" in [63, Chapter 5], with probabilities $\{p_1(x), \ldots, p_k(x)\}$, where

$$p_1(x) + \cdots + p_k(x) = 1.$$

Every point $x \in S$ of a realization of the PPP Ξ with intensity function $\lambda(x)$ is colored according to the outcome of the multinomial trial. For every color j, let Ξ_j denote the point process that corresponds to points of color j. Then Ξ_j is a PPP, and its intensity is

$$\lambda_j(x) = p_j(x)\lambda(x).$$

Poisson's gambit and Step 2 of the realization procedure shows that the PPPs independent. The intensity of their superposition is

$$\sum_{j=1}^{k} \lambda_j(x) = \sum_{j=1}^{k} p_j(x)\lambda(x) = \lambda(x),$$

which is the intensity of the original PPP.

2.9.3 Inevitability of the Poisson Distribution

If an orderly point process satisfies the independent scattering property and the number of points in any bounded set $\mathcal{R}$ is finite and not identically zero (with probability one), then the number of points of the process in a given set $\mathcal{R}$ is necessarily Poisson distributed—the Poisson distribution is inevitable (as Kingman wryly observes). This result shows that if the number points in realizations of the point process is not Poisson distributed for even one set $\mathcal{R}$, then it is not an independent scattering process, and hence not a PPP. To see this, a physics-style argument (due to Kingman [63, pp. 9–10]) is adopted.

Given a set $A \neq \varnothing$ with no "holes", or voids, define the family of sets A_t, $t \geq 0$ by

$$A_t = \cup_{a \in A} \left\{ x \in \mathbb{R}^m : \|x - a\| \leq t \right\},$$

where $\| \cdot \|$ is the usual Euclidean distance. Because A has no voids, the boundary of A_t encloses the boundary of A_s if $t > s$. Let

$$p_n(t) = \Pr[N(A_t) = n]$$

and

$$q_n(t) = \Pr[N(A_t) \leq n],$$

where $N(A_t)$ is the random variable that equals the number of points in a realization that lie in A_t. The point process is orderly, so it is assumed that the function $p_n(t)$ is differentiable. Let

$$\mu(t) \equiv E[N(A_t)].$$

Finding an explicit mathematical form for this expectation is *not* the goal here. The goal is to show that

$$p_n(t) = e^{-\mu(t)} \frac{\mu^n(t)}{n!}.$$

In words, the number $N(A_t)$ is Poisson distributed with parameter $\mu(t)$.

Since $N(A_t)$ increases with increasing t, the function $q_n(t)$ is decreasing. Similarly, $\mu(t)$ is an increasing function. For $h > 0$, the probability that $N(A_t)$ jumps from n to $n + 1$ between t and $t + h$ is $q_n(t) - q_n(t + h) \geq 0$. This is the probability that exactly one point of the realization occurs in the annular region

$$A_t^h = A_{t+h} \setminus A_t.$$

Another way to write this probability uses independent scattering. For sufficiently small $h > 0$, the probability that one point falls in A_t^h is

$$\mu(t + h) - \mu(t) = \Pr\left[N\left(A_t^h \right) = 1 \right] \geq 0.$$

This probability is independent of $N(A_t)$ since $A_t \cap A_t^h = \varnothing$, so

$$q_n(t) - q_n(t+h) = \Pr[N(A_t) = n] \Pr\left[N\left(\mathcal{A}_t^h\right) = 1\right] [\,|\, N(A_t) = n]$$
$$= \Pr[N(A_t) = n] \Pr\left[N\left(\mathcal{A}_t^h\right) = 1\right]$$
$$= p_n(t)\,(\mu(t+h) - \mu(t))\,.$$

Dividing by h and taking the limit as $h \to 0$ gives

$$-\frac{\mathrm{d}q_n(t)}{\mathrm{d}t} = p_n(t)\,\frac{\mathrm{d}\mu(t)}{\mathrm{d}t}\,. \tag{2.69}$$

For $n = 0$, $q_0(t) = p_0(t)$, so (2.69) gives

$$-\frac{\mathrm{d}p_0(t)}{\mathrm{d}t} = p_0(t)\,\frac{\mathrm{d}\mu(t)}{\mathrm{d}t} \quad \Leftrightarrow \quad \frac{\mathrm{d}}{\mathrm{d}t}\,(\mu(t) + \log p_0(t)) = 0\,.$$

Since $p_0(0) = 1$ and $\mu(0) = 0$, it follows that

$$p_0(t) = e^{-\mu(t)}\,. \tag{2.70}$$

For $n \geq 1$, from (2.69),

$$p_{n-1}(t)\,\mu'(t) = \left(-q'_{n-1}(t) - p_n(t)\,\mu'(t)\right) + p_n(t)\,\mu'(t)$$
$$= \left(-q'_{n-1}(t) + q'_n(t)\right) + p_n(t)\,\mu'(t)$$
$$= p'_n(t) + p_n(t)\,\mu'(t)\,,$$

where the last step follows from $p_n(t) = q_n(t) - q_{n-1}(t)$. Multiplying both sides by $e^{\mu(t)}$ and using the product differentiation rule gives

$$\frac{\mathrm{d}}{\mathrm{d}t}\left(p_n(t)\,e^{\mu(t)}\right) = p_{n-1}(t)\,e^{\mu(t)}\,\frac{\mathrm{d}\mu(t)}{\mathrm{d}t}\,.$$

Integrating gives the recursion

$$p_n(t) = e^{-\mu(t)} \int_0^t p_{n-1}(x)\,e^{\mu(x)}\,\frac{\mathrm{d}\mu(x)}{\mathrm{d}x}\,\mathrm{d}x\,. \tag{2.71}$$

Solving the recursion starting with (2.70) gives $p_n(t) = e^{-\mu(t)}\,\mu^n(t)/n!$, the Poisson density (2.4) with mean $\mu(t)$.

The class of sets without voids is a very large class of "test" sets. To see that the Poisson distribution is inevitable for more general sets requires more elaborate theoretical methods. Such methods are conceptually lovely and mathematically rigorous. They confirm but do not deepen the insights provided by the physics-style argument, so they are not presented here.

2.9.4 Connection to Stochastic Processes

The notion of independent increments is defined for stochastic processes. A stochastic process $X(t)$ is a family of random variables indexed by a continuous parameter $t \geq t_0$, where t_0 is an arbitrarily specified starting value. In many problems, t is identified with time. They are widely used in engineering, physics, and finance. A stochastic process is an independent increments process if for every set of ordered time indices $t_0 \leq t_1 < \cdots < t_n$, the n random variables $X(t_1)$, $X(t_2) - X(t_1)$, ..., $X(t_n) - X(t_{n-1})$ are independent. The differences $X(t_j) - X(t_{j-1})$ are called increments.

There are two different kinds of independent increments stochastic process, namely, the Poisson process and the Wiener process. Independent increments stochastic processes are linear combinations of these two processes [39, Chapter 6].

In the univariate case, the Poisson process is the counting process, often denoted by $\{N(t) : t \geq t_0\}$, of the points of a PPP with intensity $\lambda(t)$. The process $N(t)$ counts the number of points of a PPP realization in the interval $[t_0, t)$. The sample paths of $N(t)$ are therefore piecewise constant and jump in value by $+1$ at the locations of the points of the PPP realization. The CDF of the time interval τ between successive points (the interarrival time) of the PPP is

$$
\begin{aligned}
F_{t_{j-1}}[\tau] &= \Pr\left[\text{next point after } t_{j-1} \text{ is } \leq \tau + t_{j-1}\right] \\
&= 1 - \Pr\left[\text{next point after } t_{j-1} \text{ is } > \tau + t_{j-1}\right] \\
&= 1 - \Pr\left[N(\tau + t_{j-1}) - N(t_{j-1}) = 0\right] \\
&= 1 - e^{-(\Lambda(\tau + t_{j-1}) - \Lambda(t_{j-1}))},
\end{aligned}
\tag{2.72}
$$

where

$$
\Lambda(t) = \int_{t_0}^{t} \lambda(\tau)\, d\tau, \quad t \geq t_0.
$$

Differentiating (2.72) with respect to τ gives the pdf of interarrival times as

$$
p_{t_{j-1}}(\tau) = \lambda\left(\tau + t_{j-1}\right) e^{-(\Lambda(t_j) - \Lambda(t_{j-1}))}.
$$

The interarrival times are identically exponentially distributed if the PPP is homogeneous. Explicitly, for $\lambda(t) \equiv \lambda_0$,

$$
p_{t_{j-1}}(\tau) \equiv p_0(\tau) = \lambda_0\, e^{-\lambda_0 \tau}.
$$

Because of independent scattering property of PPPs, the interarrival times are also independent in this case.

In contrast to the discontinuous sample paths of the Poisson process, the sample paths of the Wiener process are continuous with probability one. For Wiener processes, the random variable $X(t_1)$ is zero mean Gaussian distributed with variance

$t_1 \Sigma$, where Σ is a positive definite matrix, and the increments $X(t_j) - X(t_{j-1})$ are zero mean Gaussian distributed with variances $(t_j - t_{j-1}) \Sigma$. The interval between zero crossings, or more generally between level crossings, of the sample paths of one dimensional Wiener processes is discussed in [93, Section 14.7] and also in [101].

2.10 Nonlinear Transformations

An important property of PPPs is that they are still PPPs after undergoing a *deterministic* nonlinear transformation. The invariance of PPPs under nonlinear mapping is important in many applications.

Let the function $f : \mathcal{S} \to \mathcal{T}$ be given, where $\mathcal{S} \subset \mathbb{R}^m$ and $\mathcal{T} \subset \mathbb{R}^\kappa$, $\kappa \geq 1$. The PPP, say $\varXi$, is transformed, or mapped, by f from $\mathcal{S}$ to $\mathcal{T}$ by mapping the realization $\xi = (n, \{x_1, \ldots, x_n\})$ to the realization $f(\xi) \equiv (n, \{f(x_1), \ldots, f(x_n)\})$. The transformed process is denoted $f(\varXi)$, and it takes realizations in the event space $\mathcal{E}(f(\mathcal{S}))$, where $f(\mathcal{S}) \equiv \{t \in \mathcal{T} : t = f(s) \text{ for some } s \in \mathcal{S}\} \subset \mathcal{T}$.

For a broad class of functions f, the transformed process is a PPP. To see this when f is a change of variables, $y = f(x)$, note that

$$\int_{\mathcal{R}} \lambda(x)\, dx = \int_{f(\mathcal{R})} \lambda\left(f^{-1}(y)\right) \left| \frac{\partial f^{-1}(y)}{\partial y} \right| dy , \qquad (2.73)$$

where $|\partial f^{-1}/\partial y|$ is the determinant of the Jacobian of the inverse of the change of variables. Since (2.73) holds for all bounded subsets $\mathcal{R}$, the intensity function of $f(\varXi)$ is

$$\nu(y) = \lambda\left(f^{-1}(y)\right) \left| \frac{\partial f^{-1}(y)}{\partial y} \right| . \qquad (2.74)$$

Orthogonal coordinate transformations are especially nice since the Jacobian is identically one.

Example 2.8 Change of Variables. From (2.74) it is a straightforward calculation to see that the linear transformation $y = Ax + b$, where the matrix $A \in \mathbb{R}^{m \times m}$ is invertible, transforms the PPP with intensity function $\lambda(x)$ into the PPP with intensity function

$$\nu(y) = \frac{1}{|A|} \lambda\left(A^{-1}(y - b)\right) , \qquad (2.75)$$

where $|A|$ is the determinant of A. What if A is singular?

Example 2.9 Mapping Nonhomogeneous to Homogeneous PPPs. On the real line, every nonhomogeneous PPP can be transformed to a homogeneous PPP

[100, Chapter 4]. Suppose that Ξ is a PPP with intensity function $\lambda(x) > 0$ for all $x \in \mathcal{S} \equiv \mathbb{R}^1$, and let

$$y = f(x) = \int_0^x \lambda(t)\,dt \quad \text{for} \quad -\infty < x < \infty. \tag{2.76}$$

The point process $f(\Xi)$ is a PPP with intensity one. To see this, use (2.74) to obtain

$$v(y) = \frac{\lambda\left(f^{-1}(y)\right)}{|\partial f(x)/\partial x|} = \frac{\lambda(x)}{\lambda(x)} = 1\,,$$

where the chain rule is used to show that $|\partial f^{-1}(x)/\partial x| = 1/|\partial f(x)/\partial x|$. An alternative, but more direct way, to see the same thing is to observe that since f is monotone, its inverse exists and the mean number of points in any bounded interval $[a, b]$ is

$$\int_{f^{-1}(a)}^{f^{-1}(b)} df(x) = \int_a^b dy = b - a\,. \tag{2.77}$$

Therefore, $f(\Xi)$ is homogeneous with intensity function $v(y) \equiv 1$. Obvious modifications are needed to make this method work for $\lambda(y) \geq 0$.

A scalar multiple of the mapping (2.76) is used in the well known algorithm for generating i.i.d. samples of a one dimensional random variable via the inverse cumulative density function. The transformation fails for $\mathbb{R}^m$, $m \geq 2$, because the inverse function is a "one to many" mapping. For the same reason, nonhomogeneous PPPs on spaces of dimension more than two do not transform to homogeneous ones of the same dimension.

Transformations may alter all the statistical properties of the original PPP, not just the PPP intensity function. For instance, in Example 2.9, because $f(\Xi)$ is a homogeneous PPP, the interval lengths between successive points of $f(\Xi)$ are independent. (see Section 2.9.4.) However, the intervals between successive points of the original nonhomogeneous PPP Ξ are not independent [63, p. 51]. In practice, it is necessary to understand how the transformation affects all the statistical properties deemed important in the application.

An important class of "many to one" mappings are the projections π from $\mathbb{R}^m$ to $\mathbb{R}^\kappa$, where $\kappa \leq m$. Let π map the point $x = (v_1, \ldots, v_m) \in \mathbb{R}^m$ to the point $y = \pi(x) = (v_1, \ldots, v_\kappa) \in \mathbb{R}^\kappa$. The set of all $x \in \mathbb{R}^m$ that map to the point y is $\pi^{-1}(y)$. This set is a continuous manifold in $\mathbb{R}^m$. Explicitly,

$$\pi^{-1}(y) = \{(v_1, \ldots, v_\kappa, v_{\kappa+1}, \ldots, v_m) : v_{\kappa+1} \in \mathbb{R}, \ldots, v_m \in \mathbb{R}\}. \tag{2.78}$$

Integrating over the manifold $\pi^{-1}(y)$ gives the intensity function

$$v(v_1, \ldots, v_\kappa) = \int_{\mathcal{R}} \cdots \int_{\mathcal{R}} \lambda(v_1, \ldots, v_\kappa, v_{\kappa+1}, \ldots, v_m) \, dv_{\kappa+1} \cdots dv_m.$$

$$(2.79)$$

This is the intensity function of a PPP on $\mathbb{R}^\kappa$ denoted by $\pi(\Xi)$.

That the projection of a PPP is still a PPP is an instance of a general nonlinear mapping property. The nonlinear mappings $y = f(x)$ for which the result holds are those for which the sets

$$\mathcal{M}(y) \equiv \left\{ f^{-1}(y) : y \in \mathbb{R}^\kappa \right\} \subset \mathbb{R}^m \qquad (2.80)$$

are all commensurate, that is, all have the same intrinsic dimension. For these functions, if Ξ is a PPP, then so is $f(\Xi)$. The intensity function of $f(\Xi)$ is

$$v(x) = \iint_{\mathcal{M}(y)} \lambda\left(f^{-1}(y)\right) \, d\mathcal{M}(y), \qquad (2.81)$$

where $d\mathcal{M}(y)$ is the differential in the tangent space at the point $f^{-1}(y)$ of the set $\mathcal{M}(y)$. The special case of projection mappings provides the basic intuitive insight into the nonlinear mapping property of PPPs. To see that the result holds requires a more careful and mathematically subtle analysis than is deemed appropriate here. See [63, Section 2.3] for further details.

In practice, the sets $\mathcal{M}(y)$ are commensurate for most nonlinear mappings. For example, it is easy to see that the projections have this property. However, some nonlinear functions do not. As the next example shows, the problem with forbidden mappings is that they lead to "intensities" that are generalized functions.

Example 2.10 A Forbidden Nonlinear Mapping. The sets $\mathcal{M}(y)$ of the function $f : \mathbb{R}^2 \to \mathbb{R}^1$ defined by

$$y = f(x_1, x_2) = \begin{cases} 0, & \text{if } x_1^2 + x_2^2 < 1 \\ \sqrt{x_1^2 + x_2^2} - 1, & \text{if } x_1^2 + x_2^2 \geq 1 \end{cases}$$

are not commensurate for all y. Clearly

$$\mathcal{M}(0) = \left\{ (x_1, x_2) : x_1^2 + x_2^2 \leq 1 \right\} \subset \mathbb{R}^2$$

is a disc of radius one and, for $y > 0$,

$$\mathcal{M}(y) = \{ ((y+1)\cos\theta, (y+1)\sin\theta) : 0 \leq \theta < 2\pi \} \subset \mathbb{R}^2$$

is a circle of radius $y + 1$. The intrinsic dimension of $f^{-1}(0)$ is two, and that of $f^{-1}(y)$ for $y > 0$ is one. Assume that Ξ is a PPP with intensity one on $\mathbb{R}^2$. Then, integrating over these sets gives the intensity

$$v(0) = \iint\limits_{\mathcal{M}(0)} 1 \, dx_1 dx_2 = \pi$$

and

$$v(y) = \int\limits_{\mathcal{M}(y)} 1 \, d\theta = 2\pi(y+1), \quad y > 0.$$

This gives

$$v(y) = \pi \, \delta(y) + 2\pi(y+1), \quad y \geq 0,$$

where $\delta(y)$ is the Dirac delta function.

Example 2.11 Polar Coordinate Projections. The change of variables from Cartesian to polar coordinates in the plane, given by

$$
\begin{aligned}
(y_1, y_2) &= f(x_1, x_2) \\
&\equiv \left(\left(x_1^2 + x_2^2\right)^{1/2}, \ \arctan(x_1, x_2) \right),
\end{aligned}
$$

maps a PPP with intensity function $\lambda(x_1, x_2)$ on $\mathbb{R}^2$ to a PPP with intensity function

$$v(y_1, y_2) = y_1 \, \lambda(y_1 \cos y_2, \ y_1 \sin y_2)$$

on the semi-infinite strip

$$\{(y_1, y_2) : y_1 > 0, \ 0 \leq y_2 < 2\pi\}. \tag{2.82}$$

If $\lambda(x_1, x_2) \equiv 1$, then $v(y_1, y_2) = y_1$. From (2.79), the projection onto the range y_1 gives a PPP on $[0, \infty) \subset \mathbb{R}^1$ with intensity function $v(y_1) = 2\pi \, y_1$, and the projection onto the angle y_2 is of infinite intensity on $[0, 2\pi]$. Alternatively, if $\lambda(x_1, x_2) = \left(x_1^2 + x_2^2\right)^{-1/2}$, then $v(y_1, y_2) \equiv 1$. The projection onto range is $v(y_1) = 2\pi$; the projection onto angle is ∞.

Historical Note. Example 2.11 is the two dimensional (cylindrical propagation) version of Olber's famous paradox (1823) in astronomy. It asks, "Why is the sky dark at night?" The argument is that if star locations form a homogeneous PPP in $\mathbb{R}^3$, at the time a seemingly reasonable model for stellar distributions, then an easy calculation shows that the polar projection onto the unit sphere is a PPP with infinite intensity. If stellar intensity falls off as the inverse square of distance (due to spherical propagation), another easy calculation shows that the polar projection still has infinite intensity. Resolving the paradox (e.g., by assuming the universe is bounded) is evidently a nontrivial exercise requiring a careful study of the structure of the universe. It is left as an exercise for the interested reader.

2.11 Stochastic Transformations

Target motion modeling and measurement are both important in many applications. Suppose the targets (i.e., the points) of a PPP realization on the space S at time t_{k-1} move to another state in S at time t_k according to a Markovian transition probability function. The point process that comprises the targets after they transition is equivalent to a realization of a PPP. The intensity function of the transitioned PPP is given in Section 2.11.1 in terms of the initial target intensity function and the transition function.

Similarly, if the errors in point measurements are distributed according to a specified probability density function conditioned on target state, then the point process comprising the measured points is a PPP on the measurement space, denoted T. The intensity function of this measurement process is given in Section 2.11.2 in terms of the target intensity and the measurement conditional pdf.

The nice thing about both these results is that they hold for nonlinear target and measurement models [118, 119]. Formulated as an input-output relationship, the input is a target PPP on the state space S, while the output is a PPP on either the target space S or the measurement space T. In this sense, the transition and measurement processes are very similar.

2.11.1 Transition Processes

A PPP that undergoes a Markovian transition remains a PPP. Let Ψ be the transition pdf, so that the likelihood that the point x in the state space S transforms to the point $y \in S$ is $\Psi(y\,|\,x)$. Let Ξ be the PPP on S with intensity function $\lambda(s)$, and let $\xi = (m, \{x_1, \ldots, x_m\})$ be a realization of Ξ. After transitioning the constituent points, this realization is $\eta \equiv (m, \{y_1, \ldots, y_m\})$, where y_j is a realization of the pdf $\Psi(\cdot\,|\,x_j)$, $j = 1, \ldots, m$. The realizations $\{y_j\}$ are independent. The transition process, denoted by $\Psi(\Xi)$, is a PPP on S with intensity function

$$v(y) = \int_S \Psi(y\,|\,x)\,\lambda(x)\,\mathrm{d}x\,. \tag{2.83}$$

To see this, let $\mathcal{R}$ be any bounded subset of S. Let $\mu = \int_{\mathcal{R}} \lambda(s)\,\mathrm{d}s$ and observe that the likelihood of the transition event η is, by construction,

$$
p(\eta) = \int_{\mathcal{R}} \cdots \int_{\mathcal{R}} \left(\prod_{j=1}^{m} \Psi(y_j \mid x_j) \right) p_{\Xi}(m, \{x_1, \ldots, x_m\}) \, dx_1 \cdots dx_m
$$

$$
= \frac{e^{-\mu}}{m!} \int_{\mathcal{R}} \cdots \int_{\mathcal{R}} \left(\prod_{j=1}^{m} \Psi(y_j \mid x_j) \right) \left(\prod_{j=1}^{m} \lambda(x_j) \right) dx_1 \cdots dx_m
$$

$$
= \frac{e^{-\mu}}{m!} \prod_{j=1}^{m} \int_{\mathcal{R}} \Psi(y_j \mid x_j) \, \lambda(x_j) \, dx_j \, .
$$

Substituting (2.83) gives

$$
p(\eta) = \frac{e^{-\mu}}{m!} \prod_{j=1}^{m} \nu(y_j) \, . \tag{2.84}
$$

Since

$$
\int_{\mathcal{R}} \nu(y) \, dy = \int_{\mathcal{R}} \int_{\mathcal{R}} \Psi(y \mid x) \, \lambda(x) \, dx \, dy
$$

$$
= \int_{\mathcal{R}} \lambda(x) \, dx = \mu \, , \tag{2.85}
$$

it follows from (2.12) that the transition Poisson process $\Psi(\Xi)$ is also a PPP.

2.11.2 Measurement Processes

The transition process result generalizes to sensor measurement processes. A sensor system, comprising a sensor together with a signal processing suite, produces target measurements. These measurements depend on the intimate details of the sensor system and on the state of the target. The specific details of the sensor system, the environment in which it is used, and the target are all built into a crucially important function called the sensor conditional pdf. This conditional pdf is assumed sufficiently accurate for the application at hand. In practice, there is almost inevitably some mismatch between the theoretical model of the sensor pdf and that of the real sensor, so the fidelity of the pdf model must be carefully examined in each application.

Let the pdf of an arbitrary measurement z conditioned on target state x be $\ell(z \mid x)$. This function includes the notion of measurement error. For example, a common nonlinear measurement equation with additive error is

$$
z = h(x) + w \, ,
$$

where $h(x)$ is the measurement the sensor produces of a target at x in the absence of noise, and the error w is a zero mean Gaussian distributed with covariance matrix Σ. The conditional pdf form of the very same equation is $\mathcal{N}(z \mid h(x),\ \Sigma)$. The pdf form is general and not limited to additive noise, so it is used here. Because $\ell(z \mid x)$ is a pdf,

$$\int_{\mathcal{T}} \ell(y \mid x)\, dy = 1$$

for every $x \in \mathcal{S}$.

Now, as in the previous section, let $\xi = (m, \{x_1,\ \ldots,\ x_m\})$ be the PPP realization and $\lambda(x)$ the PPP intensity function. Each point x_j is observed by a sensor. The sensor generates a measurement $z_j \in \mathcal{T} \equiv \mathbb{R}^\kappa$, $\kappa \geq 1$ for the target x_j. The pdf of this measurement is $\ell(y \mid x)$. In words, $\ell(z_j \mid x_j)$ is the pdf of z_j conditioned on x_j. Let $\eta = (m, \{z_1,\ \ldots,\ z_m\})$. Then η is a realization of a PPP defined on the range $\mathcal{T}$ of the pdf ℓ. To see this, it is only necessary to follow the same reasoning used to establish (2.83). The intensity function of this PPP is

$$v(y) = \int_{\mathcal{S}} \ell(y \mid x)\, \lambda(x)\, dx\,, \quad y \in \mathcal{T}\,. \tag{2.86}$$

The PPP $\ell(\Xi)$ is called a "measurement" process because it includes the effects of measurement errors. It is also an appropriate name for many applications, including tracking. (It is called a translated process in [119, Chapter 3].)

Example 2.12 PPP Target Modeling. This example is multi-purpose. At the simplest level, it is merely an example of a measurement process. Another purpose is described shortly. For concreteness, the example is presented in terms of an active sonar sensor. Such sensors generate a measurement of target location by transmitting a "ping" and detecting the same ping after it reflects off a target, e.g., a ship. The sensor estimates target direction θ from the arrival angle of the reflected ping, and it estimates range r from the travel time difference between the transmitted and reflected ping. In two dimensions, target measurements are range, $r = (x^2 + y^2)^{1/2}$, and angle, $\theta = \arctan(x,\ y)$. In the notation above,

$$h(x,\ y) = \begin{bmatrix} (x^2 + y^2)^{1/2} \\ \arctan(x,\ y) \end{bmatrix}. \tag{2.87}$$

The errors in these measurements are assumed to be additive zero mean Gaussian distributed with variances σ_r^2 and σ_θ^2, respectively. The measurement pdf conditioned on target state is therefore

$$\ell(r,\ \theta \mid x,\ y) = \mathcal{N}\left(r\,;\ (x^2 + y^2)^{1/2},\ \sigma_r^2\right) \mathcal{N}\left(\theta\,;\ \arctan(x,\ y),\ \sigma_\theta^2\right). \tag{2.88}$$

Now consider a stationary target modeled as a PPP with intensity function

$$\lambda_c(x,\, y) = c\, \mathcal{N}\left(x;\, x_0,\, \sigma_x^2\right)\, \mathcal{N}\left(y;\, y_0,\, \sigma_y^2\right), \qquad (2.89)$$

where $c = 200$, $x_0 = 6$, and $y_0 = 0$.

The other purpose of this example is to ask–but not to answer–a question: what meaning, if any, can be assigned to a PPP model for physical targets? If $\lambda_c(x,\, y)$ were an a priori pdf, the target model would be interpreted in a standard Bayesian manner. However, PPP intensity function is not a pdf. This important question is answered in Chapter 6. Given the PPP target model, the predicted measurement intensity function is, from (2.86),

$$v(r,\, \theta) = \int_{-\infty}^{\infty} \int_{-\infty}^{\infty} \ell(r,\, \theta \mid x,\, y)\lambda_c(x,\, y)\, \mathrm{d}x\, \mathrm{d}y\,. \qquad (2.90)$$

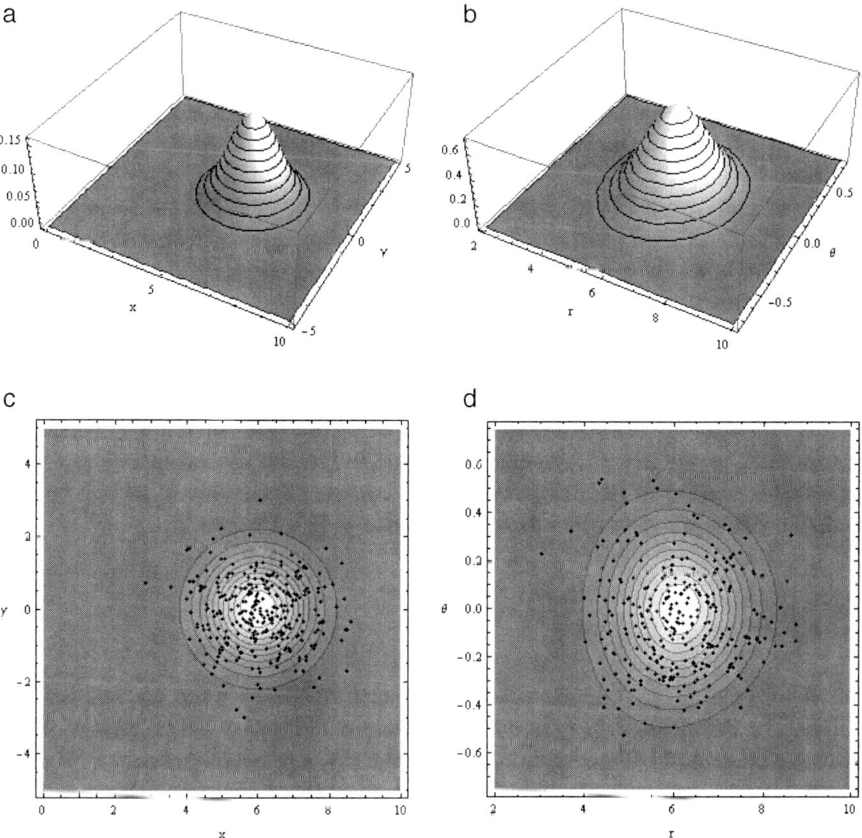

Fig. 2.4 The predicted measurement PPP intensity function in polar coordinates of a Gaussian shaped PPP intensity function in the x-y plane: $\sigma_x = \sigma_y = 1$, $\sigma_r = 0.1$, $\sigma_\theta = 0.15$ (radians), and $c = 200$, $x_0 = 6$, $y_0 = 0$

Figure 2.4a, b give the intensities (2.89) and (2.90), respectively. A realization of the PPP with intensity function $\lambda_c(x, y)$ generated by the two step procedure is given in Fig. 2.4c. Randomly perturbing each of these samples gives the realization in Fig. 2.4d. The predicted intensity $v(r, \theta)$ is nearly Gaussian in the r-θ plane. If the likelihood function (2.88) is truncated to the semi-infinite strip (2.82), the predicted intensity (2.90) is also restricted to the semi-infinite strip.

2.12 PPPs on Other Spaces

Defining PPPs on state spaces other than $\mathbb{R}^m$ enables them to model more complex phenomena. The two spaces considered in more detail in this section are discrete spaces and discrete-continuous spaces. Both are natural extensions of the underlying PPP idea set. PPPs are defined on the discrete space of countably infinite isolated points (e.g., lattices) in [17, Problem 2.4.3]. PPPs on discrete spaces are discussed in Section 2.12.1.

PPPs are defined on a discrete-continuous augmented space in Section 2.12.2. These augmented spaces are used in Chapter 6 for joint detection and tracking. The augmented space is $\mathcal{S}^+ \equiv \mathcal{S} \cup \phi$, where ϕ is an arbitrary point not in $\mathcal{S}$. Augmented spaces have been used for many years for theoretical purposes, but are not so often used in applications. The first use of $\mathcal{S}^+$ in a tracking application seems to be due to Kopec [64] in 1986.

It is straightforward to see from the discussion below that S is easily augmented with any finite or countable number of discrete points. Multiply augmented spaces are potentially useful provided the discrete points are meaningfully interpreted in the application.

PPPs are defined on locally compact, separable, Hausdorff spaces in [79, p. 1] and [57, p. 4]. Concrete examples of this general space include the spaces $\mathbb{R}^n$ and the discrete and discrete-continuous spaces. This book is not the right place to delve further into topological details (i.e., compact neighborhoods, separability, Hausdorff spaces, etc.), except to say that separability implies that the general space has at most a countable number of isolated points. A more relaxed discussion of what is needed to define PPPs on spaces other than $\mathbb{R}^m$ is found in [63, Chapter 2].

2.12.1 Discrete Spaces

Let $\Phi = \{\phi_1, \phi_2, \ldots\}$ denote a finite or countably infinite set of discrete isolated points. The definition in [17] is for homogeneous PPPs on a lattice, that is, on an equi-spaced grid of isolated points. More generally, a nonhomogeneous PPP Ξ on a countable discrete space is defined as a sequence of independent Poisson random variables $\{N_1, N_2, \ldots\}$ with (dimensionless) parameter vector

$$\lambda = \{\lambda_1, \lambda_2, \ldots\}.$$

The pdf of N_j is, from (2.4),

$$p_{N_j}(n_j) = e^{-\lambda_j} \frac{\lambda_j^{n_j}}{n_j!}, \quad n_j \geq 0. \tag{2.91}$$

The intensity of Ξ on the discrete space Φ is defined to be the intensity vector λ. With this definition, realizations of Ξ on a specified finite subset $\mathcal{R} \subset \Phi$ are generated by sampling independently each of the Poisson variates in $\mathcal{R}$. The immediate advantage of this definition is that it is very clear that PPP realizations have repeated points in $\mathcal{R}$, so it is not orderly. This contrasts sharply to PPPs on continuous spaces.

An essentially equivalent definition is consistent with the two-step generation procedure for defining PPPS on continuous spaces. Let

$$\mu(\mathcal{R}) = \sum_{j \in \mathcal{R}} \lambda_j \, .$$

In Step 1, the total number of samples n is drawn from the Poisson random variable with parameter $\mu(\mathcal{R})$. In Step 2, these n samples, denoted by ϕ_{x_j}, are i.i.d. draws from the multinomial distribution with pdf

$$\left\{ \frac{\lambda_j}{\mu(\mathcal{R})} : j \in \mathcal{R} \right\} \, .$$

The integers x_j range over the set of indices of the discrete points in $\mathcal{R}$, but they are otherwise unrestricted. The PPP realization is

$$\xi = (n, \{\phi_{x_1}, \ldots, \phi_{x_n}\}).$$

Nothing prevents the same discrete point, say $\phi_j \in \mathcal{R}$, from occurring more than once in the list $\{\phi_{x_1}, \ldots, \phi_{x_n}\}$; that is, repeated samples of the points in $\mathcal{R}$ are permitted. The number n_j of occurrences of $\phi_j \in \mathcal{R}$ as a point of the PPP realization ξ is a Poisson distributed random variable with parameter λ_j and pdf (2.91). Because of Poisson's gambit, these Poisson variates are independent. The two definitions are therefore equivalent.

The event space of PPPs on Φ is

$$\mathcal{E}(\mathcal{R}) = \{(0, \varnothing)\} \cup_{n=1}^{\infty} \left\{ (n, \{\phi_{x_1}, \ldots, \phi_{x_n}\}) : \phi_{x_j} \in \mathcal{R}, \ j = 1, \ldots, n \right\}. \tag{2.92}$$

Except for the small change in notation that highlights the indices x_j, it is identical to (2.1). The pdf of the unordered realization ξ is

$$p_{\Xi}(\xi) = e^{-\sum_{j \in \mathcal{R}} \lambda_j} \prod_{j \in \mathcal{R}} \lambda_{x_j}. \tag{2.93}$$

This is the discrete space analog of the continuous space expression (2.12). The expectation operator is changed only in that integrals are everywhere replaced by sums over the discrete points of $\mathcal{R} \subset \Phi$. The notions of superposition and thinning are also unchanged.

The intensity functions of transition and measurement processes are similar to (2.83) and (2.86), but are modified to accommodate discrete spaces. The transition pdf $\Psi(\phi_j \mid \phi_i)$ is now a transition matrix whose (i, j)-entry is the probability that the discrete state ϕ_i maps to the discrete state ϕ_j. The intensity of the transition process $\Psi(\Xi)$ is

$$v(\phi_j) = \sum_{\phi_i \in \Phi} \Psi(\phi_j \mid \phi_i) \, \lambda(\phi_i) \,, \tag{2.94}$$

where the vector λ is the intensity vector of Ξ.

Measurement processes are a little different from transition processes because the conditioning is more general. Let the point measurement be $\phi_j \in \Phi$. It is desirable (see, e.g., Section 5.2) to define the conditioning variable to take values in either discrete or continuous spaces, or both. For the conditioning variable taking values x in the continuous space $\mathcal{S}$, the measurement pdf $\ell(\phi_j \mid x)$ is probability of obtaining the measurement ϕ_j given that the underlying state is $x \in \mathcal{S}$. The measurement intensity vector is therefore

$$v(\phi_j) = \int_{\mathcal{S}} \ell(\phi_j \mid x) \, \lambda(x) \, \mathrm{d}x \,, \tag{2.95}$$

where $\lambda(x)$ is the intensity function of a PPP, say Υ, on the state space $\mathcal{S}$. If the conditioning variable takes values u in a discrete space $\mathcal{U}$, the pdf $\ell(\phi_j \mid u)$ is the probability of ϕ_j given $u \in \mathcal{U}$ and the measurement intensity vector is

$$v(\phi_j) = \sum_{u \in \mathcal{U}} \ell(\phi_j \mid u) \, \lambda(u), \tag{2.96}$$

where in this case $\lambda(u)$ is the intensity vector of the discrete PPP defined on $\mathcal{U}$. The discrete-continuous case is discussed in the next section.

Example 2.13 Histograms. The cells $\{\mathcal{R}_j\}$ of a histogram are probably the most natural example of a set of discrete isolated points. Consider a PPP Ξ defined on the underlying continuous space in which the histogram cells reside. Aggregating, or quantizing, the i.i.d. points of realizations of Ξ into the nonoverlapping cells $\{\mathcal{R}_j\}$ and reporting only the total counts in each cell yields a realization of a PPP on a discrete space with points $\phi_j \equiv \mathcal{R}_j$. The intensity vector of this discrete PPP, call it Ξ_H, are

$$\lambda_j = \int_{\mathcal{R}_j} \lambda_c(s) \, \mathrm{d}s \,,$$

where $\lambda_c(s)$ is the intensity function of Ξ. By the independent scattering, since the histogram cells $\{\mathcal{R}_j\}$ are disjoint, the number of elements in cell $\mathcal{R}_j$ is Poisson distributed with parameter λ_j. The fact that the points ϕ_j are, or can be, repeated in realizations of the discrete PPP Ξ_H hardly needs saying.

Concrete examples of discrete spaces occur in emission and transmission tomography. In these examples, the points in Φ correspond to the individual detectors in a detector array, and the number of occurrences of ϕ_j in a realization is the number of detected photons (or other particle) in the j-th detector. These topics are discussed in Chapter 5.

2.12.2 Discrete-Continuous Spaces

This subsection begins with a discussion of the "augmented" space used in Chapter 6 for joint detection and tracking. It concludes with an example showing the relationship between multiply-augmented spaces and non-orderly PPPs.

The one point augmented space is $\mathcal{S}^+ \equiv \mathcal{S} \cup \phi$, where ϕ is a discrete (isolated) point not in $\mathcal{S}$. As is seen shortly, a PPP on $\mathcal{S}^+$ is not orderly because repeated points ϕ occur with nonzero probability.

Several straightforward modifications are needed for PPPs on $\mathcal{S}^+$. The intensity is defined for all $s \in \mathcal{S}^+$. It is an intensity function on $\mathcal{S}$, and hence orderly on $\mathcal{S}$; however, it is not orderly on the full space $\mathcal{S}^+$. The number $\lambda(\phi)$ is a dimensionless quantity, unlike $\lambda(s)$ for $s \in \mathcal{S}$. The bounded sets of $\mathcal{S}^+$ are $\mathcal{R}$ and $\mathcal{R}^+ \equiv \mathcal{R} \cup \phi$, where $\mathcal{R}$ is a bounded subset of $\mathcal{S}$. Integrals of $\lambda(s)$ over bounded subsets of $\mathcal{S}^+$ must be finite; thus, the requirement (2.2) holds and is supplemented by the discrete-continuous integral

$$0 \le \int_{\mathcal{R}^+} \lambda(s)\, ds$$
$$\equiv \lambda(\phi) + \int_{\mathcal{R}} \lambda(s)\, ds < \infty. \tag{2.97}$$

The event space of a PPP on the augmented space is $\mathcal{E}(\mathcal{R}^+)$. The event space $\mathcal{E}(\mathcal{R})$ is a proper subset of $\mathcal{E}(\mathcal{R}^+)$.

Realizations are generated as before for the bounded sets $\mathcal{R}$. For bounded sets $\mathcal{R}^+$, the integrals in (2.4) are replaced by the integrals over $\mathcal{R}^+$ as defined in (2.97); otherwise, Step 1 is unchanged. Step 2 is modified slightly. If n is the outcome of Step 1, then n i.i.d. Bernoulli trials with probabilities

$$\Pr[\phi] = \frac{\lambda(\phi)}{\lambda(\phi) + \int_{\mathcal{R}} \lambda(s)\, ds}$$

$$\Pr[\mathcal{R}] = \frac{\int_{\mathcal{R}} \lambda(s)\, ds}{\lambda(\phi) + \int_{\mathcal{R}} \lambda(s)\, ds}$$

are performed. The number $n(\phi)$ is the number of occurrences of ϕ in the realization. The number of i.i.d. samples drawn from $\mathcal{R}$ is $n - n(\phi)$.

The number $n(\phi)$ is a realization of a random variable, denoted by $N(\phi)$, that is Poisson distributed with parameter $\lambda(\phi)$. This is seen from the discussion in Section 2.9.2. The expected number of occurrences of ϕ is $\lambda(\phi)$. Also, probability of repeated occurrences of ϕ is never zero. The possibility of repeated occurrences of ϕ is important to understanding augmented PPP models for applications such as multitarget tracking.

The probability that the list $\{x_1, \ldots, x_n\}$ is a set is the probability that no more than one realization of ϕ occurs in the n Bernoulli trials. Consequently, if $\lambda(\phi) > 0$, the probability that the list $\{x_1, \ldots, x_n\}$ is a set is strictly less than one. In augmented spaces, random finite sets are more accurately describes as random finite lists.

The likelihood function and expectation operator are unchanged, except that the integrals are over either $\mathcal{R}$ or $\mathcal{R}^+$, as the case may be. Superposition and thinning are unchanged. The intensity of the diffusion and prediction processes are also unchanged from (2.83) and (2.86), except that the integrals are over $\mathcal{S}^+$.

It is necessary to define the transitions $\Psi(y \mid \phi)$ and $\Psi(\phi \mid y)$ for all $y \in \mathcal{S}$, as well as $\Psi(\phi \mid \phi) = \Pr[\phi \mid \phi]$. The measurement, or data, likelihood function $L(\cdot \mid \phi)$ must also be defined. These quantities have natural interpretations in target tracking.

Example 2.14 Tracking Interpretations. A one-point augmented space is used in Chapter 6. The state ϕ is the hypothesis that no target is present in the tracking region $\mathcal{R}$, and the point $x \in \mathcal{R}$ is the hypothesis that a target is present with state x. State transitions and the measurement likelihood function are interpreted in tracking applications as follows:

- $\Psi(y \mid \phi)$ is the likelihood that the transition initiates a target at the point $y \in \mathcal{R}$.
- $\Psi(\phi \mid y)$ is the probability that the transition terminates a target at the point $y \in \mathcal{R}$.
- $\Psi(\phi \mid \phi)$ is the probability that no target is present both before or after transition.
- $L(\cdot \mid \phi)$ is the likelihood that the data are clutter-originated, i.e., the likelihood function of the data conditioned on the absence of a target in $\mathcal{R}$.

Initiation and termination of target track is therefore an intrinsic part of the tracking function when using a Bayesian tracking method (see Appendix C) on an augmented state space $\mathcal{S}^+$.

As is seen in Chapter 6, augmented spaces play an important role in simplifying difficult enumerations related to joint detection and tracking of targets. Only one state ϕ is considered here, but there is no intrinsic limitation.

Example 2.15 Non-Orderly PPPs. The intensity of general PPPs on a continuous space $\mathcal{S}$ is given in (2.3) as the sum of an ordinary function $\lambda(s)$ and a countable number of weighted Dirac delta functions located at the isolated points $\{a_j\}$. The points $\{a_j\}$ are identified with the discrete points $\Phi = \{\phi_j\}$. Let $\mathcal{S}^+ = \mathcal{S} \cup \Phi$. Realizations on the augmented space $\mathcal{S}^+$ generated in the manner outlined above for the one point augmented case map directly to realizations of the non-orderly PPP on $\mathcal{S}$ via the identification $\phi_j \leftrightarrow a_j$. Other matters are similarly handled.

Chapter 3
Intensity Estimation

How can we know the dancer from the dance?
William Butler Yeats, Among School Children, *1928*

Abstract PPPs are characterized, or parameterized, by their intensity function. When the intensity is not fully specified by application domain knowledge (e.g., the physics), it is necessary to estimate it from data. These inverse problems are addressed by the method of maximum likelihood (ML). The Expectation-Maximization method is used to obtain estimators for problems involving superposition of PPPs. Gaussian sums are developed as an important special case. Estimators are obtained for both sample and histogram data. Regularization methods are presented for stabilizing Gaussian sum algorithms.

Keywords Maximum likelihood estimation · Gaussian crosshairs · Edge effects · Expectation-Maximization (EM) · Superposed intensities · Affine Gaussian sums · Histogram data · Regularization · Parametric tying · Luginbuhl's broadband harmonic spectrum · Bayesian methods

The basic geometric and algebraic properties of PPPs are discussed in Chapter 2. These fundamental properties are exploited in this chapter to obtain algorithms for estimating the PPP intensity function from data using the method of maximum likelihood (ML). Estimation algorithms are critically important in applications in which intensity functions are not fully determined a priori. Many heuristic methods have also been proposed in the literature; though not without interest, they are not discussed here.

The fundamental notion is that the intensity function is specified parametrically, and that appropriate numerical values of one or more of the parameters are unknown and must be estimated from collected data. Two kinds of intensity models are natural for PPPs: those that do *not* involve superposition, and those that do. Moreover, two kinds of data are natural for PPPs: sample data and histogram data (see below in Section 3.1). This gives four combinations, each of which requires a different, though closely related, ML algorithm. In all cases, the parameterized intensity is written $\lambda(s; \theta)$, where the parameter vector θ is estimated from data.

The first section of this chapter discusses intensity models that do not involve superposition. The natural method of ML estimation is used: Find the right

R.L. Streit, *Poisson Point Processes*, DOI 10.1007/978-1-4419-6923-1_3,
© Springer Science+Business Media, LLC 2010

likelihood function for the available data, set the gradient with respect to the parameters to be estimated equal to zero, and solve. The likelihood functions discussed in this section correspond to PPP sample and histogram data. This takes care of two of the four combinations mentioned above.

The remaining sections are all about intensity functions that involve superposition, or linear combinations, of intensity functions. The EM method is the natural method for deriving ML estimation algorithms, or estimators, in this case. Readers with little or no familiarity with it may want to consult Appendix A or other references before reading these sections. Other readers, who wish merely to get quickly to the algorithm, are provided with two tables that outline the steps of the EM algorithm for affine Gaussian sums, perhaps the most important example. All readers, even those for whom EM may have lost some of its charm, are provided with enough details to "read along" with little need to lift a pencil. These details reside primarily in the E-step and can be skipped at the reader's pleasure with little loss of continuity or essential content.

Parametric models of superposed intensities come in at least two forms—Gaussian sums and step functions. In the latter, the steps often correspond to pixels (or voxels) in an image. With sufficiently many terms, both models can approximate any continuous intensity function arbitrarily closely. For that reason they are sometimes described as nonparametric even though they have parameters that must be either specified or estimated. To distinguish them from truly nonparametric "modern" sequential Monte Carlo (SMC) models involving particles, Gaussian sums and step functions are herein referred to as parametric models.

Gaussian sum models are the main objects of study in this chapter. Step function models are also very important, but they are discussed mostly in the Chapter 5 on medical imaging. SMC methods are discussed in the context of the tracking applications in Section 6.3.

3.1 Maximum Likelihood Algorithms

The ML estimate is, by definition, the *global* maximum. The problem in practice is that the likelihood function is often plagued by multiple local maxima. For example, when the intensity is a Gaussian shape of unknown location plus a constant (see (4.38) below), the likelihood function of the location parameter may have multiple local maxima. When it is important to find the global maximum, locally convergent algorithms are typically restarted with different initializations, and the ML estimate is taken to be the estimate with the largest likelihood found. This chapter is content to investigate algorithms that converge to a local maximum of the likelihood function.

Two important kinds of data for PPPs are considered. One kind is PPP sample data, sometimes called "count record" data, which means simply that the available data set is a realization of a PPP with intensity $\lambda(s; \theta)$. The other is histogram data, in which a realization of the PPP is represented not by the points themselves, but by the number of points that fall in a fixed number of nonoverlapping cells that partition the space $\mathcal{S}$. Such data are equivalent to a realization of a PPP on a discrete space. In practice, both kinds of data must be carefully collected to avoid unintentionally altering its essential Poisson character. The quality of the ML estimators may be adversely affected if the data do not match the underlying PPP assumption.

3.1.1 Necessary Conditions

Sample data comprise the points $x \equiv (x_1, \ldots, x_m)$, $m \geq 1$, of a PPP realization on $\mathcal{R} \subset \mathcal{S} \subset \mathbb{R}^{n_x}$, where n_x is the dimension of the points of the PPP. These data are conditionally independent given their total number m. It is implicit in this statement that the points x_j are in $\mathcal{R}$ and that the intensity of the PPP is estimated only on $\mathcal{R}$, not the full set $\mathcal{S}$. The order of the points in x is irrelevant. From (2.12), the logarithm of the pdf of x given the PPP intensity $\lambda(s\,;\,\theta)$ is

$$
\mathcal{L}_{iid}(\theta;\,x) = \log\, p\,(x\,;\,\theta)
$$

$$
= -\int_{\mathcal{R}} \lambda\,(s\,;\,\theta)\;\mathrm{d}s \;+\; \sum_{j=1}^{m} \log \lambda\left(x_j\,;\,\theta\right). \tag{3.1}
$$

The maximum likelihood estimate of θ is

$$
\hat{\theta}_{ML} \equiv \arg\max_{\theta}\, \mathcal{L}_{iid}\,(\theta\,;\,x). \tag{3.2}
$$

Assuming differentiability, the natural way to compute $\hat{\theta}_{ML}$ is to solve the necessary conditions $\nabla_\theta \mathcal{L}(\theta\,;\,x) = 0$:

$$
\sum_{j=1}^{m} \frac{1}{\lambda\left(x_j\,;\,\theta\right)}\, \nabla_\theta \lambda\left(x_j\,;\,\theta\right) = \int_{\mathcal{R}} \nabla_\theta \lambda\,(s\,;\,\theta)\;\mathrm{d}s. \tag{3.3}
$$

This system of equations may have multiple solutions, depending on the form of the intensity $\lambda(x\,;\,\theta)$ and the particular PPP sample data x.

A similar system of equations holds for histogram data. Adopting the notation of Section 2.9.1 for a K cell histogram with counts data $m_{1:K} = \{m_1, \ldots, m_K\}$, the logarithm of the pdf is, from (2.62),

$$
\mathcal{L}_{hist}\,(\theta\,;\,m_{1:K}) = -\int_{\mathcal{R}} \lambda\,(s\,;\,\theta)\;\mathrm{d}s \;+\; \sum_{j=1}^{K} \log(m_j!)
$$

$$
+ \sum_{j=1}^{K} m_j\, \log \left(\int_{\mathcal{R}_j} \lambda\,(s\,;\,\theta)\;\mathrm{d}s \right). \tag{3.4}
$$

Taking the gradient gives the necessary conditions

$$\sum_{j=1}^{K} \frac{m_j}{\int_{\mathcal{R}_j} \lambda(s \,;\, \theta)\, \mathrm{d}s} \int_{\mathcal{R}_j} \nabla_\theta \lambda(s \,;\, \theta)\, \mathrm{d}s \;=\; \int_{\mathcal{R}} \nabla_\theta \lambda(s \,;\, \theta)\, \mathrm{d}s. \qquad (3.5)$$

As for PPP sample data, the system may have multiple solutions.

It is necessary to verify that the loglikelihood function of the data is concave at the ML solution, that is, that the negative Hessian matrix of the loglikelihood function is positive definite. In practice, intuition often replaces verification.

3.1.2 Gaussian Crosshairs and Edge Effects

This example is a two dimensional problem that involves training the "crosshairs" of a receiver on a dim light source. In optical communications, this means estimating the brightest direction to the source [114, 115, 119, Section 4.5]. The receiver in this case is a photodetector with a (flat) photoemissive surface. (Photoemissive materials give off electrons when they absorb energetic photons.) Photons are detected over a specified finite time period. Recording the number and locations of each detected photon provides i.i.d. data $x = \{x_j \in \mathbb{R}^2 \,:\, j = 1, \ldots, m\}$. This kind of data may or may not be practical for large m, depending on the application. Feasible or not, it is nonetheless interesting to consider. In practice, the photodetector surface $\mathcal{R}$ is often divided into a number of disjoint regions that constitute histogram cells, and a count of the number of photons arriving in each cell is recorded. The photodetector surface is assumed to be rectangular of known size, and its center is taken as the coordinate system origin. The axes are taken parallel to the sides of $\mathcal{R}$.

Example 3.1 Sample Data. The intensity of the light distributed across the photodetector surface is proportional to a Gaussian pdf. The 2×2 covariance matrix Σ determines the elliptical shape of the "spotlight". For simplicity, suppose the shape is circular with known width ρ, so that $\Sigma = Diag\left[\rho^2, \rho^2\right]$. The light intensity is

$$\lambda(x \,;\, I_0, \mu) \;=\; I_0\, \mathcal{N}(x \,;\, \mu, \Sigma), \qquad (3.6)$$

where $I_0/(2\pi\rho^2)$ is the peak intensity and the vector $\mu = [\mu_1, \mu_2]^T \in \mathbb{R}^2$ is the location of the peak. The parameters to be estimated are $\theta = (I_0, \mu_1, \mu_2)$. The only constraint is that $I_0 > 0$.

For sample data, the necessary conditions yield three equations in three unknowns. Setting the derivative in (3.3) with respect to I_0 to zero gives the ML estimate

$$\hat{I}_0 \;=\; \frac{m}{\int_{\mathcal{R}} \mathcal{N}(s \,;\, \mu, \Sigma)\, \mathrm{d}s}, \qquad (3.7)$$

where the integral is a double integral over the photoemissive surface. The estimate $\hat{I}_0$ automatically satisfies the nonnegativity constraint. The value of μ in (3.7) is the ML estimate $\hat{\mu}$, so it is coupled to the necessary equations for μ. Setting the gradient[1] in (3.3) with respect to the vector μ equal to zero, substituting the estimate (3.7), and rearranging terms gives [114]

$$\frac{\int_{\mathcal{R}} s\,\mathcal{N}(s;\mu,\Sigma)\,ds}{\int_{\mathcal{R}} \mathcal{N}(s;\mu,\Sigma)\,ds} = \frac{1}{m}\sum_{j=1}^{m} x_j . \tag{3.8}$$

The left hand side is the mean vector of the Gaussian pdf restricted to the set $\mathcal{R}$. The equation thus says that ML estimate $\hat{\mu}$ is such that the conditional mean on $\mathcal{R}$ equals the sample mean. The conditional mean equation is uniquely solvable for rectangular domains, as shown in Appendix B.

If the bulk of the source distribution is visible on the photoemissive surface, then $\int_{\mathcal{R}} \mathcal{N}(s;\mu,\Sigma)\,ds \approx 1$. This gives the approximation $\hat{I}_0 \approx m$. Also, the left hand side of (3.8) is approximately the unconditional mean μ, so the ML estimate $\hat{\mu}$ approximates the sample mean in this case.

Example 3.2 Compensating for Edge Effects. When the peak of the light distribution lies near the edge of the photodetector, many source photons are not counted. This example shows that the estimator (3.8) automatically compensates for the missed measurements. Intuitively, this is because the sample mean in (3.8) is, by its very nature, an estimate of the conditional mean expressed analytically by the left hand side.

A realization of the PPP (3.6) with intensity $I_0 = 250$ and mean $\mu = [0.8, 0.5]^T$ and $\rho = 0.5$ is generated. Only 143 of the points in the realization fall on the photodetector surface, which is taken to be the square $(-1, 1) \times (-1, 1)$. These points are depicted in the left hand side of Figure 3.2. The ML estimated mean is computed by solving (3.8) by the method discussed in Appendix B, giving $\hat{\mu}_{ML} = [0.75007, 0.53021]^T$. The error is remarkably small, especially compared to the error in the sample mean of the 143 detected photons, namely $[0.49600, 0.37698]^T$. The estimated intensity, computed from (3.7) with μ_{ML}, gives $\hat{I}_0 = 250.72$.

The right hand side of Fig. 3.1 repeats the procedure but with the mean shifted further into the corner to $[0.8, 0.8]^T$. Only 112 of the PPP samples fall on the photodetector. The ML estimate is $\hat{\mu}_{ML} = [0.74687, 0.79664]^T$. The sample mean for the 112 detected photons is, by contrast, $[0.49445, 0.51792]^T$. The estimated intensity is $\hat{I}_0 = 245.57$.

Example 3.3 Histogram Data. ML estimates of the parameters of (3.6) are now given for histogram data. The necessary condition (3.5) for λ_0 gives

---

[1] Use the identity $\nabla_\mu \mathcal{N}(s;\mu,R) = \mathcal{N}(s;\mu,R)\,R^{-1}(s-\mu)$.

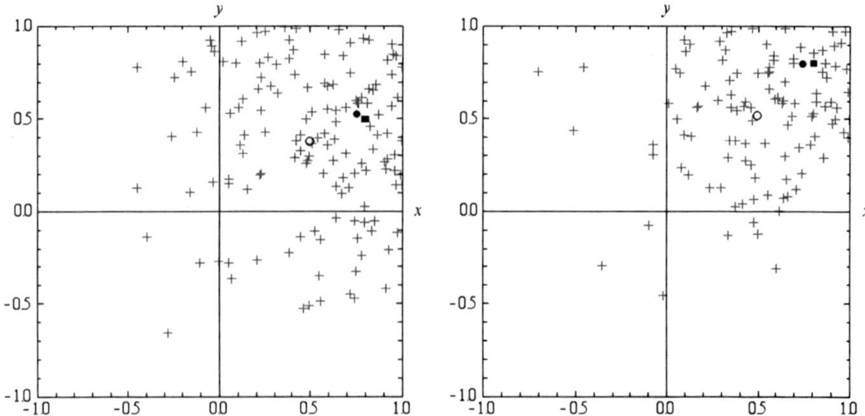

Fig. 3.1 ML mean and coefficient estimates for the model (3.6) obtained by solving equation (3.8). The mean in the left hand figure is (0.8, 0.5). The mean in the right hand figure is (0.8, 0.8). *Filled circle* = $\hat{\mu}_{ML}$; *Open circle* = uncorrected sample mean; *Square* = true mean

$$\hat{I}_0 \;=\; \frac{\sum_{j=1}^{K} m_j}{\int_{\mathcal{R}} \mathcal{N}\,(s\,;\,\mu,\,\Sigma)\,\mathrm{d}s}\,. \tag{3.9}$$

This estimator is identical to (3.7) since $m = \sum_{j=1}^{K} m_j$. It is also coupled to the estimate $\hat{\mu}$. Manipulating the necessary conditions for μ in much the same way as done in (3.8) gives

$$\frac{\int_{\mathcal{R}} s\,\mathcal{N}\,(s\,;\,\mu,\,\Sigma)\,\mathrm{d}s}{\int_{\mathcal{R}} \mathcal{N}\,(s\,;\,\mu,\,\Sigma)\,\mathrm{d}s} \;=\; \frac{1}{m} \sum_{j=1}^{K} m_j\, \frac{\int_{\mathcal{R}_j} s\,\mathcal{N}\,(s\,;\,\mu,\,\Sigma)\,\mathrm{d}s}{\int_{\mathcal{R}_j} \mathcal{N}\,(s\,;\,\mu,\,\Sigma)\,\mathrm{d}s}\,. \tag{3.10}$$

In general, this equation is solved for $\hat{\mu}$ by numerical methods. Unlike (3.8), it is not clear whether or not it has a unique solution.

As before, suppose that most of the light from the source falls on the photoemissive surface, so that $\int_{\mathcal{R}} \mathcal{N}\,(s\,;\,\mu,\,\Sigma)\,\mathrm{d}s \approx 1$. Then left hand side approximates the ML estimate $\hat{\mu}$, and the right hand side approximates the histogram mean, that is,

$$\hat{\mu} \;\approx\; \frac{1}{m} \sum_{j=1}^{K} m_j\, \gamma_j(\hat{\mu})\,,$$

where

$$\gamma_j\left(\hat{\mu}\right) \;=\; \frac{\int_{\mathcal{R}_j} s\,\mathcal{N}\left(s\,;\,\hat{\mu},\,\Sigma\right)\,\mathrm{d}s}{\int_{\mathcal{R}_j} \mathcal{N}\left(s\,;\,\hat{\mu},\,\Sigma\right)\,\mathrm{d}s}$$

is the conditional mean of cell $\mathcal{R}_j$. Since $\gamma_j\left(\hat{\mu}\right) \in \mathcal{R}_j$ regardless of $\hat{\mu}$, replace the Gaussian density in each cell with a uniform density. This final approximation, reasonably good when all histogram cells are small, gives

$$\hat{\mu} \approx \frac{1}{m} \sum_{j=1}^{K} m_j \, \bar{\gamma}_j \,,$$

where

$$\bar{\gamma}_j = \frac{1}{\mathcal{R}_j} \int_{\mathcal{R}_j} s \, ds$$

is the center of cell $\mathcal{R}_j$.

3.2 Superposed Intensities with Sample Data

Parameter estimates obtained by solving the necessary conditions (3.3) and (3.5) are coupled, as Examples 3.1 and 3.3 show. Generally, the more parameters are estimated the more difficult the necessary conditions are to solve, and the more data that are needed to obtain good parameter estimates.

An alternative to solving the ML necessary conditions directly is the EM method. Background on this method is given in Appendix A. The EM method is a hand-in-glove fit for very general applications involving superposition. In the case of PPPs, given the total number of data, the sample data are conditionally independent, and the missing data are especially simple—it consists of discrete indices that identify the components of the PPP superposition that generate the available data. (Other choices of missing data are possible, but this particular choice is by far the most common.) The EM method, by identifying an appropriate set of missing data, breaks the coupling of the necessary conditions in such a way that the parameters of each PPP in the superposition are estimated separately. These smaller systems are often—but not always—easy to solve. The resulting reduction in the number of equations that are solved simultaneously is a big advantage numerically.

Because coupling is intrinsic to the estimation problem, there is a price to be paid for the dimensional advantage that EM provides. One price for decoupling into smaller systems is iteration, that is, the smaller systems of equations are solved repeatedly in a coordinate relaxation procedure. The other price is convergence. The good news is that theory guarantees convergence to a stationary point of the likelihood function, provided the likelihood function is bounded above. For more on the importance of boundedness, see Section 3.4. In practice, if it is bounded, the EM iteration nearly always converges to a local maximum. The bad news is that the convergence rate is ultimately very slow, only linear in fact. In practice, however, the first few iterations are commonly observed to result in significant improvements in the likelihood function.

A basic review of the EM method is provided in Appendix A.2. As is discussed there, the EM method iterates over two fundamental procedures, called the E-step and the M-step. The E step is typically a symbolic step which is only performed

once, while the M-step is the parameter update step. For readers unfamiliar with
EM, there may no easier way to learn it than by first reading the appendix, or some
equivalent discussion elsewhere, and subsequently plunging headlong into the problem addressed in the next section. This problem is an excellent first application of
EM. Readers new to EM are forewarned that much of the action lies in understanding and manipulating subscripts. Further background on EM and its many variations
are given in [80, 81].

3.2.1 EM Method with Sample Data

The intensity $\lambda(x\,;\,\theta)$ is the superposition of the intensity functions of L independent PPPs:

$$\lambda(x\,;\,\theta) \;=\; \sum_{\ell=1}^{L} \lambda_\ell(x\,;\,\theta_\ell)\,, \tag{3.11}$$

where the parameter vector of the ℓ-th PPP is θ_ℓ, and $\theta \;=\; (\theta_1, \ldots, \theta_L)$ is the
parameter vector for the superposition. Different components of $\lambda(x\,;\,\theta)$ can take
different parametric forms. The parameters of the intensities $\lambda_\ell(x\,;\,\theta_\ell)$ are assumed
parametrically untied, i.e., there is no functional relationship between the vectors θ_i
and θ_j for $i \neq j$. This simplifying assumption is unnecessary theoretically, as well
as inappropriate in some applications. An example of the latter is when the centroids
of $\lambda_\ell(x\,;\,\theta_\ell)$ are required to form an equispaced grid whose orientation and spacing
are to be estimated from data.

3.2.1.1 E-step

The natural choice of the "missing data" are the conditionally independent random
indices k_j, $1 \leq k_j \leq L$, that identify which of the superposed PPPs generated the
point x_j. Let

$$x_c \;=\; \{(x_1, k_1), \ldots, (x_m, k_m)\} \tag{3.12}$$

denote the complete data. (In the language of Section 8.1, x_c is a realization of a
marked PPP.) For $\ell = 1, \ldots, L$, let

$$x_c(\ell) \;=\; \big\{(x_j, k_j) : k_j = \ell\big\}\,. \tag{3.13}$$

Let $n_c(\ell) \geq 0$ denote the number of indices j such that $k_j = \ell$, and let

$$\xi_c(\ell) \;=\; (n_c(\ell), x_c(\ell))\,. \tag{3.14}$$

It follows from the definition of k_j that $\xi_c(\ell)$ is a realization of the PPP whose intensity is $\lambda_\ell(\cdot)$. The pdf of $\xi_c(\ell)$ is (from (2.12))

$$p\left(\xi_c(\ell)\,;\,\theta\right) = \exp\left(-\int_{\mathcal{R}} \lambda_\ell\left(s\,;\,\theta_\ell\right) ds\right) \prod_{j\,:\,k_j=\ell} \lambda_\ell\left(x_j\,;\,\theta_\ell\right).$$

The superposed PPPs are independent, so

$$p\left(x_c\,;\,\theta\right) = \prod_{\ell=1}^{L} p\left(\xi_c(\ell)\,;\,\theta\right)$$

$$= \exp\left(-\int_{\mathcal{R}} \lambda(s\,;\,\theta)\,ds\right) \prod_{j=1}^{m} \lambda_{k_j}\left(x_j\,;\,\theta_{k_j}\right).$$

The loglikelihood function of θ given x_c is

$$\mathscr{L}(\theta\,;\,x_c) = \log\, p(x_c\,;\,\theta)$$

$$= -\int_{\mathcal{R}} \lambda\left(s\,;\,\theta\right) ds + \sum_{j=1}^{m} \log \lambda_{k_j}\left(x_j\,;\,\theta_{k_j}\right). \tag{3.15}$$

The conditional pdf of the missing data $(k_1,\ \ldots,\ k_L)$ is

$$p\left(k_1,\ \ldots,\ k_L\,;\,\theta\right) = \frac{p\left(x_c\,;\,\theta\right)}{p\left(x\,;\,\theta\right)} = \prod_{j=1}^{m} \frac{\lambda_{k_j}\left(x_j\,;\,\theta_{k_j}\right)}{\lambda\left(x_j\,;\,\theta\right)}, \tag{3.16}$$

where $\log\, p\left(x\,;\,\theta\right)$ is given by (3.1). From (3.11), it is easily verified that

$$\sum_{k_1,\,\ldots,\,k_n=1}^{L} \prod_{j=1}^{m} \frac{\lambda_{k_j}\left(x_j\,;\,\theta_{k_j}\right)}{\lambda\left(x_j\,;\,\theta\right)} = 1\,, \tag{3.17}$$

and

$$\sum_{k_1,\,\ldots,\,k_{j-1},\,k_{j+1},\,\ldots,\,k_m=1}^{L} \frac{\lambda_{k_j}\left(x_j\,;\,\theta_{k_j}\right)}{\lambda\left(x_j\,;\,\theta\right)} = \frac{\lambda_{k_j}\left(x_j\,;\,\theta_{k_j}\right)}{\lambda\left(x_j\,;\,\theta\right)}. \tag{3.18}$$

These identities are very useful in the E-step of the EM method.

Let $n = 0, 1, \ldots$ denote the EM iteration index, and let the initial feasible value of θ be $\theta^{(n)} = \left(\theta_1^{(n)},\ \ldots,\ \theta_L^{(n)}\right)$, the EM auxiliary function is the conditional expectation defined by

$$Q\left(\theta\,;\,\theta^{(n)}\right) = E\left[\mathcal{L}\left(\theta\,;\,x_c\right)\,;\,\theta^{(n)}\right]$$

$$\equiv \sum_{k_1=1}^{L}\cdots\sum_{k_m=1}^{L}\mathcal{L}\left(\theta\,;\,x_c\right)\prod_{j=1}^{m}\frac{\lambda_{k_j}\left(x_j\,;\,\theta_{k_j}^{(n)}\right)}{\lambda\left(x_j\,;\,\theta^{(n)}\right)}\,. \tag{3.19}$$

Substituting (3.15) gives, after interchanging the summation order and using (3.17) and (3.18),

$$Q\left(\theta\,;\,\theta^{(n)}\right) = -\int_{\mathcal{R}}\lambda(s\,;\,\theta)\,\mathrm{d}s + \sum_{j=1}^{m}\sum_{\ell=1}^{L}\frac{\lambda_\ell\left(x_j\,;\,\theta_\ell^{(n)}\right)}{\lambda\left(x_j\,;\,\theta^{(n)}\right)}\log\lambda_\ell\left(x_j\,;\,\theta_\ell\right)\,. \tag{3.20}$$

Equivalently, $Q\left(\theta\,;\,\theta^{(n)}\right)$ separates into the L term sum,

$$Q\left(\theta\,;\,\theta^{(n)}\right) = \sum_{\ell=1}^{L}Q_\ell\left(\theta_\ell\,;\,\theta^{(n)}\right)\,, \tag{3.21}$$

where

$$Q_\ell\left(\theta_\ell\,;\,\theta^{(n)}\right) = -\int_{\mathcal{R}}\lambda_\ell(s\,;\,\theta_\ell)\,\mathrm{d}s + \sum_{j=1}^{m}\frac{\lambda_\ell\left(x_j\,;\,\theta_\ell^{(n)}\right)}{\lambda\left(x_j\,;\,\theta^{(n)}\right)}\log\lambda_\ell\left(x_j\,;\,\theta_\ell\right)\,. \tag{3.22}$$

This completes the E-step.

3.2.1.2 M-step

The M-step maximizes (3.21) over all feasible θ, that is,

$$\theta^{(n+1)} = \arg\max_{\theta} Q\left(\theta\,;\,\theta^{(n)}\right)\,.$$

In general the maximization requires solving a coupled system of equations involving every parameter in θ. However, for superposed intensities, the EM method gives the separable auxiliary function (3.21). Therefore, the required M-step maximum is found by maximizing the expressions $Q_\ell\left(\theta_\ell\,;\,\theta^{(n)}\right)$ separately. Let

$$\theta_\ell^{(n+1)} = \arg\max_{\theta_\ell} Q_\ell\left(\theta_\ell\,;\,\theta^{(n)}\right)\,, \quad 1 \le \ell \le L\,. \tag{3.23}$$

Then $\theta_\ell^{(n+1)}$ satisfies the necessary conditions:

$$\sum_{j=1}^{m} w_\ell\left(x_j ; \theta^{(n)}\right) \frac{1}{\lambda_\ell\left(x_j ; \theta_\ell\right)} \nabla_{\theta_\ell} \lambda_\ell\left(x_j ; \theta_\ell\right) = \int_{\mathcal{R}} \nabla_{\theta_\ell} \lambda_\ell(s ; \theta_\ell)\, ds , \quad (3.24)$$

where the (dimensionless) weights are given by

$$w_\ell\left(x_j ; \theta^{(n)}\right) = \frac{\lambda_\ell\left(x_j ; \theta_\ell^{(n)}\right)}{\lambda\left(x_j ; \theta^{(n)}\right)} . \quad (3.25)$$

Solving the uncoupled systems (3.24) gives the recursive update for θ, namely $\theta^{(n+1)} = \left(\theta_1^{(n+1)}, \ldots, \theta_L^{(n+1)}\right)$. This completes the M-step.

3.2.2 Interpreting the Weights

The weight $w_\ell\left(x_j ; \theta^{(n)}\right)$ is the probability that the point x_j is generated by the ℓ-th PPP given the current parameter set $\theta^{(n)}$. To see this, let dx denote a multidimensional infinitesimal located at the point x_j. The probability that a point is generated by the ℓ-th PPP in the infinitesimal $(x_j, x_j + dx)$ is

$$\lambda_\ell\left(x_j ; \theta_\ell^{(n)}\right) |dx| ,$$

where $|dx|$ is the infinitesimal volume. On the other hand, the probability that a point is generated by the superposed PPP in the same infinitesimal is

$$\lambda\left(x_j ; \theta^{(n)}\right) |dx| .$$

Therefore, the probability that x_j is generated by the ℓ-th PPP *conditioned* on the event that it was generated by the superposed PPP is the ratio of these two probabilities. Cancelling $|dx|$ in the ratio gives the weight $w_\ell\left(x_j ; \theta^{(n)}\right)$. A more careful proof that avoids the use of infinitesimals is omitted.

The solution of (3.24) depends on the form of $\lambda_\ell(x; \theta_\ell)$. It differs from the direct ML necessary conditions (3.3) primarily by the presence of the weights $w_\ell\left(x_j ; \theta^{(n)}\right)$. Several illustrative examples are now given for the components of the superposed intensity $\lambda(x ; \theta)$ in (3.11).

3.2.3 Simple Examples

Example 3.1 Homogeneous Component. Let the ℓ-th PPP be homogeneous with intensity

$$\lambda_\ell\left(s ; \theta_\ell\right) \equiv I_\ell \text{ for all } s \in \mathcal{R}. \quad (3.26)$$

The only parameter is $\theta_\ell = I_\ell$. The full parameter vector θ includes not only θ_ℓ but also the parameters θ_j, $j \neq \ell$, of the other PPPs. Given $\theta^{(n)}$ and, in particular, $\theta_\ell^{(n)} = I_\ell^{(n)}$, the EM update $I_\ell^{(n+1)}$ is, from (3.24),

$$I_\ell^{(n+1)} = \frac{I_\ell^{(n)}}{|\mathcal{R}|} \sum_{j=1}^{m} \frac{1}{\lambda\left(x_j ; \theta^{(n)}\right)} , \tag{3.27}$$

where $|\mathcal{R}| = \int_\mathcal{R} ds$. The fraction

$$w_\ell\left(x_j ; \theta^{(n)}\right) = \frac{I_\ell^{(n)}}{\lambda\left(x_j ; \theta^{(n)}\right)}$$

is the probability that the point x_j is generated by the ℓ-th PPP, so the summation (3.27) is the expected number of points generated by the PPP conditioned on the current parameter set $\theta^{(n)}$. The division by $|\mathcal{R}|$ converts this number to intensity. EM updates for θ_j, $j \neq \ell$, depend on the form of the PPP intensities.

Example 3.5 Scaling a Known Component. Let the ℓ-th PPP intensity be

$$\lambda_\ell\left(s ; \theta_\ell\right) = I_\ell \, f_\ell\left(s\right) , \tag{3.28}$$

where I_ℓ is unknown and $f_\ell\left(s\right)$ is a specified intensity on $\mathcal{R}$. In contrast to Example 3.4, I_ℓ is dimensionless because the units are carried by $f_\ell(s)$. The estimated parameter is $\theta_\ell = I_\ell$. Given $\theta^{(n)}$ and thus $\theta_\ell^{(n)} = I_\ell^{(n)}$, the EM update is, from (3.24),

$$I_\ell^{(n+1)} = \frac{1}{\int_\mathcal{R} f_\ell\left(s\right) ds} \sum_{j=1}^{m} w_\ell\left(x_j ; \theta^{(n)}\right) , \tag{3.29}$$

where the weights are, from (3.25),

$$w_\ell\left(x_j ; \theta^{(n)}\right) = \frac{I_\ell^{(n)} \, f_\ell\left(x_j\right)}{\lambda\left(x_j ; \theta^{(n)}\right)} .$$

The denominator in (3.29) corrects for the intensity that lies outside of $\mathcal{R}$. Substituting into (3.29) and moving $I_\ell^{(n)}$ outside the sum gives

$$I_\ell^{(n+1)} = \frac{I_\ell^{(n)}}{\int_\mathcal{R} f_\ell\left(s\right) ds} \sum_{j=1}^{m} \frac{f_\ell\left(x_j\right)}{\lambda\left(x_j ; \theta^{(n)}\right)} , \tag{3.30}$$

an expression that reduces to (3.27) for $f_\ell(s) \equiv 1$. This expression is identical to (3.7) when $L = 1$ and $f(\cdot) = \mathcal{N}(\cdot)$.

Example 3.6 Histograms. It is amusing to examine the superposed intensity that is a sum of step functions. Let $\mathcal{R}_j \subset \mathcal{R}$, $j = 1, \ldots, K$, denote the specified cells on which the step function is defined. The step function on the cell $\mathcal{R}_\ell$ is

$$f_\ell(s) = \begin{cases} 1, & \text{if } s \in \mathcal{R}_\ell \\ 0, & \text{if } s \notin \mathcal{R}_\ell. \end{cases} \tag{3.31}$$

The EM recursion for the scaling coefficient I_ℓ is, from (3.30),

$$I_\ell^{(n+1)} = \frac{I_\ell^{(n)}}{|\mathcal{R}_\ell|} \sum_{x_j \in \mathcal{R}_\ell} \frac{f_\ell(x_j)}{\lambda\left(x_j \,;\, \theta^{(n)}\right)}. \tag{3.32}$$

It is seen through the notational fog that the EM algorithm converges in a single step to

$$\hat{I}_\ell = \frac{\sharp\{j \,:\, x_j \in \mathcal{R}_j\}}{|\mathcal{R}_\ell|},$$

where $\sharp\{\cdot\}$ is the number of points in the set. In other words, the ML estimator is proportional to the histogram if the cells are of equal size.

3.2.4 Affine Gaussian Sums

Estimating the parameters of Gaussian sums is important in widespread applications. There is no need when using the EM method to require all the components in the sum to be Gaussian shaped, i.e., proportional to a Gaussian pdf. Therefore, the notation of the previous section is continued here to preserve generality. Let the ℓ-th PPP intensity be proportional to a Gaussian pdf, that is, let

$$\lambda_\ell(s \,;\, \theta_\ell) = I_\ell \, \mathcal{N}(s \,;\, \mu_\ell, \, \Sigma_\ell), \tag{3.33}$$

where the constant I_ℓ is the total "signal level." In some applications one or more of the parameters $\{I_\ell, \, \mu, \, \Sigma\}$ are known. The more common possibilities are considered here.

Case 1. If I_ℓ is the only estimated parameter because μ_ℓ and Σ_ℓ are known, then $\theta_\ell = I_\ell$ is estimated using the recursion (3.30) with $f_\ell(x) = \mathcal{N}(x \,;\, \mu_\ell, \, \Sigma_\ell)$.

Case 2. If signal level and location vector μ_ℓ are estimated, and Σ_ℓ is known, then $\theta_\ell = (I_\ell, \mu_\ell)$. Given $\theta^{(n)}$ and thus $\theta_\ell^{(n)} = \left(I_\ell^{(n)}, \, \mu_\ell^{(n)}\right)$, the EM updates $I_\ell^{(n+1)}$ and $\mu_\ell^{(n+1)}$ are coupled. They are manipulated into a nested form in which $\mu_\ell^{(n+1)}$ is computed as the solution of a nonlinear equation, and then $I_\ell^{(n+1)}$ is computed from $\mu_\ell^{(n+1)}$. To see this, proceed in the same manner as (3.29) to obtain

$$I_\ell^{(n+1)} = \frac{1}{\int_{\mathcal{R}} \mathcal{N}(s\,;\,\mu_\ell,\,\Sigma_\ell)\,ds} \sum_{j=1}^{m} w_\ell\left(x_j\,;\,\theta^{(n)}\right), \qquad (3.34)$$

where in this equation μ_ℓ is the sought after EM update, and the weights are

$$w_\ell\left(x_j\,;\,\theta^{(n)}\right) = \frac{I_\ell^{(n)}\,\mathcal{N}\left(x_j\,;\,\mu_\ell^{(n)},\,\Sigma_\ell\right)}{\lambda\left(x_j\,;\,\theta^{(n)}\right)}. \qquad (3.35)$$

The necessary condition with respect to μ_ℓ in (3.24) is, after multiplying both sides by Σ_ℓ,

$$\sum_{j=1}^{m} w_\ell\left(x_j\,;\,\theta^{(n)}\right)(x_j - \mu_\ell) = I_\ell \int_{\mathcal{R}} \mathcal{N}(s\,;\,\mu_\ell,\,\Sigma_\ell)\,(s - \mu_\ell)\,ds. \qquad (3.36)$$

Substituting for I_ℓ the estimate $I_\ell^{(n+1)}$ from (3.34) and simplifying gives $\mu_\ell^{(n+1)}$ as the solution of a nonlinear equation:

$$\frac{\int_{\mathcal{R}} s\,\mathcal{N}(s\,;\,\mu_\ell,\,\Sigma_\ell)\,ds}{\int_{\mathcal{R}} \mathcal{N}(s\,;\,\mu_\ell,\,\Sigma_\ell)\,ds} = \frac{\sum_{j=1}^{m} w_\ell\left(x_j\,;\,\theta^{(n)}\right) x_j}{\sum_{j=1}^{m} w_\ell\left(x_j\,;\,\theta^{(n)}\right)}. \qquad (3.37)$$

(Compare this expression to the direct ML equation (3.8) to see how the EM method exploits Bayes Theorem to "split" the data into L parts, one for each component.) The right hand side of this equation is a probabilistic mean of the data, that is, a convex combination of $\{x_j\}$. In general, solving (3.37) for $\mu_\ell^{(n+1)}$ requires numerical methods. The estimate $\mu_\ell^{(n+1)}$ is then substituted into (3.34) to evaluate $I_\ell^{(n+1)}$.

If it is known that $\int_{\mathcal{R}} \mathcal{N}(s\,;\,\mu_\ell,\,\Sigma_\ell)\,ds \approx 1$, then $\int_{\mathcal{R}} s\,\mathcal{N}(s\,;\,\mu_\ell,\,\Sigma_\ell)\,ds \approx \mu_\ell$. Given this approximation, the EM updates to (3.34) and (3.37) are "self solving", so that

$$I_\ell^{(n+1)} \approx \sum_{j=1}^{m} w_\ell\left(x_j\,;\,\theta^{(n)}\right) = I_\ell^{(n)} \sum_{j=1}^{m} \frac{\mathcal{N}\left(s\,;\,\mu_\ell^{(n)},\,\Sigma_\ell\right)}{\lambda\left(x_j\,;\,\theta^{(n)}\right)} \qquad (3.38)$$

$$\mu_\ell^{(n+1)} \approx \frac{\sum_{j=1}^{m} w_\ell\left(x_j\,;\,\theta^{(n)}\right) x_j}{\sum_{j=1}^{m} w_\ell\left(x_j\,;\,\theta^{(n)}\right)}. \qquad (3.39)$$

The approximate iteration requires that $\int_{\mathcal{R}} \mathcal{N}\left(s\,;\,\mu_\ell^{(n)},\,\Sigma_\ell\right)\,ds \approx 1$ at every EM iteration, so it cannot be used if the iterates $\mu_\ell^{(n)}$ drift too close to the edge of the set $\mathcal{R}$.

Case 3. Finally, if the signal level, mean vector, and covariance matrix are all estimated, then $\theta_\ell = (I_\ell, \mu_\ell, \Sigma_\ell)$. Given $\theta^{(n)}$ and thus $\theta_\ell^{(n)} = \left(I_\ell^{(n)},\,\mu_\ell^{(n)},\,\Sigma_\ell^{(n)}\right)$, the EM updates $I_\ell^{(n+1)}$, $\mu_\ell^{(n+1)}$, and $\Sigma_\ell^{(n+1)}$ are coupled. They are manipulated so that $\mu_\ell^{(n+1)}$ and $\Sigma_\ell^{(n+1)}$ are decoupled from $I_\ell^{(n+1)}$. However, $\mu_\ell^{(n+1)}$ and $\Sigma_\ell^{(n+1)}$ remain coupled.

To see this, note that equation (3.34) holds here, but with μ_ℓ and Σ_ℓ being the desired EM updates and with the weights now given by

$$w_\ell\left(x_j\,;\,\theta^{(n)}\right) = \frac{I_\ell^{(n)}\,\mathscr{N}\left(x_j\,;\,\mu_\ell^{(n)},\,\Sigma_\ell^{(n)}\right)}{\lambda(x_j\,;\,\theta^{(n)})}. \tag{3.40}$$

The gradient in (3.24) with respect to μ_ℓ and Σ_ℓ^2 involve $I_\ell^{(n+1)}$. The gradient equation for μ is essentially identical to (3.36); the gradient for Σ is very similar but messier. Substituting the estimate $I_\ell^{(n+1)}$ for I_ℓ in both equations using (3.34) and simplifying gives $\mu_\ell^{(n+1)}$ and $\Sigma_\ell^{(n+1)}$ as the solution of a coupled system of nonlinear equations:

$$\frac{\int_\mathcal{R} s\,\mathscr{N}\,(s\,;\,\mu_\ell,\,\Sigma_\ell)\,\mathrm{d}s}{\int_\mathcal{R}\,\mathscr{N}\,(s\,;\,\mu_\ell,\,\Sigma_\ell)\,\mathrm{d}s} = \frac{\sum_{j=1}^{m} w_\ell\left(x_j\,;\,\theta^{(n)}\right)\,x_j}{\sum_{j=1}^{m} w_\ell\left(x_j\,;\,\theta^{(n)}\right)} \tag{3.41}$$

$$\frac{\int_\mathcal{R}\,\mathscr{N}\,(s;\,\mu_\ell,\,\Sigma_\ell)\,(s-\mu_\ell)\,(s-\mu_\ell)^T\,\mathrm{d}s}{\int_\mathcal{R}\,\mathscr{N}\,(s;\,\mu_\ell,\,\Sigma_\ell)\,\mathrm{d}s} = \frac{\sum_{j=1}^{m} w_\ell\left(x_j;\,\theta^{(n)}\right)(x_j-\mu_\ell)(x_j-\mu_\ell)^T}{\sum_{j=1}^{m} w_\ell\left(x_j\,;\,\theta^{(n)}\right)}. \tag{3.42}$$

The updated signal level

$$I_\ell^{(n+1)} = \frac{1}{\int_\mathcal{R}\,\mathscr{N}\left(s\,;\,\mu_\ell^{(n+1)},\,\Sigma_\ell^{(n+1)}\right)\,\mathrm{d}s}\,\sum_{j=1}^{m} w_\ell\left(x_j\,;\,\theta^{(n)}\right). \tag{3.43}$$

is evaluated after solving for $\mu_\ell^{(n+1)}$ and $\Sigma_\ell^{(n+1)}$.

Again, if $\int_\mathcal{R}\,\mathscr{N}\,(s\,;\,\mu_\ell,\,\Sigma_\ell)\,\mathrm{d}s \approx 1$, the left hand sides of the equations (3.41) and (3.42) are the mean vector μ_ℓ and covariance matrix Σ_ℓ, respectively. Given this approximation, the EM update equations are simply

---

[2] Use the matrix identity $\nabla_R\,\mathscr{N}\,(s\,;\,\mu,\,R) = \tfrac{1}{2}\,\mathscr{N}\,(s\,;\,\mu,\,R)\left[-R^{-1} + R^{-1}(s-\mu)(s-\mu)^T R^{-1}\right]$, where the gradient is written in matrix form as $\nabla_R = \left[\frac{\partial}{\partial \rho_{ij}}\right]$, where $R = [\rho_{ij}]$.

$$I_\ell^{(n+1)} \approx \sum_{j=1}^{m} w_\ell\left(x_j\,;\,\theta^{(n)}\right) \tag{3.44}$$

$$\mu_\ell^{(n+1)} \approx \frac{\sum_{j=1}^{m} w_\ell\left(x_j\,;\,\theta^{(n)}\right) x_j}{\sum_{j=1}^{m} w_\ell\left(x_j\,;\,\theta^{(n)}\right)} \tag{3.45}$$

$$\Sigma_\ell^{(n+1)} \approx \frac{\sum_{j=1}^{m} w_\ell\left(x_j\,;\,\theta^{(n)}\right) \left(x_j - \mu_\ell^{(n+1)}\right)\left(x_j - \mu_\ell^{(n+1)}\right)^T}{\sum_{j=1}^{m} w_\ell\left(x_j\,;\,\theta^{(n)}\right)}. \tag{3.46}$$

Table 3.1 EM algorithm for affine Gaussian sum with PPP sample data

---

Given data: $x_{1:m} = \{x_1, \dots, x_m\}$

Fit the intensity: $\lambda(s\,;\,\theta) = \lambda_{bgnd}(s) + \sum_{\ell=1}^{L} I_\ell \mathcal{N}(s\,;\,\mu_\ell,\,\Sigma_\ell),\ \ s \in \mathcal{R}$

Estimate the parameters: $\theta = \{(I_\ell,\,\mu_\ell,\,\Sigma_\ell) : \ell = 1, \dots, L\}$

- FOR $\ell = 1 : L$, initialize coefficients, means, and covariance matrices:
  $I_\ell(0) > 0,\quad \mu_\ell(0) \in \mathcal{R},\quad$ and $\quad \Sigma_\ell(0)$ positive definite
- END FOR
- FOR EM iteration index $n = 0, 1, 2, \dots$ until convergence:
  – FOR $j = 1 : m$ and $\ell = 1 : L$, compute:

$$w_{j\ell}(n) = \frac{I_\ell(n)\,\mathcal{N}(x_j\,;\,\mu_\ell(n),\,\Sigma_\ell(n))}{\lambda_{bgnd}(x_j) + \sum_{\ell=1}^{L} I_\ell(n)\,\mathcal{N}(x_j\,;\,\mu_\ell(n),\,\Sigma_\ell(n))}$$

  – END FOR
  – FOR $\ell = 1 : L$, compute:
  - $N_\ell(n) = \sum_{j=1}^{m} w_{j\ell}(n)$
  - $E_\ell(n) = \frac{1}{N_\ell(n)} \sum_{j=1}^{m} w_{j\ell}(n)\,x_j$
  - $V_\ell(n) = \frac{1}{N_\ell(n)} \sum_{j=1}^{m} w_{j\ell}(n)\,(x_j - \mu_\ell(n)(x_j - \mu_\ell(n))^T$
  - Solve the coupled system of equations:

$$\frac{\int_{\mathcal{R}} N(s\,;\,\mu_\ell,\,\Sigma_\ell)\,s\,ds}{\int_{\mathcal{R}} N(s\,;\,\mu_\ell,\,\Sigma_\ell)\,ds} = E_\ell(n)$$

$$\frac{\int_{\mathcal{R}} N(s\,;\,\mu_\ell,\,\Sigma_\ell)\,(s - \mu_\ell)(s - \mu_\ell)^T\,ds}{\int_{\mathcal{R}} N(s\,;\,\mu_\ell,\,\Sigma_\ell)\,ds} = V_\ell(n)$$

  for $\mu_\ell(n + 1)$ and $\Sigma_\ell(n + 1)$
  - Compute:

$$I_\ell(n + 1) = \frac{N_\ell(n)}{\int_{\mathcal{R}} N(s\,;\,\mu_\ell(n + 1),\,\Sigma_\ell(n + 1))\,ds}$$

  – END FOR
- END FOR EM iteration (Test for convergence)
- If converged: FOR $\ell = 1 : L$, $\hat{I}_\ell = I_\ell(n_{last})$, $\hat{\mu}_\ell = \mu_\ell(n_{last})$, $\hat{\Sigma}_\ell = \Sigma_\ell(n_{last})$ END FOR

The mean and covariance updates are nested in the approximate EM update, that is, the update $\mu_\ell^{(n+1)}$ is used in $\Sigma_\ell^{(n+1)}$.

Table 3.1 outlines the steps of the EM algorithm for affine Gaussian sums, that is, for a PPP that is a Gaussian sum plus an arbitrary intensity, $\lambda_{bgnd}(s)$. This intensity models the superposition of a PPP whose intensity is a Gaussian sum and a background PPP of known intensity. In many applications, the background intensity is assumed homogeneous, but this restriction is unnecessary for the estimation algorithm to converge.

3.3 Superposed Intensities with Histogram Data

The EM method is widely used for superposition (mixture) problems with either independent or conditionally independent data, but it is versatile and is applicable to any problem in which useful missing data can be found. In this section EM is used for histogram data.

3.3.1 EM Method with Histogram Data

Histogram data are difficult to treat because the loglikelihood function (3.4) involves integrals over the individual histogram cells. The key insight, for those who wish to skip the notationally burdensome details until need arises, is that the points of the PPP are the missing data. Said another way, the integrals are sums, and the variables of integration are—like the index of the summation in the superposition—a very appropriate choice for the missing data. Other choices are possible, but this is the choice made here.

3.3.1.1 E-step

The histogram notation of Section 3.1 is retained, as is the superposed intensity (3.11). In EM parlance, the incomplete data loglikelihood function is (2.62). Missing data arise from the histogram nature of the data and from the intensity superposition. For $j = 1, \ldots, K$, the count m_j in cell $\mathcal{R}_j$ corresponds to m_j points in cell $\mathcal{R}_j$, but the precise locations of these points are not observed. Let

$$\xi_j \equiv \left(x_{j1}, \ldots, x_{jm_j}\right), \quad x_{jr} \in \mathcal{R}_j, \quad r = 1, \ldots, m_j,$$

denote these locations. Denote the collection of all missing points by $\xi_{1:K} = (\xi_1, \ldots, \xi_K)$. The points in $\xi_{1:K}$ are the (missing) points of the PPP realization from which the histogram data are generated. The missing data that arise from the superposition are the same as before. The point x_{jr} is generated by one of the components of the superposition; denote the index of this component by k_{jr}, where $1 \leq k_{jr} \leq L$. Let $\mathcal{K}_j = \left(\mathcal{K}_{j1}, \ldots, \mathcal{K}_{jm_j}\right)$, and denote all missing indices by $\mathcal{K}_{1:K} = (\mathcal{K}_1 \ldots, \mathcal{K}_K)$.

The complete data are $(m_{1:K}, \xi_{1:K}, \mathcal{K}_{1:K})$. Because the points x_{jr} are equivalent to points of the PPP realization, and the indices k_{jr} indicate the components of the superposition that generated them, the *definition* of the complete data likelihood function is

$$p_{hc}\left(m_{1:K}, \xi_{1:K}, \mathcal{K}_{1:K} ; \theta\right) = \exp\left(-\int_{\mathcal{R}} \lambda\left(s ; \theta\right) ds\right) \prod_{j=1}^{K} \prod_{r=1}^{m_j} \lambda_{k_{jr}}\left(x_{jr} ; \theta\right).$$

$$(3.47)$$

Missing data do not affect the integral over all $\mathcal{R}$ because $\exp\left(-\int_{\mathcal{R}} \lambda(s ; \theta) ds\right)$ is the normalization constant that makes $p_{hc}(\cdot)$ a pdf. The conditional pdf of the missing data is, by Bayes Theorem,

$$p\left(\xi_{1:K}, \mathcal{K}_{1:K} ; \theta\right) = \frac{p_{hc}\left(m_{1:K}, \xi_{1:K}, \mathcal{K}_{1:K} ; \theta\right)}{p_{hist}\left(m_{1:K} ; \theta\right)}$$

$$= \prod_{j=1}^{K} \prod_{r=1}^{m_j} w_{k_{jr}}\left(x_{jr} ; \theta\right), \qquad (3.48)$$

where the weights are

$$w_\ell\left(x_{jr} ; \theta\right) = \frac{\lambda_\ell\left(x_{jr} ; \theta_{k_{jr}}\right)}{\int_{\mathcal{R}_j} \lambda\left(s ; \theta\right) ds}, \qquad 1 \le \ell \le L. \qquad (3.49)$$

Unlike the weights (3.25), these weights are not dimensionless; they carry the same units as the intensity. For $j = 1, \ldots, K$, let

$$\int_{\mathcal{R}_j} \cdots \int_{\mathcal{R}_j} dx_{j1} \cdots dx_{jm_j} \equiv \int_{(\mathcal{R}_j)^{m_j}} d\xi_j,$$

where $d\xi_j = dx_{j1} \cdots dx_{jm_j}$, and

$$\int_{(\mathcal{R}_1)^{m_1}} \cdots \int_{(\mathcal{R}_K)^{m_K}} d\xi_1 \cdots d\xi_K \equiv \int_{(\mathcal{R}_1)^{m_1} \times \cdots \times (\mathcal{R}_K)^{m_K}} d\xi_{1:K},$$

where $d\xi_{1:K} \equiv d\xi_1 \cdots d\xi_K$. It is easy to verify from the definition of the weights that

$$\sum_{\mathcal{K}_{1:K}} \int_{(\mathcal{R}_1)^{m_1} \times \cdots \times (\mathcal{R}_K)^{m_K}} \prod_{j=1}^{K} \prod_{r=1}^{m_j} w_{k_{jr}}\left(x_{jr} ; \theta\right) d\xi_{1:K} = 1,$$

where the sum over the indices $\mathcal{K}_{1:K}$ is

$$\sum_{\mathcal{K}_{1:K}} \equiv \left(\sum_{k_{11}=1}^{L} \cdots \sum_{k_{1m_1}=1}^{L}\right) \cdots \left(\sum_{k_{K1}=1}^{L} \cdots \sum_{k_{Km_K}=1}^{L}\right).$$

Integrating and summing over all missing data except x_{jr} and k_{jr} shows that

$$\sum_{\mathcal{K}_{1:K}\setminus k_{jr}} \int_{(\mathcal{R}_1)^{m_1}\times\cdots\times(\mathcal{R}_K)^{m_K}\setminus\mathcal{R}_j} \prod_{j'=1}^{K}\prod_{r'=1}^{m_j} w_{k_{j'r'}}\left(x_{j'r'}\,;\,\theta\right)\mathrm{d}\left(\xi_{1:K}\setminus x_{jr}\right)$$

$$= w_{k_{jr}}\left(x_{jr}\,;\,\theta\right). \tag{3.50}$$

This identity is analogous to the identity (3.18).

3.3.1.2 M-step

Let $n \geq 0$ denote the EM iteration index, and let $\theta^{(0)}$ be given. The auxiliary function is, by definition of the E-step,

$$Q\left(\theta\,;\,\theta^{(n)}\right) = E\left[\log p_{hc}\left(m_{1:K},\xi_{1:K},\mathcal{K}_{1:K}\,;\,\theta\right)\mid\theta^{(n)}\right] \tag{3.51}$$

$$\equiv \sum_{\mathcal{K}_{1:K}}\int_{(\mathcal{R}_1)^{m_1}\times\cdots\times(\mathcal{R}_K)^{m_K}} \log p_{hc}(m_{1:K},\xi_{1:K},\mathcal{K}_{1:K};\theta)\prod_{j=1}^{K}\prod_{r=1}^{m_j} w_{k_{jr}}\left(x_{jr};\theta^{(n)}\right)\mathrm{d}\xi_{1:K}.$$

Substituting the logarithm of (3.47) and paying attention to the algebra gives

$$Q\left(\theta\,;\,\theta^{(n)}\right) = -\int_{\mathcal{R}}\lambda\left(s\,;\,\theta\right)\mathrm{d}s$$

$$+ \sum_{j=1}^{K}\sum_{r=1}^{m_j}\sum_{k_{jr}=1}^{L}\int_{\mathcal{R}_j}\log\lambda_{k_{jr}}\left(x_{jr}\,;\,\theta_{k_{jr}}\right)w_{k_{jr}}\left(x_{jr}\,;\,\theta^{(n)}\right)\mathrm{d}x_{jr}$$

$$\times\left\{\sum_{\mathcal{K}_{1:K}\setminus k_{jr}}\int_{(\mathcal{R}_1)^{m_1}\times\cdots\times(\mathcal{R}_K)^{m_K}\setminus\mathcal{R}_j}\prod_{j'\neq j}\prod_{r'\neq r} w_{k_{j'r'}}\left(x_{j'r'}\,;\,\theta^{(n)}\right)\mathrm{d}\left(\xi_{1:K}\setminus x_{jr}\right)\right\}.$$

From the identity (3.50), the term in braces is identically one. Carefully examining the sum over k_{jr} shows that the index can be replaced by an index, say ℓ, that does not depend on j and r. Making the sum over ℓ the first sum, and recognizing that the integrals over x_{jr} for each index r are identical, gives the simplified auxiliary function

$$Q\left(\theta\,;\,\theta^{(n)}\right) = -\int_{\mathcal{R}}\lambda\left(s\,;\,\theta\right)\mathrm{d}s + \sum_{\ell-1}^{L}\sum_{j=1}^{K} m_j\int_{\mathcal{R}_j} w_\ell\left(s\,;\,\theta^{(n)}\right)\log\lambda_\ell\left(s\,;\,\theta_\ell\right)\mathrm{d}s \tag{3.52}$$

Finally, recalling (3.11), the auxiliary function is written

$$Q\left(\theta\,;\,\theta^{(n)}\right) = \sum_{\ell=1}^{L} Q_\ell\left(\theta_\ell\,;\,\theta^{(n)}\right),$$ (3.53)

where

$$Q_\ell\left(\theta_\ell\,;\,\theta^{(n)}\right) = -\int_{\mathcal{R}} \lambda_\ell(s\,;\,\theta_\ell)\,\mathrm{d}s + \sum_{j=1}^{K} m_j \int_{\mathcal{R}_j} w_\ell\left(s\,;\,\theta^{(n)}\right)\log \lambda_\ell(s\,;\,\theta_\ell)\,\mathrm{d}s.$$ (3.54)

This completes the E-step.

The M-step separates into L independent maximization procedures, just as it did with conditionally independent PPP data. The similarity of (3.54) and (3.4) is striking. The details of the maximization of $Q_\ell(\cdot)$ depends on the parametric form of $\lambda_\ell(\cdot\,;\,\theta_\ell)$.

3.3.2 Affine Gaussian Sums

The Gaussian case considered here is

$$\lambda_\ell(s\,;\,\theta_\ell) = I_\ell\,\mathcal{N}\left(s\,;\,\mu_\ell,\,\Sigma_\ell\right),$$ (3.55)

where I_ℓ, μ_ℓ, and Σ_ℓ are estimated. Given $I_\ell^{(n)}$, $\mu_\ell^{(n)}$, and $\Sigma_\ell^{(n)}$ for EM iteration n, the M-step necessary equation for I_ℓ is

$$I_\ell = \frac{\sum_{j=1}^{K} m_j \int_{\mathcal{R}_j} w_\ell\left(s\,;\,\theta^{(n)}\right)\mathrm{d}s}{\int_{\mathcal{R}} \mathcal{N}\left(s\,;\,\mu,\,\Sigma\right)\mathrm{d}s}.$$ (3.56)

The necessary equations for μ_ℓ and Σ_ℓ are coupled:

$$\frac{\int_{\mathcal{R}} s\,\mathcal{N}\left(s\,;\,\mu_\ell,\,\Sigma_\ell\right)\mathrm{d}s}{\int_{\mathcal{R}} \mathcal{N}\left(s\,;\,\mu_\ell,\,\Sigma_\ell\right)\mathrm{d}s} = \frac{\sum_{j=1}^{K} m_j \int_{\mathcal{R}_j} s\,w_\ell\left(s\,;\,\theta^{(n)}\right)\mathrm{d}s}{\sum_{j=1}^{K} m_j \int_{\mathcal{R}_j} w_\ell\left(s\,;\,\theta^{(n)}\right)\mathrm{d}s}$$ (3.57)

and

$$\frac{\int_{\mathcal{R}} s\,\mathcal{N}(s;\mu_\ell,\Sigma_\ell)(s-\mu_\ell)(s-\mu_\ell)^T\mathrm{d}s}{\int_{\mathcal{R}} \mathcal{N}(s;\mu_\ell,\Sigma_\ell)\,\mathrm{d}s} = \frac{\sum_{j=1}^{K} m_j \int_{\mathcal{R}_j} w_\ell(s;\theta^{(n)})(s-\mu_\ell)(s-\mu_\ell)^T\mathrm{d}s}{\sum_{j=1}^{K} m_j \int_{\mathcal{R}_j} w_\ell(s;\theta^{(n)})\,\mathrm{d}s}.$$ (3.58)

The equations (3.57) and (3.58) are solved jointly for $\mu_\ell^{(n+1)}$ and $\Sigma_\ell^{(n+1)}$. Then, $I_\ell^{(n+1)}$ is evaluated using (3.56). This completes the M-step.

If $\int_{\mathcal{R}} \mathcal{N}(s; \mu_\ell, \Sigma_\ell)\, ds \approx 1$, the updates simplify significantly. The left hand side of (3.57) is the mean, so that

$$\mu_\ell^{(n+1)} \approx \frac{\sum_{j=1}^{K} m_j \int_{\mathcal{R}_j} s\, w_\ell\left(s; \theta^{(n)}\right) ds}{\sum_{j=1}^{K} m_j \int_{\mathcal{R}_j} w_\ell\left(s; \theta^{(n)}\right) ds}. \tag{3.59}$$

Using the approximation in (3.58) gives

$$\Sigma_\ell^{(n+1)} \approx \frac{\sum_{j=1}^{K} m_j \int_{\mathcal{R}_j} w_\ell\left(s; \theta^{(n)}\right)\left(s - \mu_\ell^{(n+1)}\right)\left(s - \mu_\ell^{(n+1)}\right)^T ds}{\sum_{j=1}^{K} m_j \int_{\mathcal{R}_j} w_\ell\left(s; \theta^{(n)}\right) ds}. \tag{3.60}$$

The EM update $\mu^{(n+1)}$ is used in (3.60).

Table 3.2 EM algorithm for affine Gaussian sum with PPP histogram data

---

Given data: $m_{1:K} = \{m_1, \ldots, m_K\}$ (K is the number of histogram cells)

Fit the intensity: $\lambda(s; \theta) = \lambda_{bgnd}(s) + \sum_{\ell=1}^{L} I_\ell \mathcal{N}(s; \mu_\ell, \Sigma_\ell),\;\; s \in \mathcal{R}$

Estimate the parameters: $\theta = \{(I_\ell, \mu_\ell, \Sigma_\ell) : \ell = 1, \ldots, L\}$

- FOR $\ell = 1 : L$, initialize coefficients, means, and covariance matrices:
  - $I_\ell(0) > 0$, $\mu_\ell(0) \in \mathcal{R}$, and $\Sigma_\ell(0)$ positive definite
- END FOR
- FOR EM iteration index $n = 0, 1, 2, \ldots$ until convergence:
  - FOR $\ell = 1 : L$ and $s \in \mathcal{R}$, define (be able to evaluate):

$$w_\ell(s; n) = \frac{I_\ell(n)\, \mathcal{N}(s; \mu_\ell(n), \Sigma_\ell(n))}{\int_{\mathcal{R}_\ell} \left[\lambda_{bgnd}(s) + \sum_{\ell=1}^{L} I_\ell(n)\, \mathcal{N}(s; \mu_\ell(n), \Sigma_\ell(n))\right] ds}$$

  - END FOR
  - FOR $\ell = 1 : L$, compute:
    - $N_\ell(n) = \sum_{j=1}^{K} m_j \int_{\mathcal{R}_j} w_\ell(s; n)\, ds$
    - $E_\ell(n) = \frac{1}{N_\ell(n)} \sum_{j=1}^{K} m_j \int_{\mathcal{R}_j} s\, w_\ell(s; n)\, ds$
    - $V_\ell(n) = \frac{1}{N_\ell(n)} \sum_{j=1}^{K} m_j \int_{\mathcal{R}_j} w_\ell(s; n)\, (s - \mu_\ell)(s - \mu_\ell)^T ds$
    - Solve the coupled system of equations:

$$\frac{\int_{\mathcal{R}} s\, \mathcal{N}(s; \mu_\ell, \Sigma_\ell)\, ds}{\int_{\mathcal{R}} \mathcal{N}(s; \mu_\ell, \Sigma_\ell)\, ds} = E_\ell(n)$$

$$\frac{\int_{\mathcal{R}} \mathcal{N}(s; \mu_\ell, \Sigma_\ell)\, (s - \mu_\ell)(s - \mu_\ell)^T ds}{\int_{\mathcal{R}} \mathcal{N}(s; \mu_\ell, \Sigma_\ell)\, ds} = V_\ell(n)$$

  for $\mu_\ell(n+1)$ and $\Sigma_\ell(n+1)$
  - Compute:

$$I_\ell(n+1) = \frac{N_\ell(n)}{\int_{\mathcal{R}} N(s; \mu_\ell(n+1), \Sigma_\ell(n+1))\, ds}$$

  - END FOR
- END FOR EM Test for convergence
- If converged: FOR $\ell = 1 : L$, $\hat{I}_\ell = I_\ell(n_{last})$, $\hat{\mu}_\ell = \mu_\ell(n_{last})$, $\hat{\Sigma}_\ell = \Sigma_\ell(n_{last})$ END FOR

Table 3.2 outlines the steps of the EM algorithm for affine Gaussian sums with histogram data. It is structurally similar to Table 3.1.

3.4 Regularization

The EM method guarantees convergence only if the likelihood function is bounded above, an often overlooked fact of life. Consider, for instance, the general affine Gaussian sum

$$\lambda(s) = \lambda_{bgnd}(s) + \sum_{\ell=1}^{L} I_\ell \, \mathcal{N}\left(s\,;\,\mu_l,\,\Sigma_\ell\right). \tag{3.61}$$

These heteroscedastic sums, as they are picturesquely called in the statistics literature, have $L\left(\frac{1}{2}n_x\,(n_x+1) + n_x + 1\right)$ free parameters. There are so many parameters, in fact, that the loglikelihood function (3.1) is unbounded for conditionally independent data for $L \geq 2$. It is bounded only for $L = 1$, and then only if the data are full rank.

As often as not, the ugly fact of unboundedness intrudes during the EM iteration when a covariance matrix abruptly becomes numerically singular. What is happening is that the covariance matrix is "shrink-wrapping" itself onto some less than full n_x-dimensional data subspace. Over-fitting is a more classical name for the phenomenon. The likelihood of the corresponding Gaussian component therefore grows without bound as the EM algorithm bravely iterates toward a maximum it cannot attain. In practice, unfortunately, it is all too easy to encounter initializations that are in the domain of attraction of an unbounded point of the likelihood function. Hence, the need for regularization.

3.4.1 Parametric Tying

Initialization difficulties are greatly improved by reducing the parameter count. Parameter tying reduces the effective parameter count by placing one or more constraints on the model parameters. A constraint typically reduces the parameter count by one, so the more constraints the better, provided the fitting capability of the constrained model is not significantly reduced. Two parametric tying methods for Gaussian sums are discussed. Neither method bounds the likelihood function, but in practice both greatly reduce initialization difficulties.

By far the best known parametric tie requires the covariance matrices to be identical, that is, $\Sigma_\ell \equiv \Sigma$ for all ℓ. This kind of tied sum is called a homoscedastic sum, and its parameter count is only $\frac{1}{2}n_x\,(n_x+1) + L\,(n_x+1)$. Unfortunately, in many applications, homoscedastic Gaussian sums are good fits to the data only when the number of components L is relatively large. An ML estimator $\hat{\Sigma}$ can be developed via EM. Details are omitted.

Strophoscedastic sums fall in between homoscedastic and heteroscedastic sums in terms of parameter count. Their fitting capability is similar to that of heteroscedastic sums, and their numerical behavior is almost as good as homoscedastic sums. Both attributes are potentially significant advantages in practice. Strophoscedastic Gaussian sums are sums in which $\Sigma_\ell = U_\ell^T D U_\ell$ for some orthogonal matrix U_ℓ, where the positive diagonal matrix D is the same for all Gaussian components. The matrix D and the matrices U_ℓ are estimated by an ML algorithm that can be derived via EM (see [129]).

Other interesting parameterized models have been proposed for specialized purposes. Two are mentioned here, but the details are omitted.

Example 3.7 Contaminated Data. Tukey proposed an interesting mixture model for data contaminated with outliers. The model is sometimes called a homothetic Gaussian sum, after the Greek *thetos*, meaning "placed." The components of a homothetic Gaussian sum have the same mean vector, called the homothetic center. The covariance matrices are linearly proportional to a common covariance matrix, so the general form is

$$\lambda(s) = \sum_{\ell=1}^{L} I_\ell \, \mathcal{N}\left(s \,;\, \mu, \, \rho_\ell^2 \, \Sigma\right) . \tag{3.62}$$

The scale factors, or similitude ratios, ρ_ℓ, are specified. The estimated parameters are the mean μ and the coefficients $\{\ell\}$, as well as the matrix Σ if it is not specified in the application. In some settings, it may be useful to allow components of the homothetic model to have different covariance matrices.

Example 3.8 Luginbuhl's Broadband Harmonic Spectrum. The spectrum of a truly narrow band signal with fundamental frequency ω_0 and $L - 1$ integer harmonics is a sum of Dirac delta functions:

$$\lambda(s) = N(s) + \sum_{\ell=1}^{L} I_\ell \, \delta\left(s \,;\, \ell\omega_0\right) ,$$

where s denotes frequency and $N(s)$ is a known noise spectrum. However, if the fundamental is a randomly modulated "narrow" broadband signal with mean fundamental frequency ω_0 and spectral width σ_0, a reasonable model of the power spectrum is

$$\lambda(s) = N(s) + \sum_{\ell=1}^{L} I_\ell \, \mathcal{N}\left(s \,;\, \ell\omega_0, \, \ell^2 \sigma_0^2\right) . \tag{3.63}$$

This model differs from the homothetic model because both the locations and the widths of the harmonics are multiples of the location ω and width σ_0 of the fundamental. The objective is to estimate $\hat{\omega}_0$ and $\hat{\sigma}_0$ from DFT data. In low signal to noise ratio (SNR) applications, and in adverse environments where individual harmonics can fade in and out over time, measurements of the full harmonic structure over a sliding time window may improve the stability and accuracy of these estimates. An immediate technical problem arises—DFT data are real numbers, not integer counts. However, artificially quantizing the DFT measurements renders the data as nonnegative integers. Estimates of the fundamental frequency can be computed by thinking of the DFT cells as cells of a histogram and the integer data as histogram counts of a PPP. A partial justification for this interpretation is given in Appendix F. The spectral model (3.63) was proposed for generalized (noninteger) harmonic structure in [73].

3.4.2 Bayesian Methods

As the above discussion shows, the data likelihood function for Gaussian sums is bounded above only if the covariance matrices are bounded away from singularity. One way to do this is the force the condition number of the covariance matrices to be greater than a specified threshold. In practice, this strategy is easily implemented in every EM iteration by appropriately modifying the eigenvalues of $\hat{\Sigma}_{\ell}^{(n)}$, the EM covariance matrix estimate at iteration n. However, this abusive practice destroys the convergence properties of EM.

A principled method that avoids singularities entirely while also preserving EM convergence properties employs a Bayesian methodology. In a Bayesian method, an appropriate prior pdf is assigned to each of the parameters of the Gaussian sum. For the covariance matrices, the natural prior is the Wishart matrix-valued pdf. The hyperparameters of the Wishart density is the specified number of degrees of freedom and a positive definite target matrix. Similarly, the natural priors for the coefficients and mean vectors of the sum are Dirichlet and Gaussian densities, respectively. Incorporating these Bayesian priors into the PPP likelihood function and invoking the EM method leads to Bayesian estimators. In particular, the Bayesian estimate for the covariance matrices of the Gaussian sum are necessarily bounded away from singularity because of the Wishart target matrix is positive definite. The details for Gaussian mixtures, that is, for Gaussian sums that integrate to one, are given in [99].

The full potential of the Bayesian method is not really exploited here since the prior densities are used to avoid matrix singularities and other numerical issues. These methods add robustness to the numerical procedures, so they are certainly valuable. Nonetheless, in practice, this kind of Bayesian parameter estimate does not directly address the over-fitting problem itself. The story changes entirely however if there are application-specific justifications for invoking Bayesian priors.

Chapter 4
Cramér-Rao Bound (CRB) for Intensity Estimates

It is a dreadful thing for the inhabitants of a house not to know how it is made.

Ristoro d'Arezzo Composizione del Mondo,[1] *1282*

Abstract The quality of unbiased estimators of intensity is analyzed in terms of the Cramér-Rao Bound on the smallest possible estimation variance. No estimator is needed to find the bound. The CRB for the intensity of a PPP takes a remarkably simple explicit form. The CRB is examined for several problems. One problem analyzes the effect of gating on estimating the mean of a Gaussian signal in the presence of clutter, or outliers. Another generalizes this to find the CRB of the parameters of a Gaussian sum.

Keywords Cramer-Rao bound (CRB) · Unbiased estimators · Fisher information matrix (FIM) · Cauchy-Schwarz · CRB of intensity from sample data · CRB of intensity from histogram data · CRB of intensity on discrete space · Gating · Joint CRB for Gaussian sums · Observed information matrix (OIM)

The most common measure of estimator quality is the Cramér-Rao bound (CRB) on estimation variance. It is an important bound in both theory and practice. It is important theoretically because no unbiased estimator of θ can have a smaller variance than the CRB. Versions of the CRB are known for biased estimators, but the bound in this case depends on the specific estimator used.

An estimator whose variance equals the CRB is said to be an "efficient" estimator. Although efficient estimators do not exist for every problem, the CRB is still useful in practice because, under mild assumptions, maximum likelihood estimators are asymptotically unbiased and efficient. The CRB is also useful in another way. Approximate estimation algorithms are often implemented to reduce computational complexity or accommodate data quality, and it is highly desirable to know how close these approximate estimators are to the CRB. Such studies, performed by simulation or otherwise, may reveal that the approximate estimator is good enough,

---

[1] The oldest surviving astronomical text in *dialetto toscano* (the Italian dialect spoken in Tuscany).

R.L. Streit, *Poisson Point Processes*, DOI 10.1007/978-1-4419-6923-1_4,
© Springer Science+Business Media, LLC 2010

or that there is a performance shortfall that requires further investigation. Either outcome is important in the application.

The CRB is reviewed in a general setting in the first section. The CRB for PPP intensity function estimation is presented in the second section using both PPP sample data and histogram data. As will be seen, the CRB is surprising simple and tractable in very general problems for both kinds of data.

The CRB is especially simple, even elegant, for affine Gaussian sums. Such models are important because of their widespread use in applications. The problem of estimating the location of a Gaussian shaped intensity inside a gate, or window, with superposed clutter is analyzed explicitly in terms of the signal to noise ratio (SNR). The simplicity of the CRB for these problems is in stark contrast to the significant difficulties encountered in evaluating the CRB for closely related problems involving Gaussian mixture pdfs.

The last section discusses observed information matrices (OIMs). These matrices are widely used in difficult estimation problems as surrogates for Fisher information matrices (FIMs) when the latter are either unknown or cannot be evaluated. OIMs are the negative of the Hessian matrix of the loglikelihood function evaluated at the maximum likelihood estimate.

Applications of the CRB to PPP intensity function estimation problems are not widely discussed in the literature. This is unfortunate for possible applications. One text that does discuss the CRB for PPPs is [119]. This prescient book gives not only the theory but also many interesting and useful examples of parameterized PPP intensity functions and their CRBs. Several journal papers also discuss the topic. See, for example, [97] and the references therein.

4.1 Background

Several facts about the CRB are worth mentioning explicitly at the outset. Perhaps the most interesting to readers new to the CRB is that the CRB is determined solely by the data pdf. In other words, the CRB does *not* involve any actual data. Data influence the CRB only via the parametric form of the data pdf. For the CRB to be useful, it is imperative that the pdf accurately describe the real data.

Classical thinking says the only good estimators are the unbiased estimators which are defined in the next section. Such thinking is incorrect—there are useful biased estimators in several applications. Moreover, techniques are known for trading off variance and bias to minimize the mean squared error, a quantity of great utility in practice. Nonetheless, the classical CRB for unbiased estimators is the focus to the discussion here. The CRB for biased estimators is given in Section 4.1.4 for estimators with known bias.

For the special case of one parameter, the crucial insight that makes the CRB tick is the Cauchy-Schwarz inequality. Multiparameter problems require a modified approach because of the necessity of defining what a lower bound means in more than one dimension. Maximizing a ratio of positive definite quadratic forms (a Rayleigh quotient) replaces the Cauchy-Schwarz inequality in the multiparameter case. This approach parallels the one given in [46]. The CRB for multiple parameters is discussed in this section in a general context.

Let $p_X(x\,;\,\theta)$ denote the parameterized data pdf, where X is the random data, x is a realization of the data, and θ is a real valued parameter vector. Let

$$\theta \in \Theta \subset \mathbb{R}^{n_\theta},$$

where n_θ denotes the number of parameters and Θ is the set of all valid parameter vectors. The pdf $p_X(x\,;\,\theta)$ is assumed differentiable with respect to every component of the vector θ for all $x \in \mathcal{R} \subset \mathbb{R}^{n_x}$, where n_x is the dimension of the data space. The data space $\mathcal{R}$ is independent of θ. The gradient $\nabla_\theta\, p_X(x\,;\,\theta)$ is a column vector in $\mathbb{R}^{n_\theta}$.

4.1.1 Unbiased Estimators

Let the data X be accurately modeled by the pdf $p_X(x\,;\,\theta_0)$, where $\theta_0 \in \Theta$ is an unknown but fixed value of the parameter vector. Let $\hat{\theta}(x)$ denote an arbitrary function of the data x. The function $\hat{\theta}(x)$ is called an estimator. It is an unbiased estimator if and only if

$$E\left[\hat{\theta}(X)\right] = \int_{\mathcal{R}} \hat{\theta}(x)\, p_X(x\,;\,\theta_0)\,\mathrm{d}x \,=\, \theta_0\,. \tag{4.1}$$

The covariance matrix of an unbiased estimator $\hat{\theta}(x)$ is

$$\begin{aligned}
\mathrm{Var}\left[\hat{\theta}\right] &= E\left[\left(\hat{\theta}(X) - \theta_0\right)\left(\hat{\theta}(X) - \theta_0\right)^T\right] \\
&= \int_{\mathcal{R}} \left(\hat{\theta}(x) - \theta_0\right)\left(\hat{\theta}(x) - \theta_0\right)^T p_X(x\,;\,\theta_0)\,\mathrm{d}x\,.
\end{aligned}$$

Unbiased estimators are compared in terms of their covariance matrices. Define $\Sigma_1 \leq \Sigma_2$ if the difference $\Sigma_2 - \Sigma_1$ is positive semidefinite.[2] The CRB is based on this interpretation of matrix inequality. For $n_\theta = 1$, the definition reduces to the usual inequality for real numbers.

For biased estimators, the CRB depends on the estimator. It is given in Section 4.1.4.

4.1.2 Fisher Information Matrix and the Score Vector

For any parameter vector $\theta \in \Theta$, the Fisher Information Matrix (FIM) of θ is the $n_\theta \times n_\theta$ matrix defined by

---

[2] A square (real) matrix A is positive semidefinite if and only if $c^T A c \geq 0$ for all vectors c.

$$J(\theta) = E\left[(\nabla_\theta \log p_X(x\,;\,\theta))\,(\nabla_\theta \log p_X(x\,;\,\theta))^T\right] \tag{4.2}$$

$$\equiv \int_{\mathcal{R}} (\nabla_\theta \log p_X(x\,;\,\theta))\,(\nabla_\theta \log p_X(x\,;\,\theta))^T\, p_X(x\,;\,\theta)\,dx\,, \tag{4.3}$$

where ∇_θ is the gradient with respect to θ. The expectation operator is wonderfully succinct, and writing it explicitly as an integral in (4.3) adds clarity. If the data X are discrete or have both discrete and components, the integral in (4.3) should be replaced by the appropriate sums and integrals.

The matrix $J(\theta)$ is the covariance matrix of the score vector,

$$s(x\,;\,\theta) \equiv \nabla_\theta \log p_X(x\,;\,\theta) \in \mathbb{R}^{n_\theta}. \tag{4.4}$$

To see this, first note that $s(x\,;\,\theta)$ is zero mean since

$$E\left[s(X\,;\,\theta)\right] \equiv \int_{\mathcal{R}} (\nabla_\theta \log p_X(x\,;\,\theta))\, p_X(x\,;\,\theta)\,dx$$

$$= \int_{\mathcal{R}} \nabla_\theta\, p_X(x\,;\,\theta)\,dx \;=\; \nabla_\theta \int_{\mathcal{R}} p_X(x\,;\,\theta)\,dx \;=\; \nabla_\theta\, 1 \;=\; 0\,. \tag{4.5}$$

Therefore, by definition of covariance matrix,

$$J(\theta) = \mathrm{Var}\left[s(X\,;\,\theta)\right]. \tag{4.6}$$

In words, the FIM is the covariance matrix of the score vector.

4.1.3 CRB and the Cauchy-Schwarz Inequality

The $n_\theta \times n_\theta$ matrix $CRB(\theta_0) \equiv J^{-1}(\theta_0)$ is a lower bound on the covariance matrix of any unbiased estimator of θ_0, provided the FIM is nonsingular. Thus, for all unbiased estimators $\hat{\theta}(x)$,

$$\mathrm{Var}\left[\hat{\theta}\right] \geq CRB(\theta_0) \equiv J^{-1}(\theta_0) = (\mathrm{Var}\left[s(\theta_0)\right])^{-1}. \tag{4.7}$$

Several lower bounds on the variance of unbiased estimators are available, but the best known and most widely used is the CRB. To see that (4.7) holds, first write the equations that are equivalent to the statement that $\hat{\theta}(X)$ is unbiased:

$$\int_{\mathcal{R}} \hat{\theta}_i(x)\, p_X(x\,;\,\theta)\,dx = \theta_i\,, \quad i = 1,\ldots,n_\theta,$$

where θ_i and $\hat{\theta}_i(X)$ are the i-th components of θ and $\hat{\theta}(X)$, respectively. Differentiating each of these equations with respect to θ_j, $j = 1,\ldots,n_\theta$, gives

$$\int_{\mathcal{R}} \hat{\theta}_i(x)\, p_X(x\,;\,\theta)\, \frac{\partial}{\partial \theta_j}\left(\log\, p_X(x\,;\,\theta)\right) dx \;=\; \delta_{ij},$$

where δ_{ij} is Kronecker's delta: $\delta_{ij} = 1$ if $i = j$ and 0 if $i \neq j$. Evaluating this system of n_θ^2 equations at the point $\theta = \theta_0$ and writing it in matrix form gives

$$\int_{\mathcal{R}} \hat{\theta}(x)\, (s(x\,;\,\theta_0))^T\, p_X(x\,;\,\theta_0)\, dx \;=\; I^{n_\theta \times n_\theta}, \tag{4.8}$$

where $I^{n_\theta \times n_\theta}$ is the identity matrix.

Consider the one-dimensional random variables $U(X) = a^T \hat{\theta}(X)$ and $V(X) = b^T s(X\,;\,\theta_0)$, where the constant vectors a and b are both nonzero. Because $\hat{\theta}(X)$ is unbiased, the mean of $U(X)$ is $a^T \theta_0$. The mean of $V(X)$ is $b^T E\left[s(X\,;\,\theta)\right] = 0$, as seen from (4.5). By definition, the covariance of $U(X)$ and $V(X)$ is $\int_{\mathcal{R}} \left(U(x) - a^T \theta_0\right) V(x)\, p_X(x\,;\,\theta_0)\, dx$. The well known covariance inequality (the Cauchy-Schwarz inequality in disguise) is

$$\frac{\int_{\mathcal{R}} \left(U(x) - a^T \theta_0\right) V(x)\, p_X(x\,;\,\theta_0)\, dx}{\left[\int_{\mathcal{R}} \left(U(x) - a^T \theta_0\right)^2 p_X(x\,;\,\theta_0)\, dx\right]^{\frac{1}{2}} \left[\int_{\mathcal{R}} V^2(x)\, p_X(x\,;\,\theta_0)\, dx\right]^{\frac{1}{2}}} \;\leq\; 1. \tag{4.9}$$

Now substitute $U(x) = a^T \hat{\theta}(x)$ and $V(x) = b^T s(x\,;\,\theta_0)$. The numerator simplifies to $a^T b$ using (4.5) and (4.8). The first factor in the denominator is $a^T\, \mathrm{Var}\left[\hat{\theta}\right] a$. From (4.6), the second factor in the denominator is $b^T\, J(\theta_0)\, b$. Making these substitutions, squaring both sides of (4.9) and then multiplying through by $a^T\, \mathrm{Var}\left[\hat{\theta}\right] a$ gives

$$\frac{\left(a^T b\right)^2}{b^T\, J(\theta_0)\, b} \;\leq\; a^T\, \mathrm{Var}\left[\hat{\theta}\right] a. \tag{4.10}$$

The inequality (4.10) holds for all nonzero vectors b, so

$$\max_{b \neq 0} \frac{b^T \left(a\, a^T\right) b}{b^T\, J(\theta_0)\, b} \;\leq\; a^T\, \mathrm{Var}\left[\hat{\theta}\right] a. \tag{4.11}$$

The maximum value of the Rayleigh quotient on the left hand side is unchanged if b is multiplied by an arbitrary nonzero real number. The equality constraint $b^T\, J(\theta_0)\, b = 1$ eliminates this scale factor. The maximum value is attained at the solution of the constrained optimization problem:

$$\max_{b}\; b^T a\, a^T b \quad \text{subject to} \quad b^T\, J(\theta_0)\, b = 1.$$

The method of Lagrange multipliers gives the solution,

$$b^* \;=\; \frac{1}{a^T\, J(\theta_0)\, a}\, J^{-1}(\theta_0)\, a.$$

Substituting $b = b^*$ into (4.11) yields the inequality

$$a^T J^{-1}(\theta_0) \, a \; \leq \; a^T \, \mathrm{Var}\left[\hat{\theta}\right] a \,, \tag{4.12}$$

from which it follows immediately that

$$a^T \left(\mathrm{Var}\left[\hat{\theta}\right] - J^{-1}(\theta_0)\right) a \; \geq \; 0 \tag{4.13}$$

for all nonzero vectors a. Hence, $\mathrm{Var}\left[\hat{\theta}\right] - J^{-1}(\theta_0)$ is positive semidefinite, and this establishes (4.7).

4.1.4 Spinoffs

There is more to learn from (4.13). For all i and j, denote the (i, j)-th elements of $\mathrm{Var}\left[\hat{\theta}\right]$ and $J^{-1}(\theta_0)$ by $\mathrm{Var}_{ij}\left[\hat{\theta}\right]$ and $J_{ij}^{-1}(\theta_0)$, respectively. Suppose all the components of the vector $a \equiv (a_1, \ldots, a_{n_\theta})^T$ are zero except for $a_j \neq 0$. Then (4.13) is $a_j \left(\mathrm{Var}\left[\hat{\theta}\right] - J^{-1}(\theta_0)\right)_{jj} a_j \geq 0$. Equivalently, since $a_j \neq 0$,

$$\mathrm{Var}_{jj}\left[\hat{\theta}\right] \; \geq \; J_{jj}^{-1}(\theta_0) \,. \tag{4.14}$$

In words, this says that the smallest variance of any unbiased estimator of the j-th parameter θ_j is the (j, j)-th element of the inverse of the FIM of θ. The result is important in the many applications in which the off-diagonal elements of the CRB are of no intrinsic interest.

The inequality (4.13) yields still more. If all the components of a are zero except for the subset with indices in $M \subset \{1, \ldots, n_\theta\}$, then

$$\mathrm{Var}_{M \times M}\left[\hat{\theta}\right] \; \geq \; J_{M \times M}^{-1}(\theta_0) \,. \tag{4.15}$$

Here, $\mathrm{Var}_{M \times M}\left[\hat{\theta}\right]$ denotes the $M \times M$ submatrix of $\mathrm{Var}\left[\hat{\theta}\right]$ obtained by removing all rows and columns of $\mathrm{Var}\left[\hat{\theta}\right]$ that are not in the index set M, and similarly for $J_{M \times M}^{-1}(\theta_0)$. The inequality (4.15) allows selected off-diagonal elements of the CRB to be evaluated as needed in the application.

If $p_X(x \,; \theta)$ is twice differentiable, then it can be shown that

$$J(\theta_0) \; = \; -E\left[\nabla_\theta (\nabla_\theta)^T \log p_X(x \,; \theta_0)\right] , \tag{4.16}$$

or, written in terms of the (i, j)-th element,

$$J_{ij}(\theta_0) = -\int_{\mathcal{R}} \left(\frac{\partial^2}{\partial \theta_i \, \partial \theta_j} \log p_X(x \,;\, \theta_0) \right) p_X(x \,;\, \theta_0) \, dx \,. \tag{4.17}$$

This form of the FIM is widely known and used in many applications. It is also the inspiration behind the observed information matrix (OIM) that is often used when the FIM is unavailable. For more details in a PPP context, see Section 4.7.

An unbiased estimator is efficient if $\mathrm{Var}\left[\hat{\theta}\right] = CRB(\theta_0)$. This definition is standard in the statistical literature, but it is misleading in one respect: the *best* unbiased estimator is not necessarily efficient. Said more carefully, there are pdfs $p_X(x \,;\, \theta)$ for which the unbiased estimator with the smallest covariance matrix is known explicitly, but this estimator does not achieve the CRB.

An estimator $\hat{\theta}(x)$ is biased if

$$E\left[\hat{\theta}\right] = \theta_0 + b(\theta_0) \tag{4.18}$$

and $b(\theta_0)$ is nonzero. The nonzero term $b(\theta_0)$ is called the estimator bias. The bias clearly depends on the particular estimator, and it is often difficult to evaluate. If the form of $b(\theta_0)$ is known, and $b(\theta_0)$ is differentiable with respect to θ, then the CRB for the biased estimator $\hat{\theta}$ is

$$\mathrm{Var}\left[\hat{\theta} - \theta_0 - b(\theta_0)\right] = \left(I + \nabla_\theta \, b^T(\theta)\right)_{\theta=\theta_0} J^{-1}(\theta_0) \left(I + \nabla_\theta \, b^T(\theta)\right)^T_{\theta=\theta_0} , \tag{4.19}$$

where I is the $n_x \times n_x$ identity matrix, and the gradients are evaluated at the true value θ_0. The matrix dimensions are consistent since $b^T(\theta) = \left(b_1(\theta), \ldots, b_{n_x}(\theta)\right)$ is a row, and its gradient is the $n_x \times n_x$ matrix

$$\left(\nabla_\theta b_1(\theta), \ldots, \nabla_\theta b_{n_x}(\theta)\right) \,.$$

In other words, $\nabla_\theta \, b^T(\theta)$ is a matrix whose (i, j)-th entry is $\partial b_j(\theta)/\partial \theta_i$. Since

$$\mathrm{Var}\left[\hat{\theta} - \theta_0 - b(\theta_0)\right] = \mathrm{Var}\left[\hat{\theta} - \theta_0\right] - b(\theta_0) \, b^T(\theta_0) \,, \tag{4.20}$$

the bound (4.19) is often written

$$\mathrm{Var}\left[\hat{\theta} - \theta_0\right] = b(\theta_0) b^T(\theta_0) + \left(I + \nabla_\theta \, b^T(\theta)\right)_{\theta=\theta_0} J^{-1}(\theta_0) \left(I + \nabla_\theta \, b^T(\theta)\right)^T_{\theta=\theta_0} \,.$$

The matrix $J(\theta_0)$ is the FIM for θ_0. The bound depends on the estimator via the derivative of the bias.

A Bayesian version of the CRB called the posterior CRB (PCRB) is useful when the parameter θ is a random variable with a specified prior pdf. A good discussion of the PCRB, and in fact the very first discussion of it anywhere, is found in Van

Trees [134, pp. 66–73]. An excellent collection of papers on PCRBs is provided in the book by Van Trees and Bell [135].

Recursive PCRBs for use in tracking and filtering applications are available. The Tichavský recursion, as it is often called, is derived in [112, 133] and also in [104, Chapter 4]. The classic example in tracking applications is for linear Gaussian target motion and measurement models, in which case the Tichavský recursion is identical to the Kalman filter recursion. The recursive PCRB is potentially useful for PPP intensity estimation in filtering applications.

In the case of PPPs, the expectation in the definition of the FIM is the expectation with respect to the PPP. This expectation comprises a sum over n of n-fold integrals, so evaluating the FIM explicitly looks formidable in even the simplest cases. Fortunately, as is seen in the next section, appearances are deceiving.

4.2 CRB for PPP Intensity with Sample Data

For PPP sample data on the bounded set $\mathcal{R}$, the FIM is amazingly simple. If $\lambda(s\,;\,\theta) > 0$ for all $s \in \mathcal{R}$, then the FIM for unbiased estimators of θ is

$$J(\theta) = \int_{\mathcal{R}} \frac{1}{\lambda(s\,;\,\theta)} \left[\nabla_\theta \lambda(s\,;\,\theta)\right] \left[\nabla_\theta \lambda(s\,;\,\theta)\right]^T \, ds \,. \tag{4.21}$$

It is also delightfully straightforward to derive. From (2.12),

$$\log p_\Xi(\xi\,;\,\theta) = -\log n! - \int_{\mathcal{R}} \lambda(s\,;\,\theta)\, ds + \sum_{j=1}^{n} \log \lambda(x_j\,;\,\theta)\,. \tag{4.22}$$

The gradient of (4.22) is

$$\nabla_\theta \log p_\Xi(\xi\,;\,\theta) = -\int_{\mathcal{R}} \nabla_\theta \lambda(s\,;\,\theta)\, ds + \sum_{j=1}^{n} \frac{1}{\lambda(x_j\,;\,\theta)} \nabla_\theta \lambda(x_j\,;\,\theta)\,. \tag{4.23}$$

The FIM is the expectation of the outer product

$$\left[\nabla_\theta \log p_\Xi(\xi\,;\,\theta)\right] \left[\nabla_\theta \log p_\Xi(\xi\,;\,\theta)\right]^T$$

$$= \left(\int_{\mathcal{R}} \nabla_\theta \lambda(s\,;\,\theta)\, ds\right) \left(\int_{\mathcal{R}} \nabla_\theta \lambda(s\,;\,\theta)\, ds\right)^T$$

$$- 2 \left(\int_{\mathcal{R}} \nabla_\theta \lambda(s\,;\,\theta)\, ds\right) \left(\sum_{j=1}^{n} \frac{1}{\lambda(x_j\,;\,\theta)} \nabla_\theta \lambda(x_j\,;\,\theta)\right)^T \tag{4.24}$$

$$+ \left(\sum_{j=1}^{n} \frac{1}{\lambda(x_j\,;\,\theta)} \nabla_\theta \lambda(x_j\,;\,\theta)\right) \left(\sum_{j=1}^{n} \frac{1}{\lambda(x_j\,;\,\theta)} \nabla_\theta \lambda(x_j\,;\,\theta)\right)^T \,.$$

The FIM is the sum of the expected values of the three terms in (4.24), where the expectation is given by (2.23). The expectation of the first term is trivial—it is the expectation of a constant. The other two expectations look formidable, but they are not. 4The sums in both terms are the same form as (2.30), so the expectation of the second term is, from (2.32),

$$-2 \left(\int_{\mathcal{R}} \nabla_\theta \lambda(s ; \theta)\, ds \right) \left(\int_{\mathcal{R}} \nabla_\theta \lambda(s ; \theta)\, ds \right)^T, \tag{4.25}$$

and that of the third term is, from (2.34),

$$\left(\int_{\mathcal{R}} \nabla_\theta \lambda(s ; \theta)\, ds \right) \left(\int_{\mathcal{R}} \nabla_\theta \lambda(s ; \theta)\, ds \right)^T + J(\theta), \tag{4.26}$$

where $J(\theta)$ is given by (4.21). Upon collecting terms, everything cancels except $J(\theta)$.

Example 4.1 Scaled Intensity. Let $\lambda(x ; \theta) = I\, f(x)$, where $f(x) > 0$ is a known intensity function and the constant scale factor is estimated, so $\theta \equiv I$. From (4.21), the Fisher information is $J(I) = \left(\int_{\mathcal{R}} f(s)\, ds \right) / I$, so the CRB for unbiased estimators is the reciprocal

$$CRB(I) = \frac{I}{\int_{\mathcal{R}} f(s)\, ds}.$$

The ML estimator is

$$\hat{I}_{ML} = \frac{m}{\int_{\mathcal{R}} f(s)\, ds},$$

where m is the number of points in a realization. It is unbiased since

$$E\left[\hat{I}_{ML} \right] = \frac{E[m]}{\int_{\mathcal{R}} f(s)\, ds} = \frac{\int_{\mathcal{R}} I\, f(s)\, ds}{\int_{\mathcal{R}} f(s)\, ds} = I.$$

The variance of $\hat{I}_{ML}$ is, using (2.28),

$$\mathrm{Var}\left[\hat{I}_{ML} \right] = \frac{\mathrm{Var}[m]}{\left(\int_{\mathcal{R}} f(s)\, ds \right)^2} = \frac{\int_{\mathcal{R}} I\, f(s)\, ds}{\left(\int_{\mathcal{R}} f(s)\, ds \right)^2} = CRB(I),$$

so the ML estimator attains the lower bound and is, by definition, an efficient estimator.

Example 4.2 Linear Combination of Known Intensities. Generalizing the previous example, let $\lambda(x ; \theta) = I^T f(x)$, where the components of the vector $f(x) \equiv$

$(f_1(x), \ldots, f_L(x))^T$ are specified and the vector of constants $I = (I_1, \ldots, I_L)^T \in \mathbb{R}^L$ is estimated. Then, from (4.21), the FIM for unbiased estimators of I is

$$J(I) = \int_{\mathcal{R}} \frac{1}{I^T f(s)} \, f(s) f^T(s) \, \mathrm{d}s \; \in \mathbb{R}^{L \times L} \,.$$

Evaluating $FIM(I)$ and its inverse $CRB(I)$ requires numerical methods. The ML estimator $\hat{I}_{ML}$ is also not explicitly known, but is found numerically by solving the necessary equations (3.3), or by the EM recursion (3.29). Even in this simple example, it is not clear whether or not $\hat{I}_{ML}$ is unbiased.

Example 4.3 Oops. For the parameter vector $\theta \in \mathbb{R}^2$, define the intensity

$$\lambda(x) = (\theta - c)^T (\theta - c) = \|\theta - c\|^2, \quad x \in \mathcal{R} \subset \mathbb{R}^2, \tag{4.27}$$

where $c \in \mathbb{R}^2$ is given and $\mathcal{R} = [0, 1] \times [0, 1]$. The FIM is, from (4.21),

$$J(\theta) = \frac{(\theta - c)(\theta - c)^T}{\|\theta - c\|^2} \,.$$

The matrix $J(\theta)$ is clearly rank one, so the CRB fails to exist. The problem lies in the parameterization, not the FIM. The intensity function (4.27) is constant on $\mathcal{R}$, so the "intrinsic" dimensionality of the intensity function parameter is only one. The dimension of θ is two.

4.3 CRB for PPP Intensity with Histogram Data

The FIM for histogram data is somewhat more complicated than for PPP sample data. As in Section 3.1, the histogram cells are $\mathcal{R}_1, \ldots, \mathcal{R}_K$, where $\mathcal{R}_i \cap \mathcal{R}_j = \varnothing$ for $i \neq j$. The CRB of θ for histogram data is the inverse of the FIM given by

$$J(\theta) = \sum_{j=1}^{K} \frac{1}{\int_{\mathcal{R}_j} \lambda(s \,;\, \theta) \, \mathrm{d}s} \left[\int_{\mathcal{R}_j} \nabla_\theta \lambda(s \,;\, \theta) \, \mathrm{d}s \right] \left[\int_{\mathcal{R}_j} \nabla_\theta \lambda(s \,;\, \theta) \, \mathrm{d}s \right]^T \,. \tag{4.28}$$

This expression reduces to (4.21) in the limit as the number histogram cells goes to infinity and their size goes to zero. To see (4.28) requires nothing more than matrix algebra, but it is worth presenting nonetheless. The rest of the section up to Example 4.4 can be skipped on a first reading.

Start with the loglikelihood function of the data. Let $n_{1:K} = (n_1, \ldots, n_K)$ denote integer counts in the histogram cells. From (2.62), the pdf of the data is

$$\log p(n_{1:K} ; \theta) = -\sum_{j=1}^{K} n_j! - \int_{\mathcal{R}} \lambda(s ; \theta)\, ds + \sum_{j=1}^{K} n_j \log \left(\int_{\mathcal{R}_j} \lambda(s ; \theta)\, ds \right).$$

The gradient with respect to θ is

$$\nabla_\theta \log p(n_{1:K} ; \theta) = -\int_{\mathcal{R}} \nabla_\theta \lambda(s ; \theta)\, ds + \sum_{j=1}^{K} \frac{n_j}{\int_{\mathcal{R}_j} \lambda(s ; \theta)\, ds} \int_{\mathcal{R}_j} \nabla_\theta \lambda(s ; \theta)\, ds.$$

$$(4.29)$$

The FIM is the expectation of the outer product of this gradient with itself. As was done with conditionally independent data, the outer product is written as the sum of three terms:

$$\begin{aligned}
\left[\nabla_\theta \log p(n_{1:K} ; \theta)\right]&\left[\nabla_\theta \log p(n_{1:K} ; \theta)\right]^T \\[1mm]
= &\left(\int_{\mathcal{R}} \nabla_\theta \lambda(s ; \theta)\, ds \right) \left(\int_{\mathcal{R}} \nabla_\theta \lambda(s ; \theta)\, ds \right)^T \\[1mm]
&- 2 \left(\int_{\mathcal{R}} \nabla_\theta \lambda(s ; \theta)\, ds \right) \left(\sum_{j=1}^{K} \frac{n_j}{\int_{\mathcal{R}_j} \lambda(s ; \theta)\, ds} \int_{\mathcal{R}_j} \nabla_\theta \lambda(s ; \theta)\, ds \right)^T \qquad (4.30) \\[1mm]
&+ \left(\sum_{j=1}^{K} \frac{n_j}{\int_{\mathcal{R}_j} \lambda(s ; \theta)\, ds} \int_{\mathcal{R}_j} \nabla_\theta \lambda(s ; \theta)\, ds \right) \left(\sum_{j=1}^{K} \frac{n_j}{\int_{\mathcal{R}_j} \lambda(s ; \theta)\, ds} \int_{\mathcal{R}_j} \nabla_\theta \lambda(s ; \theta)\, ds \right)^T.
\end{aligned}$$

The first term is independent of the data, so it is identical to its expectation. The second term is a product of two factors, the first of which is independent of the data so it multiplies the expectation of the other factor, which is a sum over K terms. Since K is the number of histogram cells and is fixed, the expectation of the sum is

$$\begin{aligned}
\sum_{j=1}^{K} \frac{E\left[n_j\right]}{\int_{\mathcal{R}_j} \lambda(s ; \theta)\, ds} &\int_{\mathcal{R}_j} \nabla_\theta \lambda(s ; \theta)\, ds \\[1mm]
= \sum_{j=1}^{K} \frac{\int_{\mathcal{R}_j} \lambda(s ; \theta)\, ds}{\int_{\mathcal{R}_j} \lambda(s ; \theta)\, ds} &\int_{\mathcal{R}_j} \nabla_\theta \lambda(s ; \theta)\, ds \\[1mm]
= \sum_{j=1}^{K} \int_{\mathcal{R}_j} \nabla_\theta \lambda(s ; \theta)\, ds &= \int_{\mathcal{R}} \nabla_\theta \lambda(s ; \theta)\, ds . \qquad (4.31)
\end{aligned}$$

The expectation of the second term is therefore

$$-2 \left(\int_{\mathcal{R}} \nabla_\theta \lambda(s \,;\, \theta)\, \mathrm{d}s \right) \left(\int_{\mathcal{R}} \nabla_\theta \lambda(s \,;\, \theta)\, \mathrm{d}s \right)^T. \qquad (4.32)$$

The third term is trickier. Rewrite it as a double sum:

$$\sum_{j=1}^{K} \sum_{j'=1}^{K} \left(\frac{n_j}{\int_{\mathcal{R}_j} \lambda(s \,;\, \theta)\, \mathrm{d}s} \int_{\mathcal{R}_j} \nabla_\theta \lambda(s \,;\, \theta)\, \mathrm{d}s \right) \left(\frac{n_{j'}}{\int_{\mathcal{R}_{j'}} \lambda(s \,;\, \theta)\, \mathrm{d}s} \int_{\mathcal{R}_{j'}} \nabla_\theta \lambda(s \,;\, \theta)\, \mathrm{d}s \right)^T.$$

There are two cases. In the first case, $j \neq j'$ and the cells $\mathcal{R}_j$ and $\mathcal{R}_{j'}$ are disjoint. The summands are therefore independent and the expectation of their product is the product of their expectations. In the same manner as done in (4.31), the expectation simplifies to

$$E \left[\sum_{\substack{j,\, j'=1 \\ j \neq j'}}^{K} \right] = \sum_{\substack{j,\, j'=1 \\ j \neq j'}}^{K} \left(\int_{\mathcal{R}_j} \nabla_\theta \lambda(s \,;\, \theta)\, \mathrm{d}s \right) \left(\int_{\mathcal{R}_{j'}} \nabla_\theta \lambda(s \,;\, \theta)\, \mathrm{d}s \right)^T. \qquad (4.33)$$

In the other case, $j = j'$ and the double sum is the single sum,

$$\sum_{j=1}^{K} \left(\frac{n_j}{\int_{\mathcal{R}_j} \lambda(s \,;\, \theta)\, \mathrm{d}s} \int_{\mathcal{R}_j} \nabla_\theta \lambda(s \,;\, \theta)\, \mathrm{d}s \right) \left(\frac{n_j}{\int_{\mathcal{R}_j} \lambda(s \,;\, \theta)\, \mathrm{d}s} \int_{\mathcal{R}_j} \nabla_\theta \lambda(s \,;\, \theta)\, \mathrm{d}s \right)^T.$$

Because K is fixed, the expectation of the sum is the sum of the expectations. Denote the summand by e_{jj} and write it as a product of sums:

$$e_{jj} = \left(\sum_{\rho=1}^{n_j} \frac{\int_{\mathcal{R}_j} \nabla_\theta \lambda(s \,;\, \theta)\, \mathrm{d}s}{\int_{\mathcal{R}_j} \lambda(s \,;\, \theta)\, \mathrm{d}s} \right) \left(\sum_{\rho'=1}^{n_j} \frac{\int_{\mathcal{R}_j} \nabla_\theta \lambda(s \,;\, \theta)\, \mathrm{d}s}{\int_{\mathcal{R}_j} \lambda(s \,;\, \theta)\, \mathrm{d}s} \right)^T.$$

Using the identity (2.39) and simplifying gives

$$E\left[e_{jj}\right] = \left(\int_{\mathcal{R}_j} \nabla_\theta \lambda(s \,;\, \theta)\, \mathrm{d}s \right) \left(\int_{\mathcal{R}_j} \nabla_\theta \lambda(s \,;\, \theta)\, \mathrm{d}s \right)^T$$

$$+ \frac{1}{\int_{\mathcal{R}_j} \lambda(s \,;\, \theta)\, \mathrm{d}s} \left(\int_{\mathcal{R}_j} \nabla_\theta \lambda(s \,;\, \theta)\, \mathrm{d}s \right) \left(\int_{\mathcal{R}_j} \nabla_\theta \lambda(s \,;\, \theta)\, \mathrm{d}s \right)^T. \qquad (4.34)$$

Now add the sum over j of (4.34) to (4.33). The double sum no longer has the exception $j \neq j'$, so it becomes the product of single term sums over j. The single term sums are identical to integrals over all $\mathcal{R}$. Therefore, the expectation of the third term is

$$\left(\int_{\mathcal{R}} \nabla_{\theta}\, \lambda(s\,;\,\theta)\, ds \right) \left(\int_{\mathcal{R}} \nabla_{\theta}\, \lambda(s\,;\,\theta)\, ds \right)^{T} + J(\theta).$$

Finally, adding the expectations of the three terms gives the FIM (4.28).

Example 4.4 Scaled Intensity (continued). Let the intensity be the same as in Example 4.1. The FIM for the scale factor I using histogram data is

$$J(I) = \sum_{j=1}^{K} \frac{1}{\int_{\mathcal{R}_j} I\, f(x)\, dx} \left(\int_{\mathcal{R}_j} f(x)\, dx \right)^{2} = \frac{1}{I} \sum_{j=1}^{K} \int_{\mathcal{R}_j} f(x)\, dx = \frac{\int_{\mathcal{R}} f(x)\, dx}{I}.$$

The CRB of I is therefore the same for both conditionally independent sample data and histogram data. The ML estimator for histogram data is given by (3.6) with the Gaussian pdf replaced by $f(x)$. Its variance is

$$\begin{aligned}
\mathrm{Var}\left[\hat{I}_{ML} \right] &= \frac{\sum_{j=1}^{K} \mathrm{Var}\left[m_j \right]}{\left(\int_{\mathcal{R}} f(s)\, ds \right)^{2}} \\
&= \frac{\sum_{j=1}^{K} \int_{\mathcal{R}_j} I\, f(s)\, ds}{\left(\int_{\mathcal{R}} f(s)\, ds \right)^{2}} \\
&= I\, \frac{\sum_{j=1}^{K} \int_{\mathcal{R}_j} f(s)\, ds}{\left(\int_{\mathcal{R}} f(s)\, ds \right)^{2}} = \frac{I}{\int_{\mathcal{R}} f(s)\, ds}.
\end{aligned}$$

The ML estimator is therefore unbiased and efficient for histogram data.

4.4 CRB for PPP Intensity on Discrete Spaces

Let $\varPhi = \{\phi_1,\, \phi_2,\, \ldots\}$ be a discrete space as discussed in Section 2.12.1, and let the intensity vector $\lambda(\theta) = \{\lambda_1(\theta),\, \lambda_2(\theta),\, \ldots\} > 0$ of the PPP on $\varPhi$ depend on the parameter θ. The FIM of the PPP $\varXi$ corresponding to $\lambda(\theta)$ is

$$J(\theta) = \sum_{j \in \mathcal{R}} \frac{1}{\lambda(\phi_j\,;\,\theta)} \left[\nabla_{\theta}\, \lambda(\phi_j\,;\,\theta) \right] \left[\nabla_{\theta}\, \lambda(\phi_j\,;\,\theta) \right]^{T}. \qquad (4.35)$$

The derivatives are evaluated at the true value of θ. To see that (4.35) holds, it is only necessary to follow the steps of the proof of (4.21) for PPPs on continuous spaces, replacing integrals with the appropriate sums. This requires verifying that certain results hold in the discrete case, e.g., Campbell's Theorem. Details are omitted.

A quick intuitive way to see that the result must hold is as follows: imagine that the points of Φ are isolated points of the region over which the integral in (4.21) is performed. Let the intensity function be a test function sequence for these isolated points (of the kind used to define the Dirac delta function). Then (4.21) goes in the limit to (4.35) as the test function sequence "converges" to the Dirac delta function.

Example 4.5 Intensity Vector. Let Ξ denote a PPP on the discrete space $\Phi = \{\phi_1, \phi_2\}$ with intensity vector $\lambda(\theta) = (\lambda_1(\theta), \lambda_2(\theta))$, where $\theta = (\theta_1, \theta_2)$ and

$$\lambda_1(\theta) = \theta_1$$
$$\lambda_2(\theta) = \theta_2 . \tag{4.36}$$

For $\mathcal{R} = \Phi$, the FIM of θ is

$$J(\theta) = \frac{1}{\theta_1} \begin{bmatrix} 1 \\ 0 \end{bmatrix} \begin{bmatrix} 1 & 0 \end{bmatrix}^T + \frac{1}{\theta_2} \begin{bmatrix} 0 \\ 1 \end{bmatrix} \begin{bmatrix} 0 & 1 \end{bmatrix}^T \tag{4.37}$$
$$= \begin{bmatrix} \frac{1}{\theta_1} & 0 \\ 0 & \frac{1}{\theta_2} \end{bmatrix} .$$

The FIM $J(\theta)$ is a diagonal matrix, so the CRBs of the estimates of θ_1 and θ_2 are uncoupled, as anticipated. Their CRBs are equal to the CRBs of θ_1 and θ_2, respectively. That the CRBs are equal to the mean number of occurrences of ϕ_1 and ϕ_2 is a reflection of the fact that the mean and variance of a Poisson distributed random variable are equal.

Example 4.6 Parametrically Tied Intensity Vector. Let Ξ denote a PPP on the discrete space $\Phi = \{\phi_1, \phi_2\}$ with intensity vector

$$\lambda(\theta) = (I \cos\theta, \ I \sin\theta) ,$$

where I is known and $\theta \in \left(0, \frac{\pi}{2}\right)$. For $\mathcal{R} = \Phi$, the FIM is

$$J(\theta) = \frac{1}{I \cos\theta} (-I \sin\theta)^2 + \frac{1}{I \sin\theta} (I \cos\theta)^2$$
$$= I \frac{\sin^3\theta + \cos^3\theta}{\sin\theta \cos\theta} .$$

The CRB is the inverse of $J(\theta)$. As is easily seen, $\frac{d}{d\theta} J(\theta) = -I \cos(2\theta)$. The derivative is zero at $\theta = \pi/4$, and this corresponds to the global minimum for $J(\theta)$ of $I/2$ on the interval $\left(0, \frac{\pi}{2}\right)$. Hence, the CRB of θ is largest when the Poisson distributions for the number of occurrences of ϕ_1 and ϕ_2 are identical. The CRB of θ is $2/I$ for $\theta = \pi/4$.

4.5 Gating: Gauss on a Pedestal

A classic problem concerns estimating a Gaussian signal in noise, or clutter. For PPPs the CRB for the location of the Gaussian signal takes an especially compact and explicit form. The intensity is modeled as the sum of a nonhomogeneous signal and a nonhomogeneous noise (pedestal):

$$\lambda(x \,;\, \mu) \,=\, \lambda_N \, p_{\text{noise}}(x) \,+\, \lambda_S \, \mathcal{N}(x \,;\, \mu, \, \Sigma) \,, \tag{4.38}$$

where $\int_{\mathcal{R}} p_{\text{noise}}(x)\, dx \,=\, 1$. The intensity of the signal process is proportional to a multivariate Gaussian pdf centered at μ with spread determined by the eigenvalues and eigenvectors of Σ. The expected signal and noise counts, λ_S and λ_N, respectively, are dimensionless. Define the signal to noise ratio (SNR) by

$$\gamma_{SNR} \,=\, \frac{\lambda_S}{\lambda_N} \,.$$

The CRB for the Gaussian mean μ is given in this section.

The parameter μ is estimated from a realization of the PPP with intensity (4.38) on a bounded set $\mathcal{R}$ called the gate. The gate is an arbitrary bounded subset of $\mathbb{R}^{n_x}$, but is typically a standardized ellipsoid (as in Example 4.7 below). The gradient of (4.38) is

$$\nabla_\mu \, \lambda(x \,;\, \mu) \,=\, \lambda_S \, \mathcal{N}(x \,;\, \mu, \, \Sigma) \, \Sigma^{-1}(x - \mu) \,. \tag{4.39}$$

From (4.21), the FIM for μ using conditionally independent sample data is written in the form

$$J_{\mathcal{R}}(\mu) \,=\, \lambda_S \, \Sigma^{-1} \, W_{\mathcal{R}}(\mu) \, \Sigma^{-1} \,, \tag{4.40}$$

where the weighted covariance matrix $W_{\mathcal{R}}(\mu)$ is defined by

$$W_{\mathcal{R}}(\mu) \,=\, \gamma_{SNR} \int_{\mathcal{R}} \frac{(\mathcal{N}(x \,;\, \mu, \, \Sigma))^2}{p_{\text{noise}}(x) \,+\, \gamma_{SNR} \, \mathcal{N}(x; \mu, \Sigma)} \, (x - \mu)(x - \mu)^T \, dx \,. \tag{4.41}$$

The matrix $W_{\mathcal{R}}(\mu)$ is evaluated at the correct value of μ. The CRB is the inverse of the FIM, so

$$CRB_{\mathcal{R}}(\mu) \,=\, \frac{1}{\lambda_S} \, \Sigma \, W_{\mathcal{R}}^{-1}(\mu) \, \Sigma \,. \tag{4.42}$$

It is not required that $\mu \in \mathcal{R}$, but the CRB is large if μ is outside of $\mathcal{R}$.

A curious aspect of the CRB (4.42) is that it involves both signal strength λ_S and SNR. The signal strength λ_S governs the average *absolute* number of signal points in a realization, and thus inversely scales the variance reduction at a given SNR. On the other hand, the trade-off between signal and noise pdfs and their effect on the shape (eigenvalues and eigenvectors) of the weighted covariance matrix $W_{\mathcal{R}}(\mu)$ is determined by the average fraction of points in the realization that originate from the signal. This fraction is governed by γ_{SNR}. Good estimation therefore depends on both λ_S and γ_{SNR}.

The CRB for the covariance matrix Σ can also be found using the general result (4.21). The results seem to provide little insight and are somewhat tedious to obtain, so they are not presented here. Details for the closely related problem of evaluating the CRB of Σ for the classical multivariate Gaussian density using i.i.d. data can be found in [116].

The general result (4.21) will also give the CRB for non-Gaussian signals, that is, for data that are a realization of the PPP with intensity

$$\lambda(x \, ; \, \mu) \; = \; \lambda_N \, p_{\text{noise}}(x) \, + \, \lambda_S \, p_{\text{signal}}(x \, ; \, \mu),$$

where $p_{\text{signal}}(x \, ; \, \mu)$ is an arbitrary pdf with location parameter vector μ. These problems are application specific and are left to the reader.

A different, but related, approach to the finding CRB-like bounds for signals in clutter are the scaled "information-reduction" matrices of [89]. Comparisons between this method and the PPP approach would be interesting, but they are not given here.

Example 4.7 Effect of Gating on Estimated Mean. The structure of the CRB is explored numerically for the special case $\mathcal{R} = \mathbb{R}^1$. Let

$$\mathcal{R}(\rho) \; = \; \{x \, : \, (x - \mu)^T \Sigma^{-1}(x - \mu) \leq \rho^2\},$$

where ρ determines the standardized gate volume. In one dimension, $\Sigma \equiv \sigma^2$ and (4.42) gives

$$CRB_{\mathcal{R}(\rho)}(\mu) \; = \; \frac{\pi \, \sigma^3}{\lambda_S \, \gamma_{SNR}} \left[\int_0^\rho \frac{x^2 \, e^{-x^2}}{p_{\text{noise}}(\sigma \, x) \, + \, \frac{\gamma_{SNR}}{\sqrt{2\pi} \, \sigma} \, e^{-\frac{x^2}{2}}} \, dx \right]^{-1} . \tag{4.43}$$

The term in brackets has units of length, so the CRB has units of length squared, as required in one dimension.

The effect of SNR and gate size on the CRB is seen by plotting $CRB_{\mathcal{R}(\rho)}(\mu)$ as a function of ρ for several values of SNR and for, say, $\sigma = 1$. Let $p_{\text{noise}}(x) \equiv 1$. It is seen in Fig. 4.1 that the CRB decreases with increasing ρ and SNR. It is also seen that there is no practical reason to use gates larger than $\rho = 2.5$ in the one dimensional case, regardless of SNR.

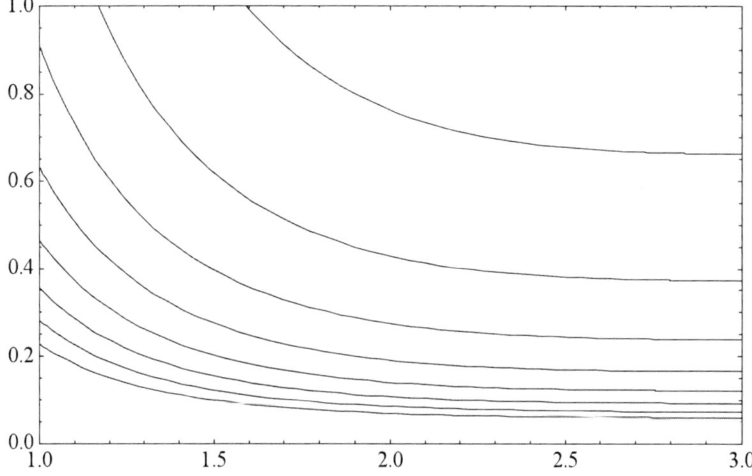

Fig. 4.1 CRB of (4.43) for μ as a function of gate size ρ: $\lambda_N = \sqrt{2\pi} \approx 2.5$ points per unit length; $SNR = 3(1)10$; $\sigma = 1$. Gates larger than $\rho = 2.5$ do little to improve estimation, at least in $\mathbb{R}^1$

4.6 Joint CRB for Gaussian Sums

The intensity of a finite superposition of noise and L independent PPPs with Gaussian shaped intensities is

$$
\lambda(x\,;\,\mu,\,\Lambda) \equiv \lambda(x\,;\,\mu_1,\ldots,\mu_L,\,\lambda_1,\ldots,\lambda_L)
$$
$$
= \lambda_N\,p_{\text{noise}}(x) + \sum_{\ell=1}^{L} \lambda_\ell\,\mathcal{N}(x\,;\,\mu_\ell,\,\Sigma_\ell), \tag{4.44}
$$

where $\mu = (\mu_1,\ldots,\mu_L)$ and $\Lambda = (\lambda_1,\ldots,\lambda_L) > 0$. The special case $L = 1$ is the model (4.38) of the previous section. If all the parameters are specified except for one of the means, say μ_j, then the CRB of μ_j is found by using the nonhomogeneous noise term

$$
\lambda_{\text{noise}(j)}(x) \equiv \lambda_N\,p_{\text{noise}}(x) + \sum_{\substack{\ell=1 \\ \ell\neq j}}^{L} \lambda_\ell\,\mathcal{N}(x\,;\,\mu_\ell,\,\Sigma_\ell) \tag{4.45}
$$

in (4.38). The FIM and CRB are then evaluated using the appropriately interpreted versions of (4.41) and (4.42).

4.6.1 Mean Vectors in a Gaussian Sum

The CRB is a joint function of all the means that are estimated. The CRB of a particular mean, say μ_j, is found from the joint FIM using the result (4.15). This CRB is very different from the one found by assuming that all the means except μ_j are known.

For simplicity, assume that all the signal mean vectors in (4.44) are estimated. Evaluating the joint FIM for the full parameter vector μ requires the gradient with respect to all the means in μ. In this case, the required gradient is the stacked vector

$$\nabla_{\mu}\,\lambda(x\,;\,\mu,\,\Lambda) = \begin{bmatrix} \nabla_{\mu_1}\,\lambda(x\,;\,\mu,\,\Lambda) \\ \vdots \\ \nabla_{\mu_L}\,\lambda(x\,;\,\mu,\,\Lambda) \end{bmatrix} = \begin{bmatrix} \lambda_1\,\mathcal{N}(x\,;\,\mu_1,\,\Sigma_1)\,\Sigma_1^{-1}\,(x-\mu_1) \\ \vdots \\ \lambda_L\,\mathcal{N}(x\,;\,\mu_L,\,\Sigma_L)\,\Sigma_L^{-1}\,(x-\mu_L) \end{bmatrix}.$$

$$(4.46)$$

The ij-th element of the outer product required by (4.41) is thus

$$\left(\lambda_i\,\mathcal{N}(x\,;\,\mu_i,\,\Sigma_i)\,\Sigma_i^{-1}\,(x-\mu_i)\right)\left(\lambda_j\,\mathcal{N}(x\,;\,\mu_j,\,\Sigma_j)\,\Sigma_j^{-1}\,(x-\mu_j)\right)^T$$
$$= \lambda_i\,\lambda_j\,\mathcal{N}(x\,;\,\mu_i,\,\Sigma_i)\,\mathcal{N}(x\,;\,\mu_j,\,\Sigma_j)\,\Sigma_i^{-1}\,(x-\mu_i)(x-\mu_j)^T\,\Sigma_j^{-1}.$$

$$(4.47)$$

The FIM, $J(\mu)$, is an $L\times L$ block matrix. Its entries are matrices of size $n_x \times n_x$, so its full dimension is $(Ln_x)\times(Ln_x)$. The ij-th block of $J(\mu)$ is the integral over $\mathcal{R}$ of (4.47) divided by $\lambda(x\,;\,\mu,\,\Lambda)$. Explicitly,

$$J_{ij}(\mu) = \lambda_i\,\lambda_j\,\Sigma_i^{-1}\,W_{\mathcal{R}}^{ij}(\mu,\,\Lambda)\,\Sigma_j^{-1},$$

$$(4.48)$$

where the $n_x \times n_x$ weighted covariance matrix is defined by

$$W_{\mathcal{R}}^{ij}(\mu,\,\Lambda) = \int_{\mathcal{R}} \frac{\mathcal{N}(x\,;\,\mu_i,\,\Sigma_i)\,\mathcal{N}(x\,;\,\mu_j,\,\Sigma_j)}{\lambda(x\,;\,\mu,\,\Lambda)}\,(x-\mu_i)(x-\mu_j)^T\,dx.$$

$$(4.49)$$

The matrix $W_{\mathcal{R}}(\mu,\,\Lambda) = \left[W_{\mathcal{R}}^{ij}(\mu,\,\Lambda)\right]$ is an $L\times L$ block matrix with $n_x\times n_x$ blocks, so its full dimension is the same as $J(\mu)$. Writing the system (4.49) in matrix form gives

$$J(\mu) = D^{-1}(\Lambda)\,W_{\mathcal{R}}(\mu,\,\Lambda)\,D^{-1}(\Lambda),$$

$$(4.50)$$

where

$$
D^{-1}(\Lambda) =
\begin{bmatrix}
\lambda_1 \Sigma_1^{-1} & 0 & \cdots & 0 \\
0 & \lambda_2 \Sigma_2^{-1} & \cdots & 0 \\
\vdots & \vdots & \ddots & \vdots \\
0 & 0 & \cdots & \lambda_L \Sigma_L^{-1}
\end{bmatrix}
\tag{4.51}
$$

is an $L \times L$ block diagonal matrix with $n_x \times n_x$ blocks. The matrix $D^{-1}(\Lambda)$ is the same overall size as the FIM. The CRB is, therefore,

$$
CRB(\mu) = D(\Lambda)\, W_{\mathcal{R}}^{-1}(\mu, \Lambda)\, D(\Lambda)\,,
\tag{4.52}
$$

where

$$
D(\Lambda) =
\begin{bmatrix}
\frac{1}{\lambda_1} \Sigma_1 & 0 & \cdots & 0 \\
0 & \frac{1}{\lambda_2} \Sigma_2 & \cdots & 0 \\
\vdots & \vdots & \ddots & \vdots \\
0 & 0 & \cdots & \frac{1}{\lambda_L} \Sigma_L
\end{bmatrix}.
\tag{4.53}
$$

The CRB depends jointly on all the means μ_j because, in general, none of the block matrix elements of $W_{\mathcal{R}}(\mu, \Lambda)$ are zero.

The joint CRB separates into CRBs for the individual PPPs in the superposition if the matrix $W_{\mathcal{R}}(\mu, \Lambda)$ is block diagonal. This happens if all the off-diagonal $n_x \times n_x$ blocks are approximately zero, that is, if

$$
W_{\mathcal{R}}^{ij}(\mu, \Lambda) \approx 0 \quad \text{for all} \quad i \neq j\,.
$$

The approximation is a good one if the coefficient of the outer product in the integral (4.49) is nearly zero for $i \neq j$. If the Gaussian pdfs are all well separated, that is, if

$$
\max_{i \neq j}\ (\mu_i - \mu_j)^T \left(\Sigma_i + \Sigma_j\right)^{-1} (\mu_i - \mu_j) \ll 1\,,
$$

then the joint CRB is block diagonal and splits into separate CRBs for each mean. More generally, the joint CRB splits the mean vectors into disjoint groups or clusters. The CRBs of the means within each cluster are coupled, but the CRBs of means in different clusters are independent.

4.6.2 Means and Coefficients in a Gaussian Sum

If the coefficients Λ and the means μ are estimated together, the CRB changes yet again. The FIM in this case, denoted by $J(\mu, \Lambda)$, is larger than $J(\mu)$ by exactly L rows and columns, so it has dimension $(L + L n_x) \times (L + L n_x)$. Partition it so that

$$J(\mu, \Lambda) = \begin{bmatrix} U_{\mathcal{R}}(\mu, \Lambda) & V_{\mathcal{R}}(\mu, \Lambda) \\ V_{\mathcal{R}}^{T}(\mu, \Lambda) & W_{\mathcal{R}}(\mu, \Lambda) \end{bmatrix}. \tag{4.54}$$

The matrix $W_{\mathcal{R}}(\mu, \Lambda)$ is the lower right $(L\,n_x) \times (L\,n_x)$ submatrix of this larger FIM. The upper left $L \times L$ submatrix, $U_{\mathcal{R}}(\mu, \Lambda)$, corresponds to the outer product $(\nabla_{\Lambda}\, \lambda(x\,;\, \mu,\, \Lambda))(\nabla_{\Lambda}\, \lambda(x\,;\, \mu,\, \Lambda))^{T}$; hence, its ij-th entry is

$$U_{\mathcal{R}}^{ij}(\mu, \Lambda) = \int_{\mathcal{R}} \frac{\mathcal{N}(x\,;\, \mu_i,\, \Sigma_i)\, \mathcal{N}(x\,;\, \mu_j,\, \Sigma_j)}{\lambda(x\,;\, \mu,\, \Lambda)}\, dx\,. \tag{4.55}$$

Finally, the $L \times (L\,n_x)$ matrix in the upper right hand corner of the partition, $V_{\mathcal{R}}(\mu, \Lambda)$, contains the cross terms, that is, the terms corresponding to the outer product

$$(\nabla_{\Lambda}\, \lambda(x\,;\, \mu,\, \Lambda))(\nabla_{\mu}\, \lambda(x\,;\, \mu,\, \Lambda))^{T}.$$

Its ij-th entry is therefore

$$V_{\mathcal{R}}^{ij}(\mu, \Lambda) = \Sigma_j^{-1} \int_{\mathcal{R}} \frac{\mathcal{N}(x\,;\, \mu_i,\, \Sigma_i)\, \mathcal{N}(x\,;\, \mu_j,\, \Sigma_j)}{\lambda(x\,;\, \mu,\, \Lambda)}\, (x - \mu_j)^{T}\, dx\,. \tag{4.56}$$

The lower left submatrix is clearly the transpose of $V_{\mathcal{R}}(\mu, \Lambda)$. The joint CRB of Λ and μ is the inverse of $J(\mu, \Lambda)$. It differs significantly from the CRB $J(\mu)$ for μ alone.

4.7 Observed Information Matrices

Evaluating the FIM requires evaluating many (multi-)dimensional integrals, and these integrals often do not have a closed form. Numerical integration is the obvious recourse, but such methods lose their appeal if they are deemed too computationally expensive in practice. This can happen if specialized numerical procedures must be tailored to the particular functions involved, if the integrals must be computed repeatedly, if there are a large number of parameters.

An alternative to the FIM is the observed information matrix (OIM). Although the OIM has no known optimality properties in general, it is nonetheless widely used in statistical practice as a surrogate for the FIM in difficult problems for which the FIM is unknown. The surrogate CRB is the inverse of the OIM. An interesting paper discussing the notion of the OIM in the single parameter case is [26].

Unlike the FIM, the OIM depends on the measured data. The sensitivity of the OIM to the actual data may be of independent interest in some applications, regardless of its utility as a surrogate to the FIM.

The OIM is the negative of the Hessian of the loglikelihood function evaluated at its maximum. For a general twice differentiable and bounded pdf, the OIM is defined by

$$OIM(\hat{\theta}_{ML}) = -\nabla_\theta (\nabla_\theta)^T \log p(x ; \hat{\theta}_{ML})$$

$$= -\nabla_\theta \left[\nabla_\theta \, p(x ; \hat{\theta}_{ML}) \right]^T , \qquad (4.57)$$

where the maximum likelihood parameter estimate is

$$\hat{\theta}_{ML} = \arg\max_\theta \log \, p(x ; \theta) .$$

The inverse of the OIM is always positive definite when $\hat{\theta}_{ML}$ is a local maximum interior to the domain of allowed parameter vectors, Θ. Intuitively, it is tempting to think of the OIM as the FIM without the expectation over the data. Unfortunately, this succinct description is slightly inaccurate technically, since the FIM is evaluated at the true parameter value while the OIM is evaluated at the ML estimate.

4.7.1 General Sums

In general, the OIM is evaluated by computing the appropriate second derivative matrix at the point $\hat{\theta}_{ML}$. This is a relatively easy calculation for PPPs whose intensity is a sum. Let

$$\lambda(x ; \theta) = \sum_{\ell=1}^{L} f_\ell(x ; \theta_\ell) , \qquad (4.58)$$

where $f_\ell(x ; \theta_\ell) > 0$ is a bounded twice differentiable function with respect to the parameter vector θ_ℓ for all $x \in \mathcal{R}$, and where $\theta = (\theta_1, \ldots, \theta_L)$ is the full parameter vector. In the most general case, the functions $f_\ell(\cdot)$ and their parameters are different; in general, θ_ℓ have different lengths. The loglikelihood of the PPP with intensity (4.58) is

$$\log p(\xi ; \theta) = -\int_{\mathcal{R}} \left(\sum_{\ell=1}^{L} f_\ell(x ; \theta_\ell) \right) dx + \sum_{j=1}^{n} \log \left(\sum_{\ell=1}^{L} f_\ell(x_j ; \theta_\ell) \right) , \qquad (4.59)$$

where $\xi = (n, \{x_1, \ldots, x_n\})$ is the given realization of the PPP. The following algebraic identity—whose terms are defined in the next paragraph—is obtained by straightforward, but tedious, differentiation and matrix algebra:

$$-\nabla_\theta [\nabla_\theta]^T \log p(\xi ; \theta) = A(\theta) + B(\theta) + C(\theta) + \sum_{j=1}^{n} S_j(\theta) \, S_j^T(\theta) . \qquad (4.60)$$

Those who wish to verify this result will find the identity

$$\nabla_{\theta_\ell} \log \left(\sum_{\ell=1}^{L} f_\ell(x \,; \theta_\ell) \right) = \frac{f_\ell(x \,; \theta_\ell)}{\lambda(x \,; \theta)} \, \nabla_{\theta_\ell} \log \, f_\ell(x \,; \theta_\ell)$$

very useful. Further details are omitted.

The stacked vector $S_j(\theta)$ has length $\dim(\theta) = \sum_{\ell=1}^{L} \dim(\theta_\ell)$ and is given by

$$S_j(\theta) \equiv \begin{bmatrix} S_{j1}(\theta) \\ \vdots \\ S_{jL}(\theta) \end{bmatrix} = \begin{bmatrix} w_1(x_1 \,; \theta) \, \nabla_{\theta_1} \log \, f_1(x_j \,; \theta_1) \\ \vdots \\ w_L(x_L \,; \theta) \, \nabla_{\theta_L} \log \, f_L(x_j \,; \theta_L) \end{bmatrix}, \quad j = 1, \ldots, n,$$

$$(4.61)$$

where the weights $w_{j\ell}(\theta)$ are

$$w_\ell(x_j \,; \theta) = \frac{f_\ell(x_j \,; \theta_\ell)}{\lambda(x_j \,; \theta)}, \quad j = 1, \ldots, n, \quad \ell = 1, \ldots, L. \qquad (4.62)$$

For $\ell = 1, \ldots, L$, the vector $S_{j\ell}(\theta)$ has length $\dim(\theta_\ell)$. The $\dim(\theta) \times \dim(\theta)$ matrices

$$A(\theta) = \mathrm{Diag}\,[A_1(\theta), \ldots, A_L(\theta)], \quad j = 1, \ldots, n,$$
$$B(\theta) = \mathrm{Diag}\,[B_1(\theta), \ldots, B_L(\theta)], \quad j = 1, \ldots, n,$$
$$C(\theta) = \mathrm{Diag}\,[C_1(\theta_1), \ldots, C_L(\theta_L)]$$

are block diagonal, and their ℓ-th diagonal blocks are size $\dim(\theta_\ell) \times \dim(\theta_\ell)$. The ℓ-th blocks are, for $\ell = 1, \ldots, L$,

$$A_\ell(\theta) = - \sum_{j=1}^{n} w_\ell(x_j \,; \theta) \left[\nabla_{\theta_\ell} \log f_\ell(x_j \,; \theta_\ell) \right] \left[\nabla_{\theta_\ell} \log f_\ell(x \,; \theta_\ell) \right]^T$$

$$B_\ell(\theta) = - \sum_{j=1}^{n} w_\ell(x_j \,; \theta) \, \nabla_{\theta_\ell} \left[\nabla_{\theta_\ell} \log f_\ell(x_j \,; \theta_\ell) \right]^T$$

$$C_\ell(\theta_\ell) = \nabla_{\theta_\ell} \left[\nabla_{\theta_\ell} \int_{\mathcal{R}} f_\ell(x \,; \theta_\ell) \, dx \right]^T .$$

The OIM is obtained by evaluating (4.60) at the ML estimate $\hat{\theta}_{ML}$.

In principle, it does not matter how the ML estimate is computed. As was first pointed out by Lewis [69], if the EM method is used, then the OIM can be computed using the weights evaluated at the last (convergent) step of the EM algorithm. The overall computational effort depends on the difficulty of evaluating the derivatives of the functions $f_\ell(x \,; \theta_\ell)$. The surrogate CRB is the inverse of the OIM, so the computational cost to find the CRB grows with cube of the total number of parameters, $\dim(\theta)$, once the OIM is known.

4.7.2 Affine Gaussian Sums

The EM approach to ML estimation is especially simple to implement for affine Gaussian sums of the form

$$\lambda(x \, ; \, \theta) = \lambda_0(x) + \sum_{\ell=1}^{L} \lambda_\ell \, \mathcal{N}(x \, ; \, \mu_\ell, \, \Sigma_\ell) \, ,$$

where $\lambda_0(x)$ is the intensity of a known PPP background process. Except for $\lambda_0(x)$, this sum is the same as (4.58) with

$$f_\ell(x \, ; \, \theta_\ell) = \lambda_\ell \, \mathcal{N}(x \, ; \, \mu_\ell, \, \Sigma_\ell) \, , \quad \ell = 1, \, \ldots, \, L \, .$$

After EM algorithm convergence, the OIM is easily evaluated using the weights that are computed during the last EM iteration. Here, for simplicity, the coefficients λ_ℓ and covariance matrices Σ_ℓ are specified, and only the mean vectors are estimated, so in the above $\theta_\ell = \mu_\ell$ and $\theta \equiv \mu = (\mu_1, \, \ldots, \, \mu_L)$. The only change in the OIM calculation required to accommodate the affine term $\lambda_0(x)$ is to adjust the weights to include it in the denominator; explicitly,

$$w_\ell(x_j \, ; \, \mu) = \frac{\lambda_\ell \, \mathcal{N}(x_j \, ; \, \mu_\ell, \, \Sigma_\ell)}{\lambda_0(x_j) + \sum_{\ell=1}^{L} \lambda_\ell \, \mathcal{N}(x_j \, ; \, \mu_\ell, \, \Sigma_\ell)} \, , \quad \begin{array}{l} j = 1, \, \ldots, \, n \, , \\ \ell = 1, \, \ldots, \, L \, . \end{array} \quad (4.63)$$

Evaluating (4.60) gives

$$A_\ell(\mu) = -\Sigma_\ell^{-1} \left[\sum_{j-1}^{n} w_\ell(x_j \, ; \, \mu) \, (x_j - \mu_\ell)(x_j - \mu_\ell)^T \right] \Sigma_\ell^{-1}$$

$$B_\ell(\mu) = \left(\sum_{j=1}^{n} w_\ell(x_j \, ; \, \mu) \right) \Sigma_\ell^{-1}$$

$$C_\ell(\mu_\ell) = \lambda_\ell \Sigma_\ell^{-1} \left\{ \int_{\mathcal{R}} \mathcal{N}(x \, ; \, \mu_\ell, \, \Sigma_\ell) \left[-\Sigma_\ell + (x - \mu_\ell)(x - \mu_\ell)^T \right] dx \right\} \Sigma_\ell^{-1}$$

$$S_j(\mu) - \begin{bmatrix} w_1(x_j \, ; \, \mu) \, \Sigma_1^{-1} \, (x_j - \mu_1) \\ \vdots \\ w_L(x_j \, ; \, \mu) \, \Sigma_L^{-1}(x_j - \mu_L) \end{bmatrix} .$$

In these equations, μ is taken to be the ML estimate $\mu_{ML} \equiv (\hat{\mu}_1, \, \ldots, \, \hat{\mu}_L)$.

The equations are more intuitive when written in a different form. Adding $A_\ell(\mu)$ and $B_\ell(\mu)$ gives

$$A_\ell(\mu) + B_\ell(\mu) = \left(\sum_{j=1}^{n} w_\ell(x_j ; \mu) \right) \Sigma_\ell^{-1} \left[\Sigma_\ell - \tilde{\Sigma}_\ell \right] \Sigma_\ell^{-1}, \qquad (4.64)$$

where the weighted covariance matrix for the ℓ-th Gaussian term is

$$\tilde{\Sigma}_\ell = \frac{\sum_{j=1}^{n} w_\ell(x_j ; \mu)\,(x_j - \mu_\ell)(x_j - \mu_\ell)^T}{\sum_{j=1}^{n} w_\ell(x_j ; \mu)}.$$

The coefficient of (4.64) is the conditional expected number of samples that originate from the ℓ-th Gaussian component. Similarly,

$$C_\ell(\mu) = - \left(\lambda_\ell \int_{\mathcal{R}} \mathcal{N}(x ; \mu_\ell, \Sigma_\ell)\,dx \right) \Sigma_\ell^{-1} \left[\Sigma_\ell - \bar{\Sigma}_\ell \right] \Sigma_\ell^{-1}, \qquad (4.65)$$

where the conditional covariance matrix is

$$\bar{\Sigma}_\ell = \frac{\int_{\mathcal{R}} \mathcal{N}(x ; \mu_\ell, \Sigma_\ell)\,(x - \mu_\ell)(x - \mu_\ell)^T\,dx}{\int_{\mathcal{R}} \mathcal{N}(x ; \mu_\ell, \Sigma_\ell)\,dx}.$$

The negative of the coefficient of (4.65) is the expected number of samples from the ℓ-th Gaussian component. If the bulk of the ℓ-th Gaussian component lies within $\mathcal{R}$, the term $C_\ell(\mu)$ is approximately zero.

With these forms it is clear that the first three terms in the OIM are comparable in some situations, in which case it is not clear whether or not their sum is positive definite. In these situations, the fourth term is an important one in the OIM, since the OIM as a whole must be positive definite at the ML estimate $\hat{\mu}_{ML}$. The (ρ, ℓ)-th block component of the sum of outer products of $S_j(\mu)$ is

$$\left[\sum_{j=1}^{n} S_j(\mu)\, S_j^T(\mu) \right]_{\rho\ell} = \Sigma_\rho^{-1} \left(\sum_{j=1}^{n} w_\rho(x_j ; \mu)\, w_\ell(x_j ; \mu)\,(x_j - \mu_\rho)(x_j - \mu_\ell)^T \right) \Sigma_\ell^{-1}.$$

$$(4.66)$$

The only off-diagonal block terms of the OIM come from (4.66). When the Gaussian components are well separated, the off-diagonal blocks are small, and the OIM is approximately block diagonal. Given good separation, then, for every ℓ the sum of the four terms is positive definite at the ML estimate $\hat{\mu}_{ML}$.

The OIM calculation requires only L numerical integrals in this case, significantly fewer that the analogous FIM calculation. Some of these integrals are unnecessary if the bulk of the Gaussian density evaluated at $\hat{\theta}_\ell$ lies inside $\mathcal{R}$, for in this case the integral in $C_\ell(\hat{\theta}_\ell)$ vanishes. Avoiding numerical integration is an advantage in practice, provided it is verified by simulation or otherwise that the OIM performs satisfactorily as a surrogate for the FIM.

Obvious modifications must be made to the algebra if the estimation problem is changed. For example, in some applications, the only estimated parameters are the coefficients λ_ℓ because both the means μ_ℓ and the covariances Σ_ℓ are known. The OIM in this case is $L \times L$. In other problems both the coefficients and the means are estimated, in which case the OIM has exactly L more rows and columns than the OIM for means alone.

Part II
Applications to Imaging, Tracking, and Distributed Sensing

Chapter 5
Tomographic Imaging

There are a thousand thousand reasons to live this life, every one of them sufficient.

Marilynne Robinson, Gilead, 2004

Abstract PPP methods for tomographic imaging are presented in this chapter. The primary emphasis is on methods for emission tomography, but transmission tomography is also included. The famous Shepp-Vardi algorithm for positron emission tomography (PET) is obtained via the EM method for time-of-flight data. Single-photon emission computed tomography (SPECT) is used in practice much more often than PET. It differs from PET in many ways, yet the models and the mathematics of the two methods are similar. (Both PET and SPECT are also closely related to multitarget tracking problems discussed in Chapter 6.) Transmission tomography is the final topic discussed. The Lange-Carson algorithm is derived via the EM method. CRBs for unbiased estimators for emission and transmission tomography are discussed. Regularization and Grenander's method of sieves are reviewed in the last section.

Keywords Positron emission tomography (PET) · Image restoration · Image reconstruction · PET with time of flight data · PET with histogram data · Shepp-Vardi algorithm · Single photon emission computed tomography (SPECT) · EM method · Miller-Snyder-Miller algorithm · Transmission tomography · Lange-Carson algorithm · CRB for emission and transmission tomography

Emission and transmission tomographic imaging methods are discussed in this chapter. The purpose here is twofold. One is to expose the nature of the problems involved in image reconstruction, so the level of engineering and medical detail is necessarily somewhat idealized from the point of view of these applications. The other purpose is to develop algorithms for emission and transmission tomography. Once the mathematical form of the relationship of the estimated parameters is written down, it is perhaps not too surprising—given the methods of Chapter 3—that both are grounded in the EM method.

Positron emission tomography (PET) is discussed first. The goal of PET is to estimate the intensity function of a physically defined PPP from sample data. The

R.L. Streit, *Poisson Point Processes*, DOI 10.1007/978-1-4419-6923-1_5, intensity function is the sought-after image. The Shepp-Vardi algorithm (or, as mentioned in the introductory chapter, the Richardson-Lucy deconvolution algorithm) produces the ML estimate of the image in a pixel-by-pixel manner, an excellent attribute for high resolution images. Both steps of the EM method are solved explicitly. The small cell limit is of particular interest for applications in tracking.

PET with histogram data is discussed next. Histogram data are harder to use than PPP sample data. Histogram data make developing an ML algorithm more elaborate, but in the end lead to an algorithm no more difficult to implement than the one for PPP sample data. The Shepp-Vardi algorithm, as first presented in [110], was designed for histogram data.

Single photon emission computed tomography (SPECT) is also discussed. For various reasons, it is used much more often diagnostically than is PET even though it provides lower resolution than PET. Algorithms for SPECT are similar to the algorithms for PET since both are based on the EM method, but there are several interesting differences between them. One is that the calculus of variations is used to solve the M-step in SPECT.

Transmission tomography is discussed in the last section. The physical nature of the problem is significantly different from PET. The goal in this instance is to estimate the attenuation coefficients of the image pixels. PPPs are involved as a known source of photons (or particles) that are subsequently attenuated by the object. Measurements comprise the numbers of photons received in an array of detectors, and therefore constitute a realization of a PPP on a discrete state space. The points of this discrete space are the individual detectors in the array. The Lange-Carson algorithm is an ML estimator of the attenuation coefficients of the pixels in the image. As in PET, it proceeds computationally in a pixel-by-pixel manner. Although it differs considerably in its development from Shepp-Vardi, both are obtained via the EM method. The most significant algorithmic difference is that the M-step is solved numerically, not explicitly. Since the equations are solved pixel-by-pixel (i.e., are in one variable only), this is not an issue in practice.

Finally, the CRBs are presented for PET and transmission tomography. They are easily obtained from the intensity function that parameterizes the PPPs.

5.1 Positron Emission Tomography

In PET imaging, a short-lived radioisotope is injected into the blood stream (attached to a sugar, e.g., fluorodesoxyglucose) and the sugar is absorbed, or metabolized. The degree of absorption varies by tissue type, but pre-cancerous cells typically show heightened metabolic activity. Pre-cancerous body tissues are imaged by estimating the spatial density of the radioisotope. This density is directly proportional to the intensity, $\lambda(x)$, of radioisotope decay at the point x. Brain imaging via PET is depicted in Fig. 5.1. The overview paper [70] discusses many issues, both theoretical and practical, not mentioned here.

The radioisotope undergoes beta decay and emits positrons. Positrons quickly encounter electrons (within a few millimeters), and they annihilate. Annihilation

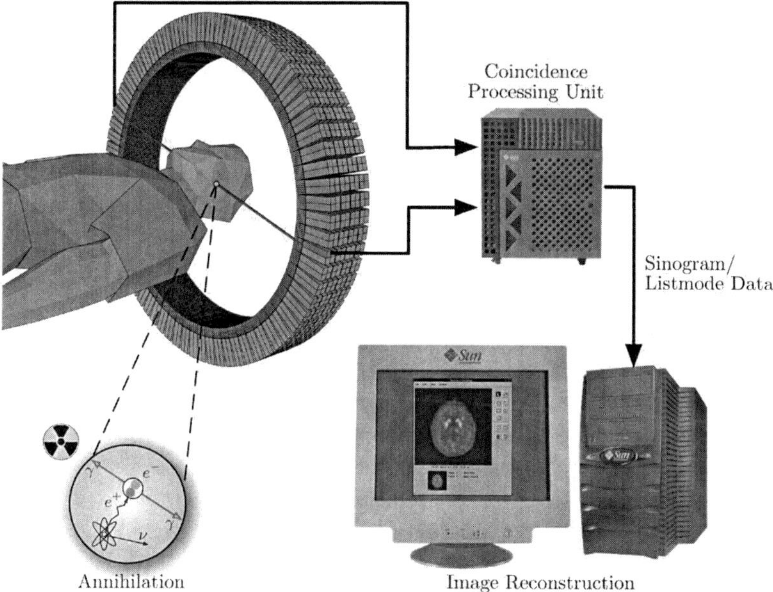

Fig. 5.1 PET processing configuration. (Image released to the Wikimedia Commons by J. Langner [66])

events emit pairs of (gamma) photons that move in opposite directions. Due to conservation of momentum, these directions are essentially collinear if the positrons and electrons are effectively zero velocity. Departures from straight line motion degrade spatial resolution, so some systems model these effects to avoid losing resolution. Straight line propagation is assumed here.

The raw measurement data are the arrival times of photons at an array of detectors that comprise scintillator crystals and photomultipliers. A pair of photons arriving within a sufficiently short time window (measured in nanoseconds, ns) at two appropriately sited detectors determine that an annihilation event occurred: the event lies on the chord segment connecting the detectors, and the specific location on the chord is determined by the time difference of arrivals. Photons without partners within the time window are discarded. The measurement procedure produces occasional spurious annihilation events.

Many subtleties attend to the data collection and preprocessing steps for PET and SPECT. An excellent review of these issues as well as the field of emission tomography as a whole is found in [70].

PET is now often part of a multisensor fusion system that combines other imaging methodologies such as CT (computed tomography) and MRI (magnetic resonance imaging). These topics are an active area of research.

Transmission tomography, more popularly known as computed tomography (CT), uses multiple sets of fan beams or cone beams to determine the spatial distribution of material density, that is, the local spatial variability of the attenuation coefficient. Both fan and cone beams employ a single source that produces radiation in many directions simultaneously. This shortens the data gathering time. Fan beams are used for two dimensional imaging problems, and their detector array is one dimensional. Similarly, cone beams are used for three dimensional problems, and they require a two dimensional detector array. In practice, CT scans are much less expensive than PET because it does not require the production of short lived radioisotopes. Contrast enhancement agents may be used to improve imaging of certain tissue types in medical applications. PET, SPECT, and CT are used diagnostically for very different purposes. PET and CT provide significantly higher resolution images than SPECT.

EM methods are discussed for PET, SPECT, and transmission tomography. Numerical issues arise with EM algorithms when the number of estimated parameters is large. For PET and SPECT, the parameters are the intensities in the array of pixels/voxels of the image; for CT, the parameters are the pixel attenuation coefficients. Good resolution therefore requires a large number of parameters. Regularization methods compatible with the EM method can alleviate many of these issues. See [119] for further discussion of this and other topics involving medical imaging applications of PPPs and related point processes.

A prominent alternative method for transmission tomography is Fourier reconstruction. This approach is fully discussed in the book [28], as well as in the paper [22]. These are Fourier analysis based methods, of which the Projection-Slice Theorem is perhaps the best known result in the field. The approach is classical analysis in the sense that it is ultimately grounded on the Radon transform (1917) and the Funk-Hecke Theorem, a lovely but little known result that stimulated Radon's interest in these problems. Fourier methods are not discussed further here.

5.2 PET: Time-of-Flight Data

The Expectation-Maximization method was first proposed for emission tomography in the 1982 paper [110]. This pioneering paper gives an excellent review of the physics of the problem and the exceptional fidelity of the model, none of which is repeated here. The objective is to estimate the (constant) intensity in an array of closely spaced detectors given data in the individual detectors. This is a nontrivial inverse problem because annihilations occurring in the volume corresponding to the same image pixel can be and are recorded in various detector pairs. The probability that any given detector pair observes photons from annihilations in any given pixel is the conditional measurement pdf. This pdf is assumed known. The measured data are conditionally independent, but the image pixel intensities are parametrically coupled via the measurement pdfs.

The great majority of current PET systems collect only the detector pairs that observe annihilation photons arriving with energy of 511 keV within a short time window (6–12 ns). The actual arrival times are not recorded. These detector pairs determine a chord on which the annihilation event occurred. Compton scattering of one or both of the gamma photons causes detections on the wrong pair of detectors, that is, the annihilation is deemed to occur on the wrong chord.

In time-of-flight (TOF) systems, the arrival times at the detectors are recorded, but with a much smaller time window of about 0.5 ns. The differential propagation time data are preprocessed to estimate the locations of every positron-electron annihilation along the detector chord. The time window of 0.5 ns corresponds to a localization uncertainty of about 7.5 cm along the chord. These TOF PET systems have met with only limited success in practice as of about 2003 [70].

The PET reconstruction problem is presented in two forms. The first uses PPP sample data that comprises the estimated locations of every annihilation event, i.e., TOF data. The algorithm given here for this data is a variant of the original Shepp-Vardi algorithm. The discussion is based on [119, Chapter 3]. Even though TOF PET is not in widespread current use, it is presented here because of the insight it provides and its intuitive mathematical appeal.

The other PET reconstruction problem uses histogram data, that is, the data comprises only the numbers of annihilation events measured by the detectors. These counts are a realization of a PPP defined on the discrete space of detectors. In the language and notation of Section 2.12.1 for PPPs on discrete spaces, the detectors are the points of the discrete space Φ. Histogram data are used in the original paper by Shepp and Vardi [110].

5.2.1 Image Reconstruction

5.2.1.1 Formulation

Given disjoint, bounded, nonempty sets $\mathcal{R}_1, \ldots, \mathcal{R}_K$, $K \geq 1$. Let $\mathcal{R} = \cup_{r=1}^{K} \mathcal{R}_r \subset \mathbb{R}^{n_x}$, where n_x is the dimension of the image space. In the simplest PET problem, $n_x = 2$. Think of these sets as a close-packed grid of small cells (pixels or voxels) in the image space.

The spatial intensity $\lambda(x)$ is assumed piecewise constant:

$$\lambda(x) = \sum_{r=1}^{K} \lambda_r \, I_r(x), \quad x \in \mathcal{R}, \tag{5.1}$$

where $\lambda_r > 0$ and

$$I_r(x) = \begin{cases} 1, & \text{if } x \in \mathcal{R}_r \\ 0, & \text{if } x \notin \mathcal{R}_r. \end{cases}$$

The parameter vector is $\Lambda = (\lambda_1, \ldots, \lambda_K)$. From the physics, the input process is a PPP on $\mathcal{R}$. Realizations of this PPP model the locations of positron-electron annihilations. An input point $x \in \mathcal{R}$ is the location of a positron-electron annihilation, and the detector array estimates that it occurred at the point y in the output (measurement) space $\mathcal{T} \subset \mathbb{R}^{n_y}$. For PET imaging, $\mathcal{T} \equiv \mathcal{R}$, but this restriction is not used here. The pdf of the measurement is $\ell(y \mid x)$, so that $\int_{\mathcal{T}} \ell(y \mid x) \, dy = 1$

for all x. This pdf is assumed to be known. From (2.86) and (5.1), the measurement point process is a PPP with intensity

$$\mu(y) = \int_{\mathcal{R}} \ell(y \mid x)\,\lambda(x)\,\mathrm{d}x \tag{5.2}$$

$$= \sum_{r=1}^{K} \lambda_r\,f_r(y), \quad y \in \mathcal{T}, \tag{5.3}$$

where

$$f_r(y) = \int_{\mathcal{R}_r} \ell(y \mid x)\,\mathrm{d}x. \tag{5.4}$$

The observed, or incomplete, data are the m output points

$$\mathcal{Y} = (y_1, \ldots, y_m), \quad y_j \in \mathcal{T}. \tag{5.5}$$

Because this is TOF data, the points correspond to the estimated locations of positron-electron annihilation events. The order of the points in $\mathcal{Y}$ is irrelevant, so the incomplete data pdf is, from (2.12),

$$p(\mathcal{Y};\,\Lambda) = e^{-\int_{\mathcal{T}} \mu(y)\,\mathrm{d}y} \prod_{j=1}^{m} \mu(y_j) \tag{5.6}$$

$$= e^{-\sum_{r=1}^{K} \lambda_r |\mathcal{R}_r|} \prod_{j=1}^{m} \left(\sum_{r=1}^{K} \lambda_r\,f_r(y_j) \right), \tag{5.7}$$

where $|\mathcal{R}_r| = \int_{\mathcal{R}_r} \mathrm{d}x < \infty$. The ML estimate of Λ is

$$\widehat{\Lambda} \equiv \arg\max_{\Lambda} \, p(\mathcal{Y};\,\Lambda). \tag{5.8}$$

The ML estimate $\widehat{\Lambda}$ is computed by solving the coupled nonlinear system of equations corresponding to the necessary conditions $\nabla_{\Lambda}\, p(\mathcal{Y};\,\Lambda) = 0$.

5.2.1.2 Uniqueness of the Estimate

The ML estimate of Λ is unique because $p(\mathcal{Y};\,\Lambda)$ is unimodal. The easy way to see this is to show that it is strictly log-concave, i.e., $\log p(\mathcal{Y};\,\Lambda)$ is concave. To see this, differentiate $\log p(\mathcal{Y};\,\Lambda)$ to find its Hessian matrix:

$$H(\Lambda) = \nabla_\Lambda \left[\nabla_\Lambda \log p(\mathcal{Y}; \Lambda) \right]^T$$

$$= -\sum_{j=1}^{m} \frac{1}{\left(\sum_{r=1}^{K} \lambda_r \, f_r(y_j) \right)^2} \begin{bmatrix} f_1(y_j) \\ \vdots \\ f_K(y_j) \end{bmatrix} \left[f_1(y_j) \cdots f_K(y_j) \right]^T . \tag{5.9}$$

The quadratic form $z^T H(\Lambda) z$ is strictly negative for nonzero vectors z if the number of data points m is at least as great as the number of pixels K, and the $m \times K$ matrix

$$\mathcal{F} = \begin{bmatrix} f_1(y_1) & f_2(y_1) & \cdots & f_K(y_1) \\ f_1(y_2) & f_2(y_2) & \cdots & f_K(y_2) \\ \vdots & \vdots & & \vdots \\ f_1(y_m) & f_2(y_m) & \cdots & f_K(y_m) \end{bmatrix} \tag{5.10}$$

is full rank. Because the Hessian matrix is negative definite for any $\Lambda \neq 0$, the function $p(\mathcal{Y}; \Lambda)$ is strictly log-concave, and hence also strictly concave.

5.2.1.3 E-step

The EM method is used to derive a recursive algorithm to compute the ML estimate $\widehat{\Lambda}$. The recursion avoids directly solving the nonlinear system $\nabla_\Lambda p(\mathcal{Y}; \Lambda) = 0$ directly. For the PET problem, EM yields a recursion for $\widehat{\Lambda}$ known as the Shepp-Vardi algorithm. Because the pdf (5.7) is unimodal, the EM method is guaranteed to converge to the ML estimate.

Let x_j be the unknown location of the annihilation event that generated the data point y_j. The question is, "Which cell/pixel/voxel contains x_j?" Since this is unknown, let the index $k_j \in \{1, \ldots, K\}$ indicate the correct cell, that is, let

$$x_j \in \mathcal{R}_{k_j}, \qquad j = 1, \ldots, m. \tag{5.11}$$

The missing data in the sense of EM are the indices $\mathcal{K} = \{k_1, \ldots, k_m\}$. The joint pdf of $(\mathcal{K}, \mathcal{Y})$ is *defined* by

$$p(\mathcal{K}, \mathcal{Y}) = e^{-\sum_{r=1}^{K} \lambda_r |\mathcal{R}_r|} \prod_{j=1}^{m} \left(\lambda_{k_j} \, f_{k_j}(y_j) \right), \tag{5.12}$$

The conditional pdf of the missing data is written

$$
\begin{aligned}
p(k_1, \ldots, k_m ; \Lambda) &\equiv p(\mathcal{K} \mid \mathcal{Y} ; \Lambda) \\
&= \frac{p(\mathcal{K}, \mathcal{Y} ; \Lambda)}{p(\mathcal{Y} ; \Lambda)} \\
&= \prod_{j=1}^{m} \frac{\lambda_{k_j} f_{k_j}(y_j)}{\sum_{r=1}^{K} \lambda_r f_r(y_j)}.
\end{aligned}
\tag{5.13}
$$

The dependence of the left hand expression on the data $\mathcal{Y}$ is suppressed to simplify the notation. The logarithm of the joint density (5.12) is

$$
\log p(\mathcal{K}, \mathcal{Y} ; \Lambda) = -\sum_{r=1}^{K} \lambda_r |\mathcal{R}_r| + \sum_{j=1}^{m} \log \left(\lambda_{k_j} f_{k_j}(y_j) \right).
$$

The term $\log f_{k_j}(y_j)$ is independent of the parameters $\{\lambda_j\}$; dropping it gives the loglikelihood function

$$
\mathcal{L}(\Lambda) = -\sum_{r=1}^{K} \lambda_r |\mathcal{R}_r| + \sum_{j=1}^{m} \log \lambda_{k_j}.
\tag{5.14}
$$

In some applications, the functions $\log f_{k_j}(y_j)$ are retained because they incorporate a Bayesian a priori pdf and depend on one or more of the estimated parameters.

5.2.1.4 M-step

Let $n \geq 0$ denote the EM recursion index, and let the initial value for the intensity parameter be $\Lambda^{(0)} = \left(\lambda_1^{(0)}, \ldots, \lambda_K^{(0)} \right)$, where $\lambda_r^{(0)} > 0$ for $r = 1, \ldots, K$. The EM auxiliary function is

$$
\begin{aligned}
Q\left(\Lambda ; \Lambda^{(n)}\right) &= E\left[\mathcal{L}(\Lambda) ; \Lambda^{(n)}\right] \\
&= \sum_{k_1=1}^{K} \cdots \sum_{k_m=1}^{K} \mathcal{L}(\Lambda)\, p\left(k_1, \ldots, k_m ; \Lambda^{(n)}\right).
\end{aligned}
\tag{5.15}
$$

Substituting (5.14) and (5.13) gives, after interchanging summations and simplifying,

$$
Q\left(\Lambda ; \Lambda^{(n)}\right) = -\sum_{r=1}^{K} \lambda_r |\mathcal{R}_r| + \sum_{k=1}^{K} \left(\sum_{j=1}^{m} \frac{\lambda_k^{(n)} f_k(y_j)}{\sum_{r=1}^{K} \lambda_r^{(n)} f_r(y_j)} \right) \log \lambda_k.
\tag{5.16}
$$

Setting $\nabla_\Lambda Q\left(\Lambda ; \Lambda^{(n)}\right) = 0$ gives the EM update:

$$\lambda_k^{(n+1)} = \frac{1}{|\mathcal{R}_k|} \sum_{j=1}^m \frac{\lambda_k^{(n)} f_k(y_j)}{\sum_{r=1}^K \lambda_r^{(n)} f_r(y_j)}, \quad k = 1, \ldots, K. \tag{5.17}$$

The Shepp-Vardi algorithm evaluates (5.17) until it satisfies the stipulated convergence criterion. An intuitive interpretation of the recursion is given below. It is summarized in Table 5.1.

Table 5.1 Shepp-Vardi algorithm for PPP sample data

Data are estimated locations of annihilation events: $y_{1:m} \equiv \{y_1, \ldots, y_m\}$
Area of pixels $|\mathcal{R}_1|, \ldots, |\mathcal{R}_K|$ of the vectorized PET image
Output: $\Lambda = \{\lambda_1, \ldots, \lambda_K\} \equiv$ vectorized PET image in pixels $\mathcal{R}_1, \ldots, \mathcal{R}_K$

- Precompute the $m \times K$ matrix of cell-level integrals, $\mathcal{F} \equiv \left[\mathcal{F}(j, r)\right]$
  - FOR annihilation event $j = 1, \ldots, m$ and pixel $r = 1, \ldots, K$, evaluate the integral

$$\mathcal{F}(j, r) = \int_{\mathcal{R}_r} \ell(y_j \mid x)\,\mathrm{d}x$$

  - END FOR
- Initialize the PET image: $\Lambda(0) = [\lambda_1(0), \ldots, \lambda_K(0)]^T$
- FOR EM iteration index $n = 0, 1, 2, \ldots$ until convergence:
  - Update the expected detector count vector, $\mathcal{D}(n) \equiv [\mathcal{D}_1(n), \ldots, \mathcal{D}_m(n)]$:

$$\mathcal{D}(n) = \mathcal{F}\,\Lambda(n) \quad \text{(matrix-vector product)}$$

  - FOR pixel $r = 1 : K$, update the intensity of r-th pixel:

$$\lambda_r(n+1) = \frac{\lambda_r(n)}{|\mathcal{R}_r|} \sum_{j=1}^m \frac{\mathcal{F}(j, r)}{\mathcal{D}_j(n)}$$

  - END FOR
  - Update vectorized PET image: $\Lambda(n+1) = [\lambda_1(n+1), \ldots, \lambda_K(n+1)]^T$
- END FOR EM iteration (Test for convergence)
- If converged: Estimated PET image is $\hat{\Lambda}_{ML} = \Lambda(n_{last}) = [\lambda_1(n_{last}), \ldots, \lambda_K(n_{last})]^T$

5.2.2 Small Cell Limit

The expression (5.17) is an ugly ducking in that it is prettier after taking the limit as $|\mathcal{R}_k| \to 0$. The limiting form for small cells (pixels or voxels) is not only insightful, but also useful in other applications. If

$$\lambda_k^{(n+1)} \to \lambda^{(n+1)}(x),$$
$$\lambda_k^{(n)} \to \lambda^{(n)}(x),$$

and $|\mathcal{R}_k| \to 0$ as $k \to 0$, then, from (5.4),

$$\frac{f_k(y_j)}{|\mathcal{R}_k|} \to \ell(y_j \mid x),$$

$$\sum_{r=1}^{K} \lambda_r^{(n)} f_r(y_j) = \sum_{r=1}^{K} \lambda_r^{(n)} \left(\frac{f_r(y_j)}{|\mathcal{R}_k|}\right) |\mathcal{R}_k| \to \int_{\mathcal{R}} \ell(y_j \mid s)\, \lambda^{(n)}(s)\, \mathrm{d}s,$$

and the limit of (5.17) is

$$\lambda^{(n+1)}(x) = \lambda^{(n)}(x) \sum_{j=1}^{m} \frac{\ell(y_j \mid x)}{\int_{\mathcal{R}} \ell(y_j \mid s)\, \lambda^{(n)}(s)\, \mathrm{d}s}, \qquad x \in \mathcal{R}. \tag{5.18}$$

The form of the algorithm used in the PET application is (5.17), not (5.18).

5.2.3 Intuitive Interpretation

The Shepp-Vardi recursion (5.18) has a simple intuitive interpretation. The data $\mathcal{y}$ are a realization of a PPP, so the best current estimate of the probability that y_j originates from an annihilation event in the multidimensional infinitesimal $(x,\, x + \mathrm{d}x)$ with volume $|\mathrm{d}x|$ is (see Section 3.2.2)

$$\frac{\ell(y_j \mid x)\, \lambda^{(n)}(x)\, |\mathrm{d}x|}{\int_{\mathcal{R}} \ell(y_j \mid s)\, \lambda^{(n)}(s)\, \mathrm{d}s}. \tag{5.19}$$

Because there is at most one annihilation event per measurement, the sum over j is the estimated number of annihilations originating in $(x,\, x + \mathrm{d}x)$ that generate a measurement. Since annihilations form a PPP, this number is equated to the updated intensity over $(x,\, x + \mathrm{d}x)$; that is, it is equated to $\lambda^{(n+1)}(x)\, |\mathrm{d}x|$. Dividing by $|\mathrm{d}x|$ gives (5.18). A similar interpretation also holds for (5.17).

5.3 PET: Histogram Data

Most current PET systems record only photon detector pairs and do not collect TOF data. In such systems, the absence of differential photon propagation times means that the measurements of the location of the annihilation event along the chord connecting the detector pairs are not available. The only change to the PET problem is that histogram data are used, not PPP sample data, but this change modifies the likelihood function of the data. As a result the EM formulation is altered, as is the reconstruction algorithm. These changes are detailed in this section.

5.3.1 Detectors as a Discrete Space

The pixels $\mathcal{R}_1, \ldots, \mathcal{R}_K$ remain the same as for TOF data, but now the sample data are accumulated by the L detectors. The available data comprise the list of integer counts $m_{1:L} = \{m_1, \ldots, m_L\}$ of the detectors. As mentioned earlier, the histogram data are a realization of a PPP on the discrete space of detectors. The measurement conditional pdf $\ell(\cdot)$ is replaced by the discrete measurement function

$$\ell(j \mid r) = \Pr\left[\text{detection in cell } \mathcal{T}_j \mid \text{annihilation event } x \in \mathcal{R}_r\right]. \tag{5.20}$$

The precise location of the event x within $\mathcal{R}_r$ is irrelevant because the intensity of annihilation events in pixel $\mathcal{R}_j$ is λ_j. The vector Λ of these parameters is estimated from the measured counts $m_{1:L}$.

5.3.2 Shepp-Vardi Algorithm

5.3.2.1 Missing and Complete Data

For $1 \le r \le K$ and $1 \le j \le L$, let $m(r, j)$ be the number of annihilation events that occur in pixel $\mathcal{R}_r$ that are subsequently detected in cell $\mathcal{T}_j$. These are the complete data, denoted by $m_{1:K, 1:L}$. These numbers are not observed, hence the index $m(r, j)$ ranges from 0 to m_j. The expected number of such events is the intensity

$$E[m(r, j)] = \ell(j \mid r)\,\lambda_r. \tag{5.21}$$

The observed data constitute constraints on the complete data:

$$m_j = \sum_{r=1}^{K} m(r, j), \quad 1 \le j \le L. \tag{5.22}$$

The intensity of the observed data in cell $\mathcal{T}_j$ is

$$E[m_j] = E\left[\sum_{r=1}^{K} m(r, j)\right] = \sum_{r=1}^{K} E\left[m(r, j)\right] = \sum_{r=1}^{K} \ell(j \mid r)\,\lambda_r. \tag{5.23}$$

The observed data are independent Poisson distributed variables, so their joint pdf is the product

$$p(m_{1:L};\, \Lambda) = \prod_{j=1}^{L} e^{-\sum_{r=1}^{K} \ell(j \mid r)\lambda_r}\,\frac{\left(\sum_{r=1}^{K} \ell(j \mid r)\lambda_r\right)^{m_j}}{m_j!}. \tag{5.24}$$

Because m_j is Poisson distributed, the complete data corresponding to it, namely $\{m(1, j), \ldots, m(K, j)\}$, are also independent and Poisson distributed. Hence the PDF of the complete data $m_{1:K, 1:L}$ is

$$p(m_{1:K, 1:L}; \Lambda) = \prod_{j=1}^{L} \prod_{r=1}^{K} e^{-\ell(j\,|\,r)\lambda_r} \frac{(\ell(j\,|\,r)\,\lambda_r)^{m(r,j)}}{m(r, j)!}. \tag{5.25}$$

Thus, by the definition of conditioning,

$$p(m_{1:K, 1:L}\,|\,m_{1:L}; \Lambda) = \prod_{j=1}^{L} \frac{\binom{m_j}{m(1, j)\ \cdots\ m(K, j)} \prod_{r=1}^{K} (\ell(j\,|\,r)\,\lambda_r)^{m(r,j)}}{\left(\sum_{r'=1}^{K} \ell(j\,|\,r')\,\lambda_{r'}\right)^{m_j}}, \tag{5.26}$$

where the multinomial coefficient is

$$\binom{m_j}{m(1, j)\ \cdots\ m(K, j)} = \frac{m_j!}{m(1, j)!\ \cdots\ m(K, j)!}. \tag{5.27}$$

Since the numerator of the ratio (5.26) is a multinomial expansion, it follows easily that

$$\sum_{\{m(r,j)\}_{r,\,j=1}^{K,\,L}} p(m_{1:K, 1:L}\,|\,m_{1:L}; \Lambda) = 1, \tag{5.28}$$

where the sum is over all indices $m(r, j)$ that satisfy the measurement constraints (5.22).

The logarithm of the complete data pdf is, from (5.25),

$$\log p(m_{1:K, 1:L}; \Lambda)$$

$$= \sum_{j=1}^{L} \sum_{r=1}^{K} \{-\ell(j\,|\,r)\,\lambda_r + m(r, j) \log(\ell(j\,|\,r)\,\lambda_r) - \log m(r, j)!\}$$

$$= c + \sum_{r=1}^{K} \left\{-\lambda_r + \left(\sum_{j=1}^{L} m(r, j)\right) \log \lambda_r\right\}, \tag{5.29}$$

where in the last equation the constant c contains only terms not involving Λ.

5.3.2.2 E-step

Let n denote the EM iteration index, and let $\Lambda^{(0)} = \left(\lambda_1^{(0)}, \ldots, \lambda_K^{(0)}\right) > 0$ be an initial set of intensities. For $n \geq 0$, the auxiliary function of the EM method is, by definition,

$$
\begin{aligned}
Q\left(\Lambda;\,\Lambda^{(n)}\right) &= E_{\Lambda^{(n)}}\left[\log p(m_{1:K,1:L};\,\Lambda)\right] \\
&= \sum_{\{m(r,j)\}_{r,\,j=1}^{K,\,L}} \left(\log p\left(m_{1:K,1:L};\,\Lambda\right)\right) p\left(m_{1:K,1:L}\,\middle|\,m_{1:L};\,\Lambda^{(n)}\right),
\end{aligned}
$$

$$(5.30)$$

where the sum in (5.30) is over all indices $m(r,\,j)$ that satisfy the L measurement constraints (5.22). Substituting (5.29), dropping the irrelevant constant c, and using the linearity of the expectation operator gives

$$
Q\left(\Lambda;\,\Lambda^{(n)}\right) = \sum_{r=1}^{K}\left\{-\lambda_r + \left(\sum_{j=1}^{L} E_{\Lambda^{(n)}}\left[m(r,\,j)\right]\right)\log \lambda_r\right\},
\qquad (5.31)
$$

where the expectation in the last step is

$$
E_{\Lambda^{(n)}}\left[m(r,\,j)\right] = m_j\,\frac{\ell(j\,|\,r)\,\lambda_r^{(n)}}{\sum_{r'=1}^{K}\ell(j\,|\,r')\,\lambda_{r'}^{(n)}}.
\qquad (5.32)
$$

The expectation (5.32) is written more intuitively as

$$
E_{\Lambda^{(n)}}\left[m(r,\,j)\right] = m_j\,\frac{E_{\Lambda^{(n)}}[m(r,\,j)]}{\sum_{r'=1}^{K} E_{\Lambda^{(n)}}\left[m(r',\,j)\right]}.
\qquad (5.33)
$$

The E-step is complete when (5.32) is verified below.

5.3.2.3 Finishing the E-step

Encountering identities like (5.32) is commonplace when simplifying expressions during the E-step. A similar, but more transparent, identity is buried in the E-step of PET between (5.15) and (5.16). Another identity, from transmission tomography, is given by (5.84) below. Yet another example is the identity (3.18) used in estimating superposed PPPs.

   The expectation (5.32) is defined for arbitrary Λ by

$$
E_{\Lambda}\left[m(r,\,j)\right] = \sum_{\{m(r,j)\}_{r,\,j=1}^{K,\,L}} m(r,\,j)\,p(m_{1:K,1:L}\,|\,m_{1:L},\,\Lambda).
$$

Substituting $p(m_{1:K,1:L}\,|\,m_{1:L};\,\Lambda)$ and summing over all indices except $m(1,\,j),\,\ldots,$ and $m(K,\,j)$ gives

$$E_\Lambda\left[m(r,j)\right]$$

$$= \frac{\sum_{m(1,j),\dots,m(K,j)=0}^{m_j} m(r,j) \binom{m_j}{m(1,j)\ \cdots\ m(K,j)} \prod_{r'=1}^{K} \left(\ell(j\mid r')\,\lambda_{r'}\right)^{m(r',j)}}{\left(\sum_{r'=1}^{K} \ell(j\mid r')\,\lambda_{r'}\right)^{m_j}}.$$

(5.34)

Making the sum on $m(r,j)$ the outermost sum gives the numerator of the ratio on the right hand side of (5.34) as

$$\sum_{m(r,j)=0}^{m_j} m(r,j)\, \frac{m_j!}{m(r,j)!\,(m_j - m(r,j))!}\, (\ell(j\mid r)\,\lambda_r)^{m(r,j)}$$

$$\times \left[\sum_{r'\neq r} \ell(j\mid r')\,\lambda_{r'}\right]^{m_j - m(r,j)}.$$

Canceling the factor $m(r,j)$ in $m(r,j)!$, changing the index of summation to $\tilde{m}(r,j) = m(r,j) - 1$, and shuffling terms yields

$$m_j\,\ell(j\mid r)\,\lambda_r \sum_{\tilde{m}(r,j)=0}^{m_j-1} \frac{(m_j - 1)!\,(\ell(j\mid r)\,\lambda_r)^{\tilde{m}(r,j)}}{\tilde{m}(r,j)!\,(m_j - 1 - \tilde{m}(r,j))!}$$

$$\times \left[\sum_{r'\neq r} \ell(j\mid r')\,\lambda_{r'}\right]^{m_j - 1 - \tilde{m}(r,j)}.$$

The sum in this last expression simplifies immediately using the binomial theorem:

$$m_j\,\ell(j\mid r)\,\lambda_r \left[\sum_{r'=1}^{K} \ell(j\mid r')\,\lambda_{r'}\right]^{m_j - 1}.$$

Substituting this form of the numerator into the expectation (5.34) and canceling terms gives the desired expectation for any Λ. The expression (5.32) is the special case $\Lambda = \Lambda^{(n)}$. This concludes the E-step.

5.3.2.4 M-step (Shepp-Vardi Algorithm)

The maximum of $Q\left(\Lambda;\Lambda^{(n)}\right)$ with respect to Λ is found by differentiation. The result is the pixel level iteration

$$\lambda_r^{(n+1)} = \lambda_r^{(n)} \sum_{j=1}^{L} \frac{m_j \, \ell(j \mid r)}{\sum_{r'=1}^{K} \ell(j \mid r') \lambda_{r'}^{(n)}}, \quad 1 \le r \le K. \tag{5.35}$$

This iteration is the original form of Shepp-Vardi algorithm. A summary is given in Table 5.2.

Table 5.2 Shepp-Vardi algorithm for histogram data

Detector Count Data: $m_{1:L} \equiv \{m_1, \ldots, m_L\}$ in detectors $1, \ldots, L$
Output: $\Lambda = \{\lambda_1, \ldots, \lambda_K\} \equiv$ vectorized PET image in pixels $1, \ldots, K$

- Precompute the $L \times K$ likelihood matrix: $\mathcal{L} \equiv [\mathcal{L}(j, r]$
  - FOR detector $j = 1, \ldots, L$ and pixel $r = 1, \ldots, K$, evaluate the likelihood function

$$\mathcal{L}(j, r) = \ell(j \mid r)$$

  - END FOR
- Initialize the vectorized PET image: $\Lambda(0) = [\lambda_1(0), \ldots, \lambda_K(0)]^T$
- FOR EM iteration index $n = 0, 1, 2, \ldots$ until convergence:
  - Update the expected detector count vector, $\mathcal{D}(n) \equiv [\mathcal{D}_1(n), \ldots, \mathcal{D}_L(n)]$:

$$\mathcal{D}(n) = \mathcal{L} \, \Lambda(n) \qquad \text{(matrix-vector product)}$$

  - FOR pixel $r = 1 : K$, update the intensity of r-th pixel:

$$\lambda_r(n + 1) = \lambda_r(n) \sum_{j=1}^{L} m_j \, \frac{\mathcal{L}(j, r)}{\mathcal{D}_j(n)}$$

  - END FOR
  - Update vectorized PET image: $\Lambda(n + 1) = [\lambda_1(n + 1), \ldots, \lambda_K(n + 1)]^T$
- END FOR EM iteration (Test for convergence)
- If converged: Estimated PET Image is $\Lambda(n_{last}) = [\lambda_1(n_{last}), \ldots, \lambda_K(n_{last})]^T$

5.3.2.5 Convergence

General EM theory guarantees convergence of the iteration (5.35) to a stationary point of the likelihood function. The strict concavity of the loglikelihood function guarantees that it converges to the global ML estimate, $\hat{\Lambda}_{ML}$. To see that the observed data pdf $p(\mathcal{Y}; \Lambda)$ is strictly logconcave, differentiate (5.25) to find the Hessian matrix of its logarithm:

$$H(\Lambda) = \nabla_\Lambda \left[\nabla_\Lambda \log p(m_{1:L}; \Lambda) \right]^T$$

$$= -\sum_{j=1}^{L} \frac{m_j}{\left(\sum_{r=1}^{K} \ell(j \mid r) \lambda_j \right)^2} \begin{bmatrix} \ell(j \mid 1) \\ \vdots \\ \ell(j \mid K) \end{bmatrix} \left[\ell(j \mid 1) \cdots \ell(j \mid K) \right]^T. \tag{5.36}$$

The quadratic form $z^T H(\Lambda) z$ is strictly negative for nonzero vectors z if the $L \times K$ likelihood matrix

$$\mathcal{L} = \begin{bmatrix} \ell(1 \mid 1) & \ell(1 \mid 2) & \cdots & \ell(1 \mid K) \\ \ell(2 \mid 1) & \ell(2 \mid 2) & \cdots & \ell(2 \mid K) \\ \vdots & \vdots & & \vdots \\ \ell(L \mid K) & \ell(L \mid K) & \cdots & \ell(L \mid K) \end{bmatrix} \tag{5.37}$$

is full rank. Because the Hessian matrix is negative definite for any $\Lambda \neq 0$, the observed data pdf is strictly concave.

5.4 Single-Photon Computed Emission Tomography (SPECT)

A older and lower resolution imaging procedure is called SPECT (single photon emission computed tomography). In SPECT, a radioisotope (most commonly, technetium 99) is introduced into the body. As the isotope decays, gamma photons are emitted in all directions. A gamma camera (also often called an Anger camera after its developer, Hal Anger, in 1957) takes a "snapshot" of the photons emitted in the direction of the camera. Unlike PET which can be treated as a stack of two dimensional slice problems, SPECT requires solving an inherently a three dimensional reconstruction problem.

5.4.1 Gamma Cameras

A simplistic depiction of a gamma camera is given in Fig. 5.2. The camera comprises several parts:

- The emitted gamma photons are collimated using one of several methods (a lead plate with parallel drilled holes is common). Fewer than 1% of the incident gamma photons emerge from the collimator.
- Photons that emerge enter a thallium-activated sodium iodide (NaI(T1)) crystal. The NaI(T1) crystal is a flat circular plate approximately 1 cm thick and 30 cm in diameter. (The thinner the crystal, the better the resolution but the less the efficiency.) If a gamma photon is fully absorbed by an atom in the crystal (photoelectric effect), the atom ejects an electron; if it is partially absorbed (Compton effect), the atom ejects an electron and another gamma photon.
- Ejected electrons encounter other atoms in the crystalline lattice and produce visible light in a physical process called scintillation. The number scintillated light photons is proportional to the energy incident on the crystal.
- The scintillated photons are few in number, so an array of hexagonally-packed photo-multiplier tubes (PMTs) are affixed to the back of the crystal to boost the count (without also boosting the noise level). Typical gamma cameras use

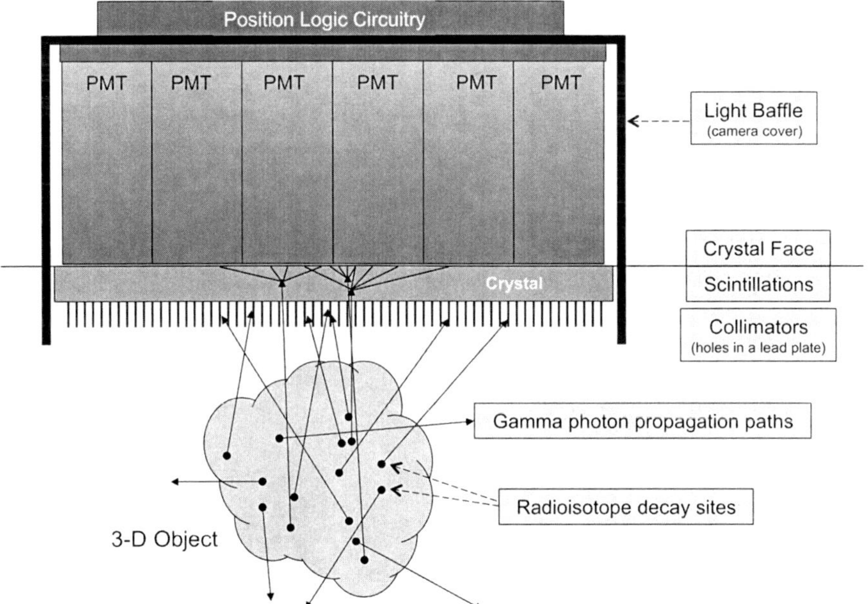

Fig. 5.2 Sketch of the basic components of a gamma (Anger) camera used in SPECT

between 37 and 120 PMTs. The face of a PMT ranges in size from 5 to 7 cm across.

- Position logic circuits are used to estimate the locations of the crystal atoms that absorb gamma photons. An estimation procedure is necessary because several PMTs detect light from the same event. The position estimate is a convex combination of the PMT data.

The output is a two dimensional "snapshot" of the estimated locations of the atoms that absorb gamma photons in the NaI(Tl) crystal. The estimated locations depend on the location of the camera.

In a typical SPECT imaging procedure, the gamma camera is moved to a fixed number of different viewing positions around the object (and, naturally, is never in physical contact with it). A snapshot is made at each camera position. The multiple view snapshots are used to reconstruct the image. The reconstructed SPECT image is the estimated intensity of radioisotope decay within the three dimensional volume of the imaged object.

The clinical use of SPECT is well established. A common use of SPECT is cardiac blood perfusion studies, in which an ECG/EKG (electrocardiogram/elektrokardiogram) acquires data from a beating heart and the heart rhythm is used to time gate the SPECT data collection procedure.

SPECT is much more widely used than PET. There are several reasons for this difference. One is that SPECT and PET are used for diagnostically different purposes. Many diagnostic needs do not require the high resolution provided by PET.

Another is that the radioisotopes needed for SPECT are readily accessible compared to those for PET. Yet another reason is simply the cost—SPECT procedures are relatively inexpensive. In 2009, SPECT procedures cost on the order of US\$500 and were often performed in physician offices and walk-in medical facilities. In contrast, PET cost US\$30,000 or more and required specialized hospital-based equipment and staff.

5.4.2 Image Reconstruction

A relatively recent overview of methods for SPECT, as well as PET, is given in [70]. The discussion here follows the outline of the method first used in [82], with appropriate modifications.

5.4.2.1 Problem Formulation and Data

The data used by SPECT are the set of snapshots made by the gamma camera. Each snapshot comprises a list of the position logic circuit estimates of the locations of atoms that absorb gamma photons. A Cartesian coordinate system $x = (x_1, x_2, x_3)$ is adopted for the imaged object, where x_3 is the horizontal axis. For simplicity, the origin is taken interior to the object. The goal of SPECT is to compute the ML estimate of the intensity of radioisotope decay at the location x, denoted $\lambda(x)$, $x \in \mathbb{R}^3$. The ML estimate is found using an algorithm derived using the EM method. A pleasing aspect of the derivation is that the M-step is solved by the calculus of variations.

The camera is rotated about the x_3-axis. The face (the NaI(T1) crystal) is kept parallel to the x_3-axis, and the locus of a specified point P on the camera face (e.g., the center) is constrained to lie a circle in the x_1–x_2 plane centered at the origin. The camera viewing positions correspond to the angles

$$\Theta \equiv \left\{\theta_0, \theta_1, \ldots, \theta_{n_\Theta - 1}\right\},$$

where n_Θ is the fixed number of camera viewing positions. See Fig. 5.3. Let $\mathbb{F}(\theta_j)$ denote the two dimensional plane of the camera face at view angle θ_j. An arbitrary point in $\mathbb{F}(\theta_j)$ is denoted by (y, θ_j), where $y \equiv (y_1, y_2) \in \mathbb{R}^2$ is a two-dimensional coordinate system. The data comprising the snapshot at view angle θ_j are the events in the list

$$S_j = \left\{\left(y_{j1}, \theta_j\right), \left(y_{j2}, \theta_j\right), \ldots, \left(y_{j,n_j}, \theta_j\right)\right\} \subset \mathbb{R}^3, \tag{5.38}$$

where $n_j \geq 1$ is the number of position estimates for the snapshot at camera angle θ_j. Let $\mathbb{S} \equiv \left\{S_0, S_1, , \ldots, S_{n_\Theta - 1}\right\}$. The intensity function $\lambda(x)$ is estimated from the data $\mathbb{S}$.

The intensity functions needed to formulate the SPECT likelihood function are defined on the Cartesian product $\mathbb{R}^3 \times \mathbb{F}(\theta_j)$ of decay points x in the imaged object

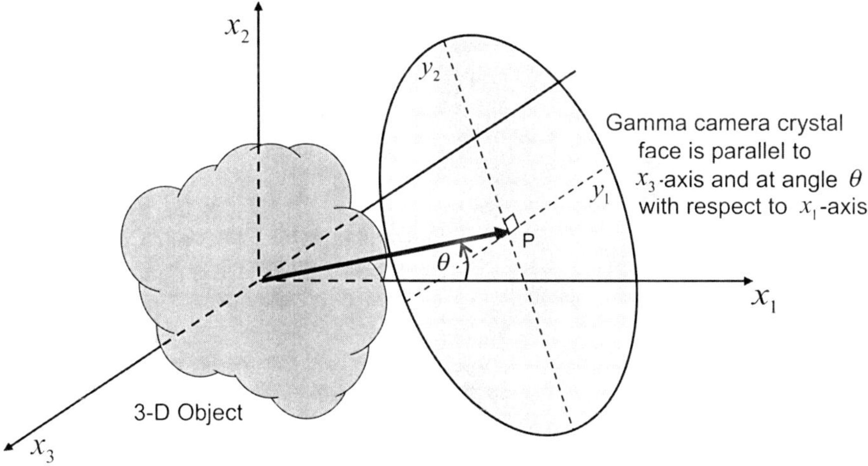

Fig. 5.3 Geometry and coordinates for SPECT imaging of a three dimensional object for a gamma camera with view angle θ. The center P of the camera face lies in the x_1–x_2 plane for all camera view angles θ

and observed gamma photon absorption points y on the crystal face of the camera at view angle θ_j. Let $\mu(x, y, \theta_j)$ denote the intensity function of the j-th PPP. There are n_Θ of these PPPs. In terms of these intensity functions, the intensity to be estimated is

$$\lambda(x) = \sum_{j=0}^{n_\Theta-1} \int_{\mathbb{F}(\theta_j)} \mu(x, y, \theta_j)\,\mathrm{d}y. \tag{5.39}$$

The j-th integral in (5.39) is a double integral over the coordinates $y \in \mathbb{F}(\theta_j)$ that correspond to points on the camera face.

The intensity function $\mu(x, y, \theta_j)$ is the superposition of the intensity functions of two independent PPPs. One is determined by the detected photons. Its intensity function is denoted by $\mu_0(x, y, \theta_j)$. The other is determined by the photons that arrive at the camera face but are not detected. The intensity function of the undetected photons is denoted by $\mu_1(x, y, \theta_j)$. Thus,

$$\mu(x, y, \theta_j) = \mu_0(x, y, \theta_j) + \mu_1(x, y, \theta_j) \tag{5.40}$$

Both $\mu_0(\cdot)$ and $\mu_1(\cdot)$ are expressed in terms of three input functions. These inputs are assumed known. They are discussed next.

5.4.2.2   Specified Functions

Several functions are required inputs to SPECT imaging. One is the pdf of the position estimate y. This density is known, or in principle knowable, from the physics and the engineering details of the system design. Its detailed mathematical form depends on where the decay x occurred and the camera position θ. It is denoted by $p_{Y|X\Theta}(y \,|\, x, \theta)$. A significant difference between this pdf and the analogous pdf for PET is that it is depth dependent—it broadens with increasing distance between the decay point x and the camera face. This pdf is assumed known.

Another function is the survival probability function

$$\beta(x, y, \theta) = \Pr \left[\begin{array}{l} \text{Photon emitted at decay point } x \text{ moving} \\ \text{toward location } y \text{ on the camera face at} \\ \text{view angle } \theta \text{ arrives at the camera face} \end{array} \right]. \qquad (5.41)$$

The survival function depends on many factors. These factors include the efficiency of the detector and the three dimensional attenuation density of the object being imaged. The efficiency is determined as part of a calibration procedure. The attenuation is clearly a complex issue since it depends on the object being imaged. It is, moreover, important for successful imaging. Methods for estimating it are discussed in some detail in [82] and the references therein. The function $\beta(x, y, \theta)$ is assumed known.

The third required function is the conditional pdf $p_{\Theta|X}\left(\theta_j \,|\, x\right)$. It is the fraction of photons emanating from x that propagate toward the camera at view angle θ_j. This fraction is dependent on geometric factors such as the solid angle subtended by the camera. It is also weighted by the relative length of the "dwell times" of the camera at the camera angles Θ. It is assumed known.

5.4.2.3 Likelihood Function

The intensity functions of the undetected and detected decay PPPs are given by

$$\mu_0(x, y, \theta_j) = p_{Y|X\Theta}\left(y \,|\, x, \theta_j\right) p_{\Theta|X}\left(\theta_j \,|\, x\right) \lambda(x) \left(1 - \beta(x, y, \theta_j)\right) \qquad (5.42)$$

and

$$\mu_1(x, y, \theta_j) = p_{Y|X\Theta}\left(y \,|\, x, \theta_j\right) p_{\Theta|X}\left(\theta_j \,|\, x\right) \lambda(x) \beta(x, y, \theta_j), \qquad (5.43)$$

respectively. These PPPs are independent because they result from an independent thinning process determined by $\beta(x, y, \theta_j)$ (*cf.* Examples 2.6 and 2.7). Their sum satisfies (5.40). The identity

$$\sum_{j=0}^{n_\Theta-1} \int_{\mathbb{R}^3 \times \mathbb{F}(\theta_j)} \mu(x, y, \theta_j) \, dx \, dy = \int_{\mathbb{R}^3} \lambda(x) \, dx \qquad (5.44)$$

is used shortly.

The j-th snapshot S_j is a realization of a PPP on the two dimensional camera face, that is, on the plane of the camera face at view angle θ_j. The photons arriving at the camera face form a PPP. The intensity of this PPP is

$$\mu(y, \theta_j) = \int_{\mathbb{R}^3} \mu(x, y, \theta_j)\, dx. \tag{5.45}$$

These intensities are defined only for values of $y \in \mathbb{F}(\theta_j) \subset \mathbb{R}^3$. In light of the relationship (5.39), the function

$$
\begin{aligned}
\mathscr{L}_{S_j}(\lambda) &= \exp\left\{-\int_{\mathbb{F}(\theta_j)} \mu(y, \theta_j)\, dy\right\} \prod_{r=1}^{n_j-1} \mu(y_{jr}, \theta_j) \\
&= \exp\left\{-\int_{\mathbb{R}^3 \times \mathbb{F}(\theta_j)} \mu(x, y, \theta_j)\, dx\, dy\right\} \prod_{r=1}^{n_j-1} \int_{\mathbb{R}^3} \mu(x, y_{jr}, \theta_j)\, dx
\end{aligned}
\tag{5.46}
$$

is the likelihood function for $\lambda(x)$ given the data S_j. The PPPs for different camera view angles are independent, so the product

$$
\begin{aligned}
\mathscr{L}_{\mathbb{S}}(\lambda) &= \prod_{j=0}^{n_\Theta-1} \mathscr{L}_{S_j}(\lambda) \\
&= \exp\left\{-\sum_{j=0}^{n_\Theta-1} \int_{\mathbb{R}^3 \times \mathbb{F}(\theta_j)} \mu(x, y, \theta_j)\, dx\, dy\right\} \prod_{j=0}^{n_\Theta-1} \prod_{r=1}^{n_j} \mu(y_{jr}, \theta_j) \\
&= \exp\left\{-\int_{\mathbb{R}^3} \lambda(x)\, dx\right\} \prod_{j=0}^{n_\Theta-1} \prod_{r=1}^{n_j} \int_{\mathbb{R}^3} \mu(x, y_{jr}, \theta_j)\, dx
\end{aligned}
\tag{5.47}
$$

is the likelihood function of $\lambda(x)$ given all the snapshots.

5.4.2.4 Missing Data

Missing data (in the EM sense) can be defined in many ways. The choice that seems most appropriate in this problem are defined, for each view angle θ_j, as follows:

- The number $N_j(0) > 0$ of gamma photons that reach the camera but are not detected;
- The locations $\{y_{jr} : r = n_j + 1, \ldots, n_j + N_j(0)\}$ at which the undetected gamma photons exit the crystal face of the camera;
- The locations $\{x_{jr} : r = 1, \ldots, n_j + N_j(0)\}$ of the decays that generated the detected and undetected gamma photons.

The complete data (in the EM sense) for the j-th snapshot are

$$S'_j = \left\{ (x_{j1},\, y_{j1},\, \theta_j),\, (x_{j2},\, y_{j2},\, \theta_j),\, \ldots,\, (x_{j,\,n_j+N_j},\, y_{j,\,n_j+N_j},\, \theta_j) \right\},$$

$$(5.48)$$

where $x_{jr} \in \mathbb{R}^3$. The points $(y_{jr},\, \theta_j)$ correspond to undetected decays for $r > n_j$.

The likelihood function of $\lambda(x)$ given the complete data $\mathbb{S}' \equiv \left\{ S'_j \right\}$ for all snapshots is, using (5.47),

$$\mathscr{L}^{\mathrm{com}}_{\mathbb{S}'}(\lambda) = \exp\left\{ -\int_{\mathbb{R}^3} \lambda(x)\,\mathrm{d}x \right\}$$

$$\times \prod_{j=0}^{n_\Theta-1} \left\{ \prod_{r=1}^{n_j} \mu_1(x_{jr},\, y_{jr},\, \theta_j) \prod_{r'=n_j+1}^{n_j+N_j(0)} \mu_0(x_{jr},\, y_{jr'},\, \theta_j) \right\}. \quad (5.49)$$

The logarithm is

$$\log \mathscr{L}^{\mathrm{com}}_{\mathbb{S}'}(\lambda) = -\int_{\mathbb{R}^3} \lambda(x)\,\mathrm{d}x + \sum_{j=0}^{n_\Theta-1} \left\{ \sum_{r=1}^{n_j} \log \mu_1(x_{jr},\, y_{jr},\, \theta_j) \right.$$

$$\left. + \sum_{r'=n_j+1}^{n_j+N_j(0)} \log \mu_0(x_{jr},\, y_{jr'},\, \theta_j) \right\}. \quad (5.50)$$

Using (5.42) and (5.43) gives the loglikelihood function of $\lambda(x)$ as

$$\log \mathscr{L}^{\mathrm{com}}_{\mathbb{S}'}(\lambda) = c - \int_{\mathbb{R}^3} \lambda(x)\mathrm{d}x + \sum_{j=0}^{n_\Theta-1} \left\{ \sum_{r=1}^{n_j} \log \lambda(x_{jr}) + \sum_{r'=n_j+1}^{n_j+N_j(0)} \log \lambda(x_{jr}) \right\},$$

$$(5.51)$$

where c contains terms independent of the intensity function $\lambda(x)$.

5.4.2.5 E-step

Let $m \geq 0$ denote the EM iteration index, and let $\lambda^{(0)}(x) > 0$ be specified. The auxiliary function of the EM method is defined as the conditional expectation of (5.50):

$$Q\left(\lambda \,|\, \lambda^{(m)}\right) = E_{\lambda^{(m)}}\left[\log \mathscr{L}^{\mathrm{com}}_{\mathbb{S}'}(\lambda) \right]$$

$$= c - \int_{\mathbb{R}^3} \lambda(x)\,\mathrm{d}x + A + B, \quad (5.52)$$

where

$$A = \sum_{j=0}^{n_\Theta-1} E_{\lambda^{(m)}} \left[\sum_{r'=n_j+1}^{n_j+N_j(0)} \log \lambda(x_{jr'}) \right] \tag{5.53}$$

$$B = \sum_{j=0}^{n_\Theta-1} E_{\lambda^{(m)}} \left[\sum_{r=1}^{n_j} \log \lambda(x_{jr}) \right]. \tag{5.54}$$

The expectations A and B are evaluated differently because the number of terms in the r' sum is both random *and* missing, while the number of terms in the r sum is specified.

To evaluate A, note that the j-th expectation in (5.53) is the expectation of a random sum, all of whose summands are $\lambda(x)$, with respect to the undetected target PPP on the camera face with view angle θ_j. The number of terms in the sum is Poisson distributed; therefore, replacing $\lambda(x)$ in (5.42) with $\lambda^{(m)}(x)$ and using Campbell's Theorem gives

$$A = \sum_{j=0}^{n_\Theta-1} \int_{\mathbb{R}^3 \times \mathbb{F}(\theta_j)} \log \lambda(x)$$

$$\times\ p_{Y|X\Theta}\left(y \mid x,\ \theta_j\right) p_{\Theta|X}\left(\theta_j \mid x\right) \lambda^{(m)}(x) \left(1 - \beta(x,\ y,\ \theta_j)\right) dx\,dy$$

$$= \int_{\mathbb{R}^3} \lambda^{(m)}(x) \log \lambda(x)$$

$$\times \left\{ \sum_{j=0}^{n_\Theta-1} \int_{\mathbb{F}(\theta_j)} p_{Y|X\Theta}(y \mid x,\ \theta_j) p_{\Theta|X}\left(\theta_j \mid x\right) \left(1 - \beta(x,\ y,\ \theta_j)\right) dy \right\} dx$$

$$= \int_{\mathbb{R}^3} \left(1 - \bar{\beta}(x)\right) \lambda^{(m)}(x) \log \lambda(x)\, dx\,, \tag{5.55}$$

where

$$\bar{\beta}(x) = \sum_{j=0}^{n_\Theta-1} \int_{\mathbb{F}(\theta_j)} p_{Y|X\Theta}\left(y \mid x,\ \theta_j\right) p_{\Theta|X}\left(\theta_j \mid x\right) \beta(x,\ y,\ \theta_j)\,dy \tag{5.56}$$

is the mean survival probability of detected gamma photons originating from decays at x. Equivalently,

$$\bar{\beta}(x) = \sum_{j=0}^{n_\Theta-1} p_{\Theta|X}\left(\theta_j \mid x\right) \beta_j(x)\,, \tag{5.57}$$

where

$$\beta_j(x) = \int_{\mathbb{F}(\theta_j)} p_{Y|X\Theta}\left(y \mid x, \theta_j\right) \beta(x, y, \theta_j)\, dy \tag{5.58}$$

is the probability that gamma photons generated from decays at x are detected in the camera at view angle θ_j.

Evaluating B requires using methods akin to those used in Chapter 3. From (5.47) and (5.49), the conditional pdf of the missing points x_{jr} for $r \leq n_j$ and all j is

$$\prod_{j=0}^{n_\Theta-1} \prod_{r=1}^{n_j} w_j(x_{jr} \mid y_{jr}, \theta_j), \tag{5.59}$$

where the weights are, for $j = 0, 1, \ldots, n_\Theta - 1$,

$$w_j(x \mid y, \theta_j)$$

$$= \frac{p_{Y|X\Theta}\left(y \mid x, \theta_j\right) p_{\Theta|X}\left(\theta_j \mid x\right) \beta(x, y, \theta_j) \lambda^{(m)}(x)}{\int_{\mathbb{R}^3} p_{Y|X\Theta}\left(y \mid x, \theta_j\right) p_{\Theta|X}\left(\theta_j \mid x\right) \beta(x, y, \theta_j) \lambda^{(m)}(x)\, dx}, \quad y \in \mathbb{F}(\theta_j). \tag{5.60}$$

It is straightforward to verify that the conditional expectation is

$$B = \sum_{j=0}^{n_\Theta-1} \left\{ \int_{\mathbb{R}^3} \cdots \int_{\mathbb{R}^3} \left[\sum_{r=1}^{n_j} \log \lambda(x_{jr}) \right] \left[\prod_{r=1}^{n_j} w_j(x_{jr} \mid y_{jr}, \theta_j) \right] dx_{j1} \cdots dx_{j,n_j} \right\}$$

$$= \sum_{j=0}^{n_\Theta-1} \sum_{r=1}^{n_j} \int_{\mathbb{R}^3} w_j(x \mid y_{jr}, \theta_j) \log \lambda(x)\, dx$$

$$= \int_{\mathbb{R}^3} f^{(m)}(x) \log \lambda(x)\, dx, \tag{5.61}$$

where

$$f^{(m)}(x) = \sum_{j=0}^{n_\Theta-1} \sum_{r=1}^{n_j} w_j(x \mid y_{jr}, \theta_j). \tag{5.62}$$

The intuitive interpretation of $f^{(m)}(x)$ is that it is the expected number of decays at the point x given the data from all camera view angles.

Substituting the expressions for A and B into (5.52) gives the final form of the auxiliary function as

$$Q\left(\lambda \,\middle|\, \lambda^{(m)}\right)$$
$$= c + \int_{\mathbb{R}^3} \left\{ -\lambda(x) + \left[\left(1 - \bar{\beta}(x)\right) \lambda^{(m)}(x) + f^{(m)}(x) \right] \log \lambda(x) \right\} dx. \tag{5.63}$$

This completes the E-step.

5.4.2.6 M-step

The EM update is the solution of the maximization problem defined by

$$\lambda^{(m+1)}(x) = \arg\max_{\lambda > 0} Q\left(\lambda \,\middle|\, \lambda^{(m)}\right) \tag{5.64}$$

Maximizing Q with respect to $\lambda(x)$ is a straightforward problem in the calculus of variations. Let $\delta(x)$ be an arbitrary "variation" about the solution $\lambda^{(m+1)}(x)$ such that $\lambda^{(m+1)}(x) + \epsilon\, \delta(x) > 0$ for sufficiently small ϵ. Then

$$Q\left(\lambda^{(m+1)} + \epsilon\, \delta(x) \,\middle|\, \lambda^{(m)}\right) = c + \int_{\mathbb{R}^3} \left\{ -\lambda^{(m+1)}(x) - \epsilon\, \delta(x) \right.$$
$$\left. + \left[\left(1 - \bar{\beta}(x)\right) \lambda^{(m)}(x) + f^{(m)}(x) \right] \log\left(\lambda^{(m+1)}(x) + \epsilon\, \delta(x)\right) \right\} dx. \tag{5.65}$$

For a specified variation $\delta(x)$, the maximum of (5.65) with respect to ϵ must occur at $\epsilon = 0$, for otherwise $\lambda^{(m+1)}(x)$ is not the solution of the M-step. Evaluating the derivative of (5.65) with respect to ϵ at zero gives

$$\int_{\mathbb{R}^3} \delta(x) \left[-1 + \frac{\left(1 - \bar{\beta}(x)\right) \lambda^{(m)}(x) + f^{(m)}(x)}{\lambda^{(m+1)}(x)} \right] dx = 0. \tag{5.66}$$

Since (5.66) must hold for all variations $\delta(x)$, it follows that

$$-1 + \frac{\left(1 - \bar{\beta}(x)\right) \lambda^{(m)}(x) + f^{(m)}(x)}{\lambda^{(m+1)}(x)} \equiv 0 \quad \text{for all } x. \tag{5.67}$$

The EM update for the SPECT intensity function estimate is

$$\lambda^{(m+1)}(x) = \left(1 - \bar{\beta}(x)\right) \lambda^{(m)}(x) + f^{(m)}(x). \tag{5.68}$$

To facilitate later discussion (in Section 6.5), the recursion (5.68) is rewritten as a sum over the camera view angles:

$$\lambda^{(m+1)}(x) = \lambda^{(m)}(x) \sum_{j=0}^{n_\Theta - 1} p_{\Theta|X}\left(\theta_j \mid x\right) \left[1 - \beta_j(x) \right.$$

$$\left. + \sum_{r=1}^{n_j} \frac{p_{Y|X\Theta}\left(y_{jr} \mid x, \theta_j\right) \beta(x, y_{jr}, \theta_j)}{\int_{\mathbb{R}^3} p_{Y|X\Theta}\left(y_{jr} \mid x, \theta_j\right) p_{\Theta|X}\left(\theta_j \mid x\right) \beta(x, y_{jr}, \theta_j) \lambda^{(m)}(x)\, \mathrm{d}x} \right].$$

$$(5.69)$$

This completes the M-step.

The SPECT recursion is similar to estimators that are discussed elsewhere for other applications. One is the Shepp-Vardi recursion for PET. The similarity is especially evident for $n_\Theta = 1$. Another is the information update of a multisensor multitarget tracking filter. This application is discussed in Chapter 6. The similarity arises because the different camera positions are analogous to multiple sensors.

5.5 Transmission Tomography

5.5.1 Background

In transmission tomography the photon source is external to the tissue or other object being imaged, so the photon count can be much higher than in PET. It is desirable to minimize patient radiation exposure, so there are limits on how much the photon count can be increased.

The estimated parameters are no longer the PPP intensities in a pixel grid, but rather the attenuation coefficients of the pixels in the grid. PET measures radioactivity concentration and, thus, preferentially images tissues with higher metabolic rates. Transmission tomography measures x-ray absorption, which is related to the average electron density (or effective atomic number) of the tissue. Readable accounts of the physics involved are found in the book by Ter-Pogossian [131]. A more recent discussion of effective atomic numbers and electron density is given in [141].

As in PET, the image pixels are denoted by $\mathcal{R}_r$, $r = 1, \ldots, K$. Let μ_r be the attenuation coefficient in $\mathcal{R}_r$. The probability that a photon is absorbed while traversing a line segment of length Δl lying inside $\mathcal{R}_r$ is $1 - e^{-\mu_r \Delta l}$, so μ_r has units of inverse length. Hence, the probability that a photon is *not* absorbed while traversing an interval of length Δl inside $\mathcal{R}_r$ is $e^{-\mu_r \Delta l}$. Imaging in transmission tomography is equivalent to estimating the attenuation coefficients $\mu = (\mu_1, \ldots, \mu_K)$.

The EM method is used to estimate the attenuation coefficients. For various reasons this problem is significantly more difficult than the PET algorithm. The approach given here parallels that of Lange and Carson [65], who were first use of the EM method for transmission tomography. In any event, the method here is very interesting and potentially of utility in applications other than transmission tomography. The word photon is used throughout the discussion because it matches the attenuation assumptions; however, photons are only placeholders for any quantity of interest that attenuates linearly.

5.5.2 Lange-Carson Algorithm

5.5.2.1 Attenuation and the Likelihood Function

Let L be the number of detectors. A detector is assumed to receive photons from only one source. This assumption is a limiting factor in some applications since it implies that the photon beam is very narrow and does not spread across multiple detectors. The source intensity typically varies across the emitted fan or cone beam. Let $\alpha_j > 0$ denote the intensity emitted in the direction of detector j. These intensities are assumed known since they can be measured before inserting the object to be imaged into the viewing field. The expected number of photons emitted toward detector j in the time interval Δt over which the data are aggregated in the detectors is

$$d_j = \alpha_j \, \Delta t,$$

so α_j has units of number per unit time. The number of photons emitted by source j is assumed to be Poisson distributed with parameter d_j.

The number of photons arriving at the detectors in the array is denoted by

$$m_{1:L} = (m_1, \ldots, m_L).$$

A photon emitted by source j traverses a number of pixels on its way to detector j. Denote the indices of these pixels by $\mathscr{I}_j \subset \{1, \ldots, K\}$. The line of flight of the photon from source j to detector j is a straight line intersecting every pixel in $\mathscr{I}_j$. Let l_{jr}, $r \in \mathscr{I}_j$, denote the length of the traversal in pixel $\mathcal{R}_r$ for source-detector pair j. Let $l_{jr} = 0$ for $r \notin \mathscr{I}_j$. The probability that a photon from source j reaches detector j is the product of probabilities that it successfully traversed each of the cells in its flight path:

$$\prod_{r \in \mathscr{I}_j} e^{-l_{jr}\,\mu_r} = \exp\left(- \sum_{r \in \mathscr{I}_j} l_{jr}\,\mu_r \right).$$

Attenuation is a thinning process, so the number of photons arriving at detector j is Poisson distributed with parameter

$$d_j \, \exp\left(- \sum_{r \in \mathscr{I}_j} l_{jr}\,\mu_r \right). \tag{5.70}$$

The detector counts in $m_{1:K}$ are independent, so the loglikelihood of $m_{1:K}$ is

$$\log p(m_{1:L}\,;\,\Lambda) = \sum_{j=1}^{L}\left\{-d_j\,\exp\left(-\sum_{r\in\mathscr{I}_j} l_{jr}\,\mu_r\right) - m_j\sum_{r\in\mathscr{I}_j} l_{jr}\,\mu_r\right\}, \tag{5.71}$$

where the constants $m_j\,\log d_j\,-\,\log m_j!$ are omitted because they do not depend on Λ.

5.5.2.2 EM Convergence

The loglikelihood function (5.71) is strictly concave. To see this, differentiate to find the (u, v)-th entry of the $K \times K$ Hessian matrix:

$$\frac{\partial^2}{\partial\mu_u\,\partial\mu_v}\log p(m_{1:L}\,;\,\Lambda) = -\sum_{j=1}^{L} a_{ju}\,d_j\,\exp\left(-\sum_{r\in\mathscr{I}_j} l_{jr}\,\mu_r\right) a_{jv}, \tag{5.72}$$

where $a_{jk} = l_{jk}$ if $k \in \mathscr{I}_j$ and $a_{jk} = 0$ otherwise. The quadratic form of this matrix is strictly negative definite for all Λ if the $L \times K$ matrix of cell traversal lengths

$$\mathbb{L} = \begin{bmatrix} l_{11} & l_{12} & \cdots & l_{1K} \\ l_{21} & l_{22} & \cdots & l_{2K} \\ \vdots & \vdots & & \vdots \\ l_{L1} & l_{L2} & \cdots & l_{LK} \end{bmatrix} \tag{5.73}$$

is full rank. Many entries of the traversal length matrix are zero, since the indices of the nonzero entries of row j are identical to the indices in $\mathscr{I}_j$. The matrix is full rank if the configuration of pixels and source-detector pairs is properly designed. The full rank condition is assumed to be incorporated into the system. Consequently, the EM method is guaranteed to converge to the global ML estimate.

5.5.2.3 E-step: Missing Data Likelihood Function

The indices in $\mathscr{I}_j$ are ordered arbitrarily via the labeling assigned to the pixels of the object being imaged. This is not the same as the order in which the pixels are traversed by photons propagating from source j to detector j. The traversal order is denoted by

$$\mathscr{I}_j(1) < \mathscr{I}_j(2) < \cdots < \mathscr{I}_j(T_j - 1), \tag{5.74}$$

where $T_j - 1$ is the total number of indices in $\mathscr{I}_j$. To be clear, the pixel adjacent to the source is $\mathcal{R}_{\mathscr{I}_j(1)}$, and the pixel adjacent to the detector is $\mathcal{R}_{\mathscr{I}_j(T_j-1)}$.

Although other choices for the missing data are possible, the natural choice here is the set of numbers $m_j(t)$, $t = 1, \ldots, T_j - 1$, of photons entering pixel $\mathcal{R}_{\mathscr{I}_j(t)}$. The number $m_j(1)$ is the number of photons emitted by the source. The number $m_j(T_j)$ is identified with the number of detected photons. The numbers $m_j(t)$, $t = 1, \ldots, T_j$, $j = 1, \ldots, L$, are random variables. Their conditioning on the observed measurements $m_j(T_j) = m_j$ is discussed in the next subsection. Because of attenuation,

$$m_j(1) \le m_j(2) \le \cdots \le m_j(T_j - 1) \le m_j(T_j). \tag{5.75}$$

The probability that the j-th source emits $m_j(1)$ photons is

$$\frac{d_j^{m_j(1)}}{m_j(1)!} e^{-d_j}.$$

A photon successfully transits pixel $\mathcal{R}_t$ with probability $e^{-l_{jt}\mu_t}$, so the probability that $m_j(t+1)$ photons are not absorbed and that $m_j(t) - m_j(t+1)$ are absorbed is the binomial term:

$$\binom{m_j(t)}{m_j(t+1)} \left(e^{-l_{jt}\mu_t}\right)^{m_j(t+1)} \left(1 - e^{-l_{jt}\mu_t}\right)^{m_j(t)-m_j(t+1)}.$$

The likelihood functions of the missing data of the j-th source-detector pair is therefore

$$p_j = \frac{d_j^{m_j(1)}}{m_j(1)!} e^{-d_j} \prod_{t=1}^{T_j-1} \binom{m_j(t)}{m_j(t+1)} \left(e^{-l_{jt}\mu_t}\right)^{m_j(t+1)} \left(1 - e^{-l_{jt}\mu_t}\right)^{m_j(t)-m_j(t+1)}. \tag{5.76}$$

The source-detector pairs are independent, so the joint likelihood function is the product:

$$p = \prod_{j-1}^{L} p_j. \tag{5.77}$$

Retaining only terms that depend on the attenuation coefficients μ gives the complete data loglikelihood function in the form

$$\mathcal{L}(\mu) = \sum_{j=1}^{L} \left\{ \sum_{t=1}^{T_j-1} m_j(t+1) \log\left(e^{-l_{jt}\mu_t}\right) \right.$$

$$\left. + (m_j(t) - m_j(t+1)) \log\left(1 - e^{-l_{jt}\mu_t}\right) \right\}. \tag{5.78}$$

The auxiliary function of the EM method is the conditional expectation of (5.78). Evaluating this expectation requires an explicit expression for the conditional expectations $E[m_j(t) \mid m_j(T_j) = m_j]$. This result is obtained in the next subsection.

5.5.2.4 E-step: Conditional Expectation Identity

The joint probability of $m_j(t)$ and $m_j(T_j)$ is, by the definition of conditioning,

$$\Pr[m_j(t), m_j(T_j); \mu] = \Pr[m_j(t); \mu] \Pr[m_j(T_j) \mid m_j(t); \mu]. \tag{5.79}$$

The random variable $m_j(t)$ is Poisson distributed with mean

$$E[m_j(t); \mu] = d_j \exp\left(-\sum_{t'=1}^{t-1} l_{j,\mathscr{I}_j(t')} \mu_{\mathscr{I}_j(t')}\right)$$

$$= \gamma_{jt} \equiv \gamma_{jt}(\mu). \tag{5.80}$$

The observed photon count $m_j(T_j)$ conditioned on $m_j(t)$ is binomially distributed because

$$\exp\left(-\sum_{t'=t}^{T_j-1} l_{j,\mathscr{I}_j(t')} \mu_{\mathscr{I}_j(t')}\right) = \frac{\gamma_{jT_j}}{\gamma_{jt}}. \tag{5.81}$$

is the probability of a photon successfully traversing pixel $\mathcal{R}_{\mathscr{I}_j(t)}$ and also all pixels it encounters on the way to the j-th detector. The joint pdf of $m_j(T_j)$ and $m_j(t)$ is, from (5.79),

$$\Pr[m_j(t), m_j(T_j); \mu]$$

$$= \left\{ e^{-\gamma_{jt}} \frac{(\gamma_{jt})^{m_j(t)}}{m_j(t)!} \right\} \left\{ \binom{m_j(t)}{m_j(T_j)} \left(\frac{\gamma_{jT_j}}{\gamma_{jt}}\right)^{m_j} \left(1 - \frac{\gamma_{jT_j}}{\gamma_{jt}}\right)^{m_j(t)-m_j(T_j)} \right\}. \tag{5.82}$$

The conditional pdf given the measurement $m_j(T_j) = m_j$ is, using (5.82),

$$\Pr[m_j(t) \mid m_j(T_j) = m_j ; \mu] = \frac{\Pr[m_j(t), m_j(T_j) = m_j ; \mu]}{\Pr[m_j(T_j) = m_j ; \mu]}$$

$$= \frac{\left\{ e^{-\gamma_{jt}} \frac{(\gamma_{jt})^{m_j(t)}}{m_j(t)!} \right\} \left\{ \frac{m_j(t)!}{m_j! (m_j(t) - m_j)!} \left(\frac{\gamma_{jT_j}}{\gamma_{jt}} \right)^{m_j} \left(1 - \frac{\gamma_{jT_j}}{\gamma_{jt}} \right)^{m_j(t) - m_j} \right\}}{e^{-\gamma_{jT_j}} \frac{\left(\gamma_{jT_j} \right)^{m_j}}{m_j!}}$$

$$= e^{-(\gamma_{jt} - \gamma_{jT_j})} \frac{\left\{ \gamma_{jt} - \gamma_j \right\}^{m_j(t) - m_j}}{(m_j(t) - m_j)!}. \tag{5.83}$$

Since $m_j(t) \geq m_j$, the expected value of (5.83) is

$$E\left[m_j(t) \mid m_j ; \mu \right] = \sum_{m_j(t) = m_j}^{\infty} m_j(t) \Pr[m_j(t) \mid m_j ; \mu]$$

$$= m_j + \gamma_{jt}(\mu) - \gamma_{jT_j}(\mu). \tag{5.84}$$

To see this, substitute (5.83) and simplify. The details are straightforward and are omitted.

5.5.2.5 E-step: Auxiliary Function

Let $n \geq 0$ denote the EM iteration index, and let $\mu^{(0)} = \left(\mu^{(0)}, \ldots, \mu_K^{(0)} \right) > 0$ be an initial value for the attenuation coefficients. The EM auxiliary function is defined as

$$Q\left(\mu ; \mu^{(n)} \right) = E_{\mu^{(n)}} \left[\mathscr{L}(\mu) \right].$$

Substituting (5.78) and using the conditional expectations

$$\tilde{M}_{jt}\left(\mu^{(n)} \right) = m_j - \gamma_{jT_j}\left(\mu^{(n)} \right) + \gamma_{jt}\left(\mu^{(n)} \right)$$

$$\tilde{N}_{jt}\left(\mu^{(n)} \right) = m_j - \gamma_{jT_j}\left(\mu^{(n)} \right) + \gamma_{j,t+1}\left(\mu^{(n)} \right),$$

gives

$$Q\left(\mu ; \mu^{(n)} \right) = \sum_{j=1}^{L} \sum_{t=1}^{T_j - 1} \left\{ \tilde{N}_{jt}\left(\mu^{(n)} \right) \log \left(e^{-l_{jt} \mu_t} \right) \right.$$

$$\left. + \left(\tilde{M}_{jt}\left(\mu^{(n)} \right) - \tilde{N}_{jt}\left(\mu^{(n)} \right) \right) \log \left(1 - e^{-l_{jt} \mu_t} \right) \right\}. \tag{5.85}$$

This form is awkward because the traversal ordering (5.74) does not naturally aggregate the parameter μ_t into separate parts of the expression. Rewriting

$Q\left(\mu;\mu^{(n)}\right)$ in the original pixel ordering is more convenient for this purpose, and is straightforward.

For $r = 1,\ldots,K$, let $M_{jr}\left(\mu^{(n)}\right)$ and $N_{jr}\left(\mu^{(n)}\right)$ denote the expected numbers of photons entering and exiting pixel r for source-detector pair j, conditioned on the measurements $m_{1:K}$ and given the parameter vector $\mu^{(n)}$. These expectations are evaluated by appropriately permuting the expectations $\left\{\tilde{M}_{jt}\left(\mu^{(n)}\right)\right\}$ and $\left\{\tilde{N}_{jt}\left(\mu^{(n)}\right)\right\}$.

Now let $\mathscr{J}_r$ denote the indices j of the source-detector pairs whose photons traverse pixel r. Then (5.85) becomes

$$Q\left(\mu;\mu^{(n)}\right) = \sum_{r=1}^{K}\sum_{j\in\mathscr{J}_r}\left\{N_{jr}\left(\mu^{(n)}\right)\log\left(e^{-l_{jr}\mu_r}\right)\right.$$
$$\left.+\left(M_{jr}\left(\mu^{(n)}\right)-N_{jr}\left(\mu^{(n)}\right)\right)\log\left(1-e^{-l_{jr}\mu_r}\right)\right\}.\quad(5.86)$$

This completes the E-step.

5.5.2.6 M-step

The EM recursive update is found by setting the gradient of (5.86) with respect to μ equal to zero. EM works like magic here—the gradient equations decouple completely into K equations, one for each attenuation coefficient. The equation for the update of μ_r is

$$\sum_{j\in\mathscr{J}_r}\left(M_{jr}\left(\mu^{(n)}\right)-N_{jr}\left(\mu^{(n)}\right)\right)\frac{l_{jr}}{e^{l_{jr}\mu_r}-1} = \sum_{j\in\mathscr{J}_r}N_{jr}\left(\mu^{(n)}\right)l_{jr}.\quad(5.87)$$

The solution of this transcendental equation is the EM update $\mu_r^{(n+1)}$. The solution of this equation is unique. To see this, observe that the left hand side of (5.87) decreases monotonically from $+\infty$ to zero as μ goes from zero to $+\infty$. Since the right hand side is a positive constant for every EM iteration n, the equation (5.87) has a unique solution.

An explicit analytical solution of (5.87) is not available, but an explicit solution is unnecessary for EM theory to work. Given the monotonicity of the left hand side of the equation, any number of numerical methods will provide the solution to arbitrary accuracy.

General EM theory guarantees that the iterates converge to a stationary point of the likelihood function. Because of the strict concavity of the loglikelihood function, this stationary point is the global ML estimate, $\hat{\mu}_{ML}$.

Solving the M-step numerically to arbitrary precision makes the Lange-Carson algorithm an exact EM algorithm. Lange and Carson in their original 1984 paper

[65] introduce an approximate solution to (5.87), but there is no longer any practical reason to use it.

Improvements to the Lange-Carson algorithm are discussed in [91] and reviewed in [34]. These include numerical algorithms to solve (5.87), methods to increase the convergence rate, and regularization to improve image quality.

A summary of the Lange-Carson algorithm is given in Table 5.3.

Table 5.3 Lange-Carson algorithm for transmission tomography

Data:

- Counts of detected events in L detectors: $m_{1:L} \equiv \{m_1, \ldots, m_L\}$
- Number of photons emitted by source j aimed at detector j: $\{d_1, \ldots, d_L\}$
- Traversal length matrix $\mathbb{L} = [l_{jr}]$: See (5.73).
- Traverse order: For $j = 1, \ldots, L, \mathscr{I}_j = (\mathscr{I}_j(1), \ldots, \mathscr{I}_j(T_j - 1))$: See (5.74).

Output:

- $\hat{\mu}_{ML} = (\hat{\mu}_1, \ldots, \hat{\mu}_K) \equiv$ vectorized image of attenuation coefficients in pixels $\mathcal{R}_1, \ldots, \mathcal{R}_K$

Algorithm:

- Initialize the attenuation coefficient image: $\mu(0) = (\mu_1(0), \ldots, \mu_K(0))$
- FOR EM iteration index $n = 0, 1, 2, \ldots$ until convergence:
  - FOR $j = 1 : L$
    - $\gamma(j, 1) = d_j$
    - FOR $t = 1 : T_j - 1$

$$\gamma(j, t + 1) = \gamma(j, t) \exp\left(-l_{j,\mathscr{I}_j(t)} \mu_{\mathscr{I}_j(t)}\right)$$

    - END FOR
    - FOR $r = 1 : K$

$$M(j, r) = N(j, r) = 0$$

    - END FOR
    - FOR $t = 1 : T_j - 1$

$$M\left(j, \mathscr{I}_j(t)\right) = m_j - \gamma\left(j, \mathscr{I}_j(T_j)\right) + \gamma\left(j, \mathscr{I}_j(t)\right)$$
$$N\left(j, \mathscr{I}_j(t)\right) = m_j - \gamma\left(j, \mathscr{I}_j(T_j)\right) + \gamma\left(j, \mathscr{I}_j(t + 1)\right)$$

    - END FOR
  - END FOR
  - FOR $r = 1 : K$, solve for the update $\mu_r(n + 1)$ of the r-th attenuation coefficient

$$\sum_{j=1}^{L} (M(i, r) - N(j, r)) \frac{l_{jr}}{e^{l_{jr}\mu_r} - 1} = \sum_{j=1}^{L} N(j, r) l_{jr}$$

  - END FOR
  - Update vectorized attenuation image: $\mu(n + 1) = (\mu_1(n + 1), \ldots, \mu_K(n + 1))$
- END FOR EM iteration (Test for convergence)
- If converged: Estimated attenuation coefficient image is $\hat{\mu}_{ML} = (\mu_1(n_{last}), \ldots, \mu_K(n_{last}))$

5.6 CRBs for Emission and Transmission Tomography

The FIM and CRB are probably most useful during system design phase of imaging system development, or perhaps for the control and analysis of alternative deployable system configurations. In any event, showing that the CRB matrix of a proposed system (or system configuration) is diagonally dominant with small variances establishes that it has the potential to estimate high quality images. This statement assumes that an efficient estimator (one that achieves the CRB) exists and is known.

The expressions above yield explicit expressions for the FIM for the PET problem for both sample data and histogram data. For sample data, the intensity is given by (5.3). Hence, using (4.21), the (i, j)-th entry of the $K \times K$ FIM is

$$J_{ij}(\Lambda) = \int_{\mathcal{R}} \frac{f_i(y) f_j(y)}{\sum_{r=1}^{K} \lambda_r f_r(y)} \, \mathrm{d}y. \tag{5.88}$$

The function $f_r(y)$ is given by (5.4). The FIM is determined by the likelihood function $\ell(\cdot)$ and by the sizes and locations of the pixels. Similarly, for histogram data, the piecewise constant intensity is such that

$$\int_{\mathcal{R}_j} \lambda(s) \, \mathrm{d}s = \sum_{r=1}^{K} \ell(j \mid r) \lambda_r. \tag{5.89}$$

Using (4.28) gives the $K \times K$ FIM as

$$J_{hist}(\Lambda) = \sum_{j=1}^{L} \frac{1}{\sum_{r=1}^{K} \ell(j \mid r) \lambda_r} \begin{bmatrix} \ell(j \mid 1) \\ \vdots \\ \ell(j \mid K) \end{bmatrix} \left[\ell(j \mid 1) \cdots \ell(j \mid K) \right]^T. \tag{5.90}$$

The (i, j)-th entry of $J_{hist}(\Lambda)$ is closely related to (5.88).

In the case of transmission tomography, the PPP is defined on the discrete space of all source-detector pairs. The appropriate FIM is therefore (4.35). It follows from (5.70) that the (u, v)-th entry of the FIM is

$$J_{tran}(\mu) = \sum_{j=1}^{L} a_{ju} d_j \exp\left(- \sum_{r \in \mathscr{I}_j} l_{jr} \mu_r \right) a_{jv}, \tag{5.91}$$

where $a_{jk} = l_{jk}$ if $k \in \mathscr{I}_j$ and $a_{jk} = 0$ otherwise. This expression is no surprise since it is identical to the Hessian matrix used in (5.72) to establish the concavity of the likelihood function.

In all cases, inverting the full FIM gives the CRB. The diagonal of the CRB is the variance of the pixel intensity, or attenuation coefficient, as the case may be, so the diagonal is often the primary quantity of interest in practice. The diagonal is of length K, so it corresponds to a vectorized image. If the CRB is computed, the diagonal image reveals which pixels may have unreliable estimates. The off-diagonal entries of the CRB are of interest, but are not as easily displayed intuitively. Such entries are potentially useful because they may reveal undesirable long distance spatial correlations between pixel estimates.

The drawback to the FIM for tomography is that practical imaging problems require high resolution. If the field of view is large and the scale of the objects of interest within the image is small, the number of pixels K rapidly becomes very large. For example, a 100×150 pixilated image gives $K = 15,000$ pixels. Large K causes two difficulties in applications. The first is that numerically evaluating any one of the entries of the FIM is straightforward, but evaluating all $K(K+1)/2$ of them can be time consuming and require more care than is desirable for a development effort. In many applications this problem can probably be circumvented in various ways. The other is more daunting—inverting the full FIM to find the CRB is infeasible for sufficiently large K. A more practical alternative is provided by Hero and Fessler [50, 51], who propose a recursive algorithm to compute the CRB for a subset of the intensities in a region of interest within the larger image.

5.7 Regularization

Tomographic imaging is known to require regularization to reduce noise artifacts that are visually similar to the Gibbs phenomenon. These artifacts are not caused by the EM method, but arise in any algorithm which estimates an infinite dimensional quantity, in this case the intensity, from finite dimensional measured data.

Regularization methods for transmission tomography are reviewed in [34]. Several interesting methods for regularizing PET are described in detail in [119]:

- Grenander's Method of Sieves.
- Penalty Functions.
- Markov Random Fields.
- Resolution Kernels.
- ML Estimator.
- Penalized maximum likelihood (ridge estimation).

These methods differ in important ways, but they have much in common. Perhaps the simplest to use that preserves the EM form of the intensity estimator is Grenander's method.

5.7.1 Grenander's Method of Sieves

Intensity estimation sometimes requires regularization to reduce noise artifacts. These artifacts, when they arise, are not caused by the EM method since the EM method is merely a method for computing the ML estimate. The fault typically is due to inadequate data size compared to the number of parameters being estimated.

Grenander's Method of Sieves is especially useful because it preserves the EM form of the intensity estimator. A sieve restricts the range of possible intensities $\lambda(x)$ to a smooth subspace, Λ, of the collection of all nonnegative functions on $\mathcal{R}$. The kernel of the sieve, denoted $k(x \mid z)$, is a specified pdf for each $z \in \mathcal{Z}$, so that

$$\int_{\mathcal{R}} k(x \mid \cdot)\, dx = 1.$$

The kernel and the space $\mathcal{Z}$ are arbitrary and can be chosen in any convenient manner. For example, the kernel is often taken to be an appropriately dimensioned Gaussian pdf. The sieve restricts $\lambda(x)$ to the collection of all functions of the form

$$\Lambda \equiv \left\{ \lambda(x) = \int_{\mathcal{Z}} k(x \mid z)\, \zeta(z)\, dz, \quad \text{for some } \zeta(z) \right\}, \tag{5.92}$$

where $\zeta(z) \geq 0$ and $\int_{\mathcal{Z}} \zeta(z)\, dz < \infty$. In effect, the integral (5.92) is a low pass filter applied to the intensity $\zeta(z)$ to produce a smoothed estimate of $\lambda(x)$.

The basic idea is to compute the ML estimate $\widehat{\zeta}(z)$ from the data, and subsequently compute the ML intensity

$$\widehat{\lambda}_{ML}(x) = \int_{\mathcal{Z}} k(x \mid z)\, \widehat{\zeta}(z)\, dz. \tag{5.93}$$

The estimate $\widehat{\zeta}(z)$ is computed by the EM method.

The likelihood function for the i.i.d. data (5.5) is given by (5.6), where the intensity is given by (5.2). Substituting the smoothed form of $\lambda(x)$ in (5.92) into (5.2) gives the measurement intensity

$$\mu(y) = \int_{\mathcal{Z}} \widetilde{\ell}(y \mid z)\, \zeta(z)\, dz, \tag{5.94}$$

where the modified measurement likelihood function is

$$\widetilde{\ell}(y \mid z) = \int_{\mathcal{R}} \ell(y \mid x)\, k(x \mid z)\, dx. \tag{5.95}$$

The likelihood function (5.6) changes slightly to

$$p(\mathcal{Y};\, \Lambda) = e^{-\int_{\mathcal{Z}} \mu(y)\, dy} \prod_{j=1}^{m} \mu(y_j).$$

If $\mathcal{Z} = \mathcal{R}$, the change in domain is essentially inconsequential if the kernel is unimodal and peaks at $x = z$. In this case, the Shepp-Vardi iteration proceeds as before to estimate $\widehat{\zeta}(z)$.

If $\mathcal{Z} \neq \mathcal{R}$, it is necessary to devise new pixels in $\mathcal{Z}$ on which to perform the EM / Shepp-Vardi iterations. In any event, the end result is an estimate of $\widehat{\zeta}(z)$. This estimate in turn generates the smoothed, or regularized, estimate $\widehat{\lambda}_{ML}(x)$ that satisfies (5.93).

Chapter 6
Multiple Target Tracking

> *Although this may seem a paradox, all exact science is*
> *dominated by the idea of approximation.*
> Bertrand Russell, The Scientific Outlook, *1931*

Abstract Multitarget tracking intensity filters are closely related to imaging problems, especially PET imaging. The intensity filter is obtained by three different methods. One is a Bayesian derivation involving target prediction and information updating. The second approach is a simple, compelling, and insightful intuitive argument. The third is a straightforward application of the Shepp-Vardi algorithm. The intensity filter is developed on an augmented target state space. The PHD filter is obtained from the intensity filter by substituting assumed known target birth and measurement clutter intensities for the intensity filter's predicted target birth and clutter intensities. To accommodate heterogeneous targets and sensor measurement models, a parameterized intensity filter is developed using a marked PPP Gaussian sum model. Particle and Gaussian sum implementations of intensity filters are reviewed. Mean-shift algorithms are discussed as a way to extract target state estimates. Grenander's method of sieves is discussed for regularization of the multitarget intensity filter estimates. Sources of error in the estimated target count are discussed. Finally, the multisensor intensity filter is developed using the same PPP target models as in the single sensor filter. It is closely related to the SPECT medical imaging problem. Both homogeneous and heterogeneous multisensor fields are discussed. Multisensor intensity filters reduce the variance of estimated target count by averaging.

Keywords Intensity filter · PHD filter · Derivation via Bayes Theorem · Derivation via expected target count · Derivation via Shepp-Vardi · Marked multisensor intensity filter · Particle implementation · Gaussian sum implementation · Mean-shift algorithm · Surrogate CRB · Regularization · Sources of target count error · Multisensor intensity filter · Variance reduction · Heterogeneous multisensor intensity filter

Multitarget tracking in clutter is a joint detection and estimation problem. It comprises two important inter-related tasks. One initiates and terminates targets, and the other associates, or assigns, data to specific targets and to clutter. The MHT (Multiple Hypothesis Tracking) formulation treats both tasks. It is well matched to the target physics and to the sensor signal processors of most radar and sonar systems. Unfortunately, exact MHT algorithms are intractable because the number of measurement assignment hypotheses grows exponentially with the number of measurements. These problems are aggravated when multiple sensors are used. Circumventing the computational difficulties of MHT requires approximation.

Approximate tracking methods based on PPP models are the topics of this chapter. They show much promise in difficult problems with high target and clutter densities. The key insight is to model the distribution of targets in state space as a PPP, and then use a filter to update the defining parameter of the PPP—its intensity. To update the intensity is to update the PPP. The intensity function of the PPP approximation characterizes the multiple target tracking model. This important point is discussed further in Section 6.1.1.

The PPP intensity model uses an augmented state space, $\mathcal{S}^+$. This enables it to estimate target birth and measurement clutter processes on-line as part of the filtering algorithm.

Three approaches to the intensity filter are provided. The first is a Bayesian derivation given in Appendix D. This approach relies on a "mean field" approximation of the posterior point process. The relationship between mean field approach and the "first moment intensity" approximation of the posterior point process used by Mahler [74] is discussed. The second approach is a short but extraordinarily insightful derivation that is ideal for readers who wish to avoid the Bayesian analysis, at least on a first reading. The third approach is based on the connection to PET and the Shepp-Vardi algorithm. The PET interpretation contributes significantly to understanding the PPP target model. A special case of the intensity filter is the well known PHD filter. It is obtained by assuming a known target birth-death process, a known measurement clutter process, and restricting the intensity filter to the non-augmented target state space $\mathcal{S}$.

Implementation issues are discussed in Section 6.3. Current approaches use either particle or Gaussian sum methods. An image processing method called the mean shift algorithm is well suited to point target estimation, especially for particle methods. Observed information matrices (cf. Section 4.7) are proposed as surrogates for the error covariance matrices widely used in single target Bayesian filters. The underlying statistical meaning of OIM estimates is as yet unresolved.

Other topics discussed include a Gaussian sum PPP filter that enables heterogeneous target motion and sensor measurement models to be used in an intensity filter setting. See Section 6.2.2 for details. Sources of error in the estimated target count are discussed in Section 6.4. Target count estimates are sensitive to the probability of target detection. This function, $P_k^D(x)$, depends on target state x at measurement time t_k. It varies over time because of slowly changing sensor characteristics and environmental conditions. Monitoring these and other factors that affect target detection probability is not typically considered part of the tracking problem.

Another topic is the multiple sensor intensity filter described in Section 6.5. This topic is the subject of some debate. The filter presented here relies on the validity of the target PPP model for every sensor. It is also closely related to the imaging problem SPECT, just as the intensity and PHD filters are related to PET. The variance of the target count is reduced by averaging over the sensors, so that the variance of estimated target count decreases with the number of sensors.

Several areas of on-going research are not discussed here. Much recent work in the area is focused on the cardinalized PHD (CPHD) filter recently proposed by Mahler [41]. This interesting filter does not assume a Poisson distributed number of targets, so that the posterior finite point process is not an independent scattering process (see Section 2.9.3). This contributes to what has been appropriately called [37] "spooky action at a distance." An even more recent topic of interest is that of smoothing PHD filters [47, 86, 91]. The intention is to reduce variance by introducing time lag into the intensity function estimates.

6.1 Intensity Filters

6.1.1 PPP Model Interpretation

The points of a realization of a PPP on the target state space are a poor representation of the physical reality of a multiple target state. This is especially easy to see when exactly one target is present, for then ideally

$$\int_{\mathcal{S}} \lambda(x)\,\mathrm{d}x \;=\; 1\,. \tag{6.1}$$

From (2.4), the probability that a realization of the PPP has *exactly* one point target is

$$p_N(n \;=\; 1) \;=\; e^{-1} \;\approx\; 37\%\,.$$

Hence, 63% of all realizations have either no target or two or more targets. Evidently, realizations of the PPP seriously mismodel this simple tracking problem. The problem worsens with increasing the target count: if exactly n targets are present, then the probability that a realization has exactly n points is $e^{-n} n^n / n! \approx (2\pi n)^{-1/2} \to 0$ as $n \to \infty$. Evidently, PPP realizations are poor models of real targets.

One interpretation of the PPP approximation is that the critical element of the multitarget model is the intensity function, *not* the PPP realizations. The shift of perspective means that the integral (6.1) is the more physically meaningful quantity. Said another way, the concept of expectation, or ensemble average over realizations, corresponds more closely to the physical target reality than do the realizations themselves.

A huge benefit comes from accepting the PPP approximation to the multiple target state—exponential numbers of assignments are completely eliminated. The PPP approximation finesses the data assignment problem by replacing it with a stochastic imaging problem, and the imaging problem is easier to solve. It is fortuitous that the imaging problem is mathematically the same problem that arises in PET; see Section 5.2. The "at most one measurement per target" rule for tracking corresponds in PET to the physics—there is at most one measurement per positron-electron annihilation.

Analogies are sometimes misleading, but consider this one: In the language of thermodynamics, the points of PPP realizations are microstates. Microstates obey the laws of physics, but are not directly observable without disturbing the system state. Physically meaningful quantities (such as temperature, etc.) are ensemble averages over the microstates. In the PPP target model, the points of a realization are thus "microtargets", and microtargets obey the same target motion and measurement models as real targets. The ensemble average over the PPP microtargets yields the target intensity function. The language of microtargets is helpful in Section 6.5 on multisensor intensity filtering, but it is otherwise eschewed in this chapter.

6.1.2 Predicted Target and Measurement Processes

6.1.2.1 Formulation

Standard filtering notation is adopted, but modified to accommodate PPPs. The general Bayesian filtering problem is reviewed in Appendix C. Let $\mathcal{S} = \mathbb{R}^{n_x}$ denote the n_x-dimensional single target state space. The augmented space is $\mathcal{S}^+ \equiv \mathcal{S} \cup \phi$, where ϕ represents the "target absent" hypothesis.

The single target transition function from time t_{k-1} to time t_k, denoted by $\Psi_{k-1}(y \,|\, x) \equiv p_{\Xi_k | \Xi_{k-1}}(y \,|\, x)$, is assumed known for all $x, y \in \mathcal{S}^+$. The augmented state space enables both target initiation and termination to be incorporated directly into Ψ_{k-1} as specialized kinds of state transitions. Novel aspects of the transition function are:

- $\Psi_{k-1}(\phi \,|\, x)$ is the probability that a target at $x \in \mathcal{S}$ terminates;
- $\Psi_{k-1}(y \,|\, \phi)$ is the likelihood that a new target initiates at $y \in \mathcal{S}$; and
- $\Psi_{k-1}(\phi \,|\, \phi)$ is the probability that the ϕ hypothesis is unchanged.

The augmented space $\mathcal{S}^+$ is discussed in Section 2.12.

The multitarget state at time t_k is Ξ_k. It is a point process on $\mathcal{S}^+$, but it is not a PPP on $\mathcal{S}^+$. Nonetheless, Ξ_k is approximated by a PPP to "close the loop" after the Bayesian information update. The multitarget state is a realization ξ_k of Ξ_k. If $\xi_k = (n, \{x_1, \ldots, x_n\})$, then every point x_j is either a point in $\mathcal{S}$ or is the hypothesis ϕ. It is stressed that repeated occurrences of ϕ are allowed in the list $\{x_1, \ldots, x_n\}$ to account for clutter measurements.

The measurement at time t_k is Υ_k. It is a point process on the (nonaugmented) space $\mathcal{T} \equiv \mathbb{R}^{n_z}$, where n_z is the dimension of a sensor measurement. The measurement data set

$$\upsilon_k = (m, \{z_1, \ldots, z_m\}) \in \mathcal{E}(\mathcal{T})$$

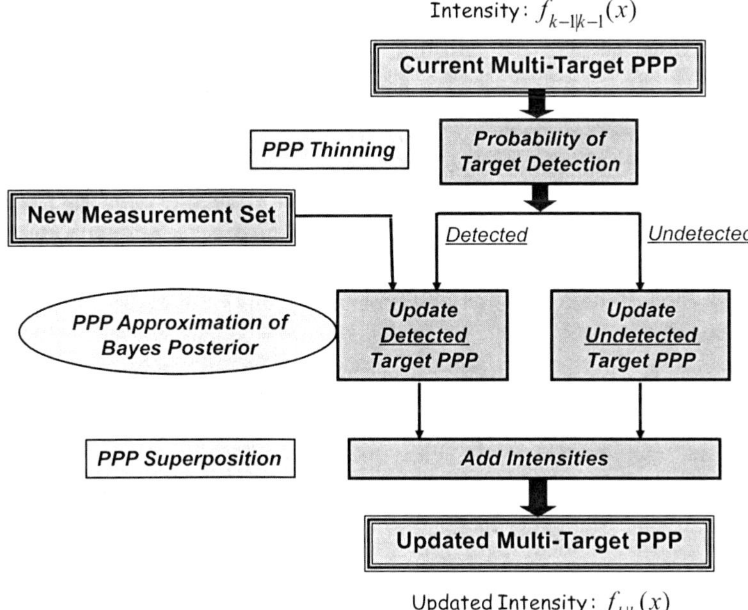

Fig. 6.1 Block diagram of the Bayes update of the intensity filter on the augmented target state space $\mathcal{S}^+$. Because the null state ϕ is part of the state space, target birth and measurement clutter estimates are intrinsic to the predicted target and predicted measurement steps. The same block diagram holds for the PHD filter on the nonaugmented space $\mathcal{S}$

is a realization of Υ_k. The pdf of a point measurement $z \in \mathcal{T}$ conditioned on a target in state $x \in \mathcal{S}^+$ at time t_k is the measurement pdf $p_k(z \mid x)$. The only novel aspect of this pdf is that $p_k(z \mid \phi)$ is the pdf that z is a clutter measurement.

The Bayesian posterior multitarget state point process conditioned on the data $\upsilon_1, \ldots, \upsilon_{k-1}$ is approximated by a PPP. Denote this PPP by $\Xi_{k-1|k-1}$. The intensity of $\Xi_{k-1|k-1}$ is $f_{k-1|k-1}(x)$, $x \in \mathcal{S}^+$. Let $\Xi_{k|k-1}$ denote the predicted PPP at time t_k. Its intensity is denoted by $f_{k|k-1}(x)$, and it is the integral of the intensity $f_{k-1|k-1}(x)$ of $\Xi_{k-1|k-1}$, as seen in Section 2.11.2, (2.86).

The goal is to update the predicted PPP, $\Xi_{k|k-1}$, with the measurement data υ_k. The information updated point process is not a PPP, so it is approximated by a PPP. Let $\Xi_{k|k}$ denote the approximating PPP, and let its intensity be $f_{k|k}(x)$.

Figure 6.1 outlines the steps of the intensity filter. The discussion below walks through the steps in the order outlined.

6.1.2.2   Target Motion and the Bernoulli Split

The first of these steps accounts for target motion and predicts the intensity at the next time step. The input is a PPP with intensity $f_{k-1|k-1}(x)$, so the transition procedure yields an output process that is also a PPP, as is seen in Section 2.11.1. Let

$f_{k|k-1}(x)$ denote the intensity of the output PPP. Adapting (2.83) to $\mathcal{S}^+$ gives

$$f_{k|k-1}(x) = \int_{\mathcal{S}^+} \Psi(x \mid y) \, f_{k-1|k-1}(y) \, dy \,, \qquad (6.2)$$

where the integral over $\mathcal{S}^+$ is defined as in (2.97).

Target motion is followed by a Bernoulli thinning procedure using the probability of the sensor detecting a target at the point x, denoted $P_k^D(x)$. This probability is state dependent and assumed known. The input PPP intensity is $f_{k|k-1}(x)$. As seen in Section 2.9.2, thinning splits it into two PPPs – one for detected targets and the other for undetected targets, denoted by $f_{k|k-1}^D(x)$ and $f_{k|k-1}^U(x)$, respectively. These PPPs are independent (see Section 2.9.2) and, from (2.56), and their intensities are

$$f_{k|k-1}^D(x) = P_k^D(x) \, f_{k|k-1}(x)$$

and

$$f_{k|k-1}^U(x) = \left(1 - P_k^D(x)\right) f_{k|k-1}(x) \,.$$

Both branches in Fig. 6.1 are now subjected to an information update.

6.1.2.3 Predicted Measurement PPP, and Why It Is Important

As seen in Section 2.11.2, the predicted measurement process is a PPP. Its intensity is

$$\lambda_{k|k-1}(z) = \int_{\mathcal{S}^+} p_k(z \mid x) \, P_k^D(x) \, f_{k|k-1}(x) \, dx \,, \qquad \text{for } z \in \mathcal{T}, \qquad (6.3)$$

as is seen from (2.86). The measurement PPP is a critical component of the intensity filter because, as is seen in (6.9), it weights the individual terms in the sum that comprises the filter.

Another way to see the importance of $\lambda_{k|k-1}(z)$ is to recall the classical single-target Bayesian tracking problem. The standard Bayesian formulation gives

$$p(x \mid z) = \frac{p(z \mid x) \, p(x)}{p(z)} \,, \qquad (6.4)$$

where the denominator is a scale factor that makes the left hand side a true pdf. It is very easy to ignore $p(z)$ in practice because the numerator is obtained by multiplication and the product scaled so that it is a pdf. When multiple conditionally independent measurements are available, the conditional likelihood is a product and it is the same story again for the scale factor. However, *if the pdfs are summed, not multiplied*, the scale factor must be included for the individual terms to be comparable. Such is the case with the intensity filter: the PPP model justifies adding Bayesian pdfs instead of multiplying them, and the scale factors are crucial to making the sum meaningful.

The scale factor clearly deserves a respectable name, and it has one. It is called the partition function in statistical physics and the machine learning communities.

6.1.3 Information Updates

It is seen in Appendix C that the mathematically correct information update procedure is to apply the Bayesian method to both the detected and undetected target PPPs to evaluate their posterior pdfs. These pdfs are defined on the event space $\mathscr{E}(\mathcal{S}^+)$. If the posterior pdfs have the proper form, then the posterior point processes are PPPs and are characterized by their intensity functions on $\mathcal{S}^+$.

The information update of the undetected target PPP is the Bayesian updated process conditioned on no target detection. The posterior point process is identical to the predicted target point process. It is therefore a PPP whose intensity, denoted by $f_{k|k}^U(x)$, is given by

$$\begin{aligned}
f_{k|k}^U(x) &= f_{k|k-1}^U(x) \\
&= \left(1 - P_k^D(x)\right) f_{k|k-1}(x).
\end{aligned} \tag{6.5}$$

This brings the right hand branch of Fig. 6.1 to the superposition stage, i.e., the block that says "Add Intensities".

The left hand branch is more difficult because, as it turns out, the information updated detected target point process is not a PPP. This is a serious dilemma since it is highly desirable both theoretically and computationally for the filter recursion to remain a closed loop. The posterior point process of the detected targets is therefore approximated by a PPP. Three methods are given for obtaining this approximation.

The first method is a Bayesian derivation of the posterior density of the point process on the event space $\mathscr{E}(\mathcal{S}^+)$ followed by a "mean field" approximation. Details about Bayesian filters in general are given in Appendix C. The Bayesian derivation is mathematically rigorous, but not particularly insightful. The situation is improved considerably by showing the close connections between the mean field approximation and the "first moment" approximation of the posterior point process.

To gain intuition, one need look no further than to the second method. While not rigorous, it is intuitively appealing and very convincing. The perspective is further enriched by the third method. It shows a direct connection between the information update of the Bayesian filter and the Shepp-Vardi algorithm for PET imaging. This third method also poses an interesting question about iterative updates of a Bayesian posterior density.

6.1.3.1 First Method: Bayesian Derivation and Mean Field Approximation

The pdf of the Bayesian posterior point process for detected targets is defined on the event space $\mathcal{E}(\mathcal{S}^+)$. The Bayesian posterior, or information updated, pdf is defined on this complex event space. The derivation is straightforward and a delight for Bayesians. It is relatively long and interferes with the flow of the discussion, so it is given in Appendix D where it can be read at leisure. The main points are outlined here, so readers suffer little or no loss of insight by skipping it on a first reading. Specific equation references are provided here so the precision of the mathematics is not lost in the flow of words.

The Bayesian derivation explicitly incorporates the "at most one measurement per target" rule into the measurement likelihood function. It imposes this constraint via the measurement conditional pdf (cf. (D.5)). This pdf sums over all possible assignments of the given data to targets. Because of the augmented space $\mathcal{S}^+$, a clutter measurement is accounted for by assigning it to an absent target with state ϕ.

The usual Bayesian update (cf. (D.6)) leads to the pdf of the Bayesian posterior point process (cf. (D.10)). This pdf uses the facts that the a priori target process is a PPP with intensity $f^D_{k|k-1}(x)$ and that the predicted measurement process is a PPP with intensity $\lambda_{k|k-1}(z)$ given by (6.3). The posterior pdf is computationally intractable except in reasonably small problems. In any event, inspection of the posterior pdf clearly reveals that it does not have the form of a PPP pdf. Approximating the posterior point process with a PPP is the next step.

The Bayesian posterior pdf is replaced by a mean field approximation, a widely used method of approximation in machine learning and statistical physics problems. This approximation is the product of the one dimensional marginal pdfs of the posterior pdf. The marginal pdfs are identical. The intensity function of the approximating PPP is therefore taken to be proportional to the marginal pdf (cf. (D.13)). The appropriate scale factor is a constant determined by maximum likelihood. In essence, the mean field approximation (cf. (D.14)) is proportional to the intensity function of the approximating PPP.

The mean field method just outlined is closely related to the method originally used by Mahler [74, 76] to derive the PHD filter via the first moment of the posterior point process. The first moment of a general finite point process is evaluated via the expected value of a sum of the Janossy densities of the process, where the Janossy densities are the joint pdfs of the points of the process (conditioned on their number). Happily, in the tracking application discussed here, only one Janossy density is nonzero, and it turns out to be the Bayes information update. The details needed to see this are straightforward and are given in the last section of Appendix D. The connections between the mean field approximation and the first moment approximation provides considerable insight into both.

6.1.3.2 Second Method: Expected Target Count

Let $p_k(\,\cdot\mid x)$, $x \in S^+$, denote the conditional pdf of a measurement in the measurement space $\mathcal{T}$, so that

$$\int_{\mathcal{T}} p_k(z \mid x)\, dz = 1, \quad \text{for all } x \in S^+. \tag{6.6}$$

The special case $p_k(z \mid \phi)$ is the pdf of a data point z conditioned on absent target state ϕ, that is, the pdf of z given that it is clutter. The predicted intensity at time t_k of the target point process is a PPP with intensity $P_k^D(x)\, f_{k|k-1}(x)$. The intensity $f_{k|k}^D(x)$ is the intensity of a PPP that approximates the information updated, or Bayes posterior, detected target process.

The measured data at time t_k are m_k points in a measurement space $\mathcal{T}$. Denote these data points by $Z_k = (z_1, \ldots, z_{m_k})$.

The information update of the detected target PPP is obtained intuitively as follows. The best current estimate of the probability that the point measurement z_j originated from a physical target with state $x \in S$ in the infinitesimal dx is (see Section 3.2.2 and also (5.19))

$$\frac{p_k(z_j \mid x)\, P_k^D(x)\, f_{k|k-1}(x)\, |dx|}{\lambda_{k|k-1}(z_j)}, \tag{6.7}$$

where the denominator is found using (6.3). Similarly, the probability that z_j originated from a target with state ϕ is

$$\frac{p_k(z_j \mid \phi)\, P_k^D(\phi)\, f_{k|k-1}(\phi)}{\lambda_{k|k-1}(z_j)}. \tag{6.8}$$

Because of the "at most one measurement per target" rule, the sum of the ratios over all measurements z_j is the estimated number of targets at x, or targets in ϕ, that generated a measurement.

The estimated number of targets at $x \in S$ is set equal to the expected number of targets conditioned on the data υ_k, namely $f_{k|k}^D(x)\, |dx|$. Cancelling $|dx|$ gives

$$f_{k|k}^D(x) = \sum_{j=1}^{m_k} \frac{p_k(z_j \mid x)\, P_k^D(x)\, f_{k|k-1}(x)}{\lambda_{k|k-1}(z_j)}. \tag{6.9}$$

Equation (6.9) holds for all $x \in S^+$, not just for $x \in S$.

The expected target count method makes it clear that the expected number of detected targets in any given set $\mathcal{R} \subset S$ is simply the integral of the posterior pdf

$$E\left[\text{Number of detected targets in } \mathcal{R}\right] = \int_{\mathcal{R}} f_{k|k}^D(x)\, dx. \tag{6.10}$$

Similarly, the expected number of targets in state ϕ is the posterior intensity evaluated at ϕ, namely, $f_{k|k}^D(\phi)$. The predicted measurement process with intensity $\lambda_{k|k-1}(z)$ is a vital part of the intensity filter.

6.1.3.3 Third Method: Shepp-Vardi Iteration

The PET model is interesting here. The measurement data and multiple target models are interpreted analogously so that:

- The target state space $\mathcal{S}^+$ corresponds to the space in which the radioisotope is absorbed.
- The measured data $Z_k \subset \mathcal{T}$ correspond to the measured locations of the annihilation events. As noted in the derivation of Shepp-Vardi, the measurement space $\mathcal{T}$ need not be the same as the state space.
- The posterior target intensity $f_{k|k}^D(x)$ corresponds to the annihilation event intensity.

The analogy makes the targets mathematically equivalent to the distribution of (hypothetical) positron-electron annihilation events in the state space.

Under the annihilation event interpretation, the information update (6.9) of the detected target process is given by the Shepp-Vardi algorithm for PET using PPP sample data. The EM derivation needs only small modifications to accommodate the augmented state space. Details are left to the reader. The n-th iteration of the Shepp-Vardi algorithm is, from (5.18),

$$f_{k|k}^D(x)^{(n+1)} = f_{k|k}^D(x)^{(n)} \sum_{j=1}^{m_k} \frac{p_k(z_j \mid x)}{\int_{\mathcal{S}^+} p_k(z_j \mid s)\, f_{k|k}^D(s)^{(n)}\, ds}, \tag{6.11}$$

where the predicted intensity $f_{k|k}^D(x)^{(0)} \equiv f_{k|k-1}^D(x) = P_k^D(x)\, f_{k|k-1}(x)$ initializes the algorithm. The first iteration of this version of the Shepp-Vardi algorithm is clearly identical to the Bayesian information update (6.9) of the detected target process.

The Shepp-Vardi iteration converges to an ML estimate of the target state intensity given only data at time t_k. It is independent of the data at times $t_1, \ldots, t_{k-1}$ except insofar as the initialization influences the ML estimate. In other words, the iteration leads to an ML estimate of an intensity that does *not* include the effect of a Bayesian prior. The problem lies not in the PET interpretation but in the pdf of the data. To see this it suffices to observe that the parameters of the pdf (5.7) are not constrained by a Bayesian prior and, consequently, the Shepp-Vardi algorithm converges to an estimate that is similarly unconstrained. It is, moreover, not obvious how to impose a Bayesian prior on the PET parameters that does not disappear in the small cell limit.

6.1.4 The Final Filter

Superposing the PPP approximation of the detected target process and the undetected target PPP gives

$$f_{k|k}(x) = f_{k|k}^{U}(x) + f_{k|k}^{D}(x), \qquad x \in \mathcal{S}^{+}$$

$$= \left[1 - P_{k}^{D}(x) + \sum_{j=1}^{m} \frac{p_{k}(z_{j} \mid x) \, P_{k}^{D}(x)}{\lambda_{k|k-1}(z_{j})} \right] f_{k|k-1}(x) \qquad (6.12)$$

as the updated intensity of the PPP approximation to $\Xi_{k|k}$.

The intensity filter comprises equations (6.2), (6.3), and (6.12). The first two equations are more insightful when written in traditional notation. Expanding the discrete-continuous integral (6.2) gives

$$f_{k|k-1}(x) = \hat{b}_{k}(x) + \int_{\mathcal{S}} \Psi_{k-1}(x \mid y) \, f_{k-1|k-1}(y) \, \mathrm{d}y, \qquad (6.13)$$

where the predicted target birth intensity is

$$\hat{b}_{k}(x) = \Psi_{k-1}(x \mid \phi) \, f_{k-1|k-1}(\phi). \qquad (6.14)$$

Also, from (6.3),

$$\lambda_{k|k-1}(z) = \hat{\lambda}_{k}(z) + \int_{\mathcal{S}} p_{k}(z \mid x) \, P_{k}^{D}(x) \, f_{k|k-1}(x) \, \mathrm{d}x, \qquad (6.15)$$

where

$$\hat{\lambda}_{k}(z) = p_{k}(z \mid \phi) \, P_{k}^{D}(\phi) \, f_{k|k-1}(\phi) \qquad (6.16)$$

is the predicted measurement clutter intensity. The probability $P_{k}^{D}(\phi)$ in (6.16) is the probability that a ϕ hypothesis generates a measurement at time t_{k}.

The computational parts of the intensity filter are outlined in Table 6.1. The table clarifies certain interpretive issues that are glossed over in the discussion. Implementation methods are discussed elsewhere in this chapter.

6.1.4.1 Likelihood Function of the Data Set

Since $f_{k|k}(x)$ is the intensity of a PPP, it is reasonable to inquire about the pdf of the data $\eta_{k} = \left(m_{k}, \{ z_{1}, \ldots, z_{m_{k}} \} \right)$. The measurement intensity after the information update of the target state intensity is, applying (2.86) on the augmented space $\mathcal{S}^{+}$,

$$\lambda_{k|k}(z) = \int_{\mathcal{S}^{+}} p_{k}(z \mid x) \, f_{k|k}(x) \, \mathrm{d}x . \qquad (6.17)$$

Table 6.1 Intensity filter on the state space $\mathcal{S}^+ = S \cup \phi$

INPUTS:

Data: $z_{1:m} \equiv \{z_1, \ldots, z_m\} \subset \mathcal{T}$ at time t_k

Probability of target detection: $P_k^D(x)$ at time t_k

OUTPUT: $\{f_{k|k}(x),\ f_{k|k}(\phi)\} = \text{IntensityFilter}\left[\ f_{k-1|k-1}(x),\ f_{k-1|k-1}(\phi),\ P_k^D(x),\ z_{1:m}\right]$

- Predicted target intensity : For $x \in \mathcal{S}$,
    - Newly born targets (target initiations): $\hat{b}_k(x) = \Psi_{k-1}(x \mid \phi)\, f_{k-1|k-1}(\phi)$
    - Moving targets: $\hat{d}_k(x) = \int_{\mathcal{S}} \Psi_{k-1}(x \mid y)\, f_{k-1|k-1}(y)\, \mathrm{d}y$
    - Target intensity: $f_{k|k-1}(x) = \hat{b}_k(x) + \hat{d}_k(x)$
- Predicted number of ϕ hypotheses:
    - Persistent: $\hat{b}_k(\phi) = \Psi_{k-1}(\phi \mid \phi)\, f_{k-1|k-1}(\phi)$
    - Newly absent (target terminations): $\hat{d}_k(\phi) = \int_{\mathcal{S}} \Psi_{k-1}(\phi \mid y)\, f_{k-1|k-1}(y)\, \mathrm{d}y$
    - Number of ϕ hypotheses: $f_{k|k-1}(\phi) = \hat{b}_k(\phi) + \hat{d}_k(\phi)$
- Predicted measurement intensity: FOR $j = 1 : m$,
    - Generated by targets: $\hat{v}_k(z_j) = \int_{\mathcal{S}} p_k(z_j \mid x)\, P_k^D(x)\, f_{k|k-1}(x)\, \mathrm{d}x$
    - Generated by ϕ hypotheses: $\hat{\lambda}_k(z_j) = p_k(z_j \mid \phi)\, P_k^D(\phi)\, f_{k|k-1}(\phi)$
    - Measurement intensity: $\lambda_{k|k-1}(z_j) = \hat{\lambda}_k(z_j) + \hat{v}_k(z_j)$
- Information updated target intensity : For $x \in \mathcal{S}$,
    - Undetected targets: $f_{k|k}^U(x) = \left(1 - P_k^D(x)\right) f_{k|k-1}(x)$
    - Detected targets: $f_{k|k}^D(x) = \left[\sum_{j=1}^m \dfrac{p_k(z_j \mid x)\, P_k^D(x)}{\lambda_{k|k-1}(z_j)}\right] f_{k|k-1}(x)$
    - Target intensity: $f_{k|k}(x) = f_{k|k}^U(x) + f_{k|k}^D(x)$
- Information updated number of ϕ hypotheses:
    - Number of ϕ hypotheses that generate no data: $f_{k|k}^U(\phi) = \left(1 - P_k^D(\phi)\right) f_{k|k-1}(\phi)$
    - Number of ϕ hypotheses that generate data: $f_{k|k}^D(\phi) = \left[\sum_{j=1}^m \dfrac{p_k(z_j \mid \phi)\, P_k^D(\phi)}{\lambda_{k|k-1}(z_j)}\right] f_{k|k-1}(\phi)$
    - Number of ϕ hypotheses: $f_{k|k}(\phi) = f_{k|k}^U(\phi) + f_{k|k}^D(\phi)$

Therefore the pdf of the data η_k is

$$p(\eta_k) = e^{-\int_{\mathbb{R}^{n_z}} \lambda_{k|k}(z)\, \mathrm{d}z} \prod_{j=1}^{m_k} \lambda_{k|k}(z_j)$$

$$= e^{-N_{k|k}} \prod_{j=1}^{m_k} \lambda_{k|k}(z_j), \tag{6.18}$$

where

$$N_{k|k} = \int_{\mathbb{R}^{n_z}} \lambda_{k|k}(z)\, \mathrm{d}z$$

$$= \int_{\mathcal{S}^+} \left(\int_{\mathbb{R}^{n_z}} p_k(z \mid x)\, \mathrm{d}z\right) f_{k|k}(x)\, \mathrm{d}x$$

$$= \int_{\mathbb{R}^{n_x}} f_{k|k}(x)\, \mathrm{d}x \tag{6.19}$$

is the estimated mean number of targets. The pdf of η_k is approximate because $f_{k|k}(x)$ is an approximation.

6.2 Relationship to Other Filters

The modeling assumptions of the intensity filter are very general, and specializations are possible. The most important is the PHD filter discussed in Section 6.2.1. By assuming certain kinds of a priori knowledge concerning target birth and measurement clutter, and adjusting the filter appropriately, the intensity filter reduces to the PHD filter. The differences between the intensity and PHD filters are nearly all attributable to the augmented state space $\mathcal{S}^+$. That is, the intensity filter uses the augmented single target state space $\mathcal{S}^+ = \mathcal{S} \cup \phi$, while the PHD filter uses only the single target space $\mathcal{S}$. Using $\mathcal{S}$ practically forces the PHD filter to employ target birth and death processes to model initiation and termination of targets.

A different kind of specialization is the marked multitarget intensity filter. This is a parameterized linear Gaussian sum intensity filter that interprets measurements as target marks. This interpretation is interesting in the context of PPP target models because it implies that joint measurement-target point process is a PPP. Details are discussed in Section 6.2.2.

6.2.1 Probability Hypothesis Density (PHD) Filter

The state ϕ is the basis for the on-line estimates of the intensities of the target birth and measurement clutter PPPs given by (6.13) and 6.15), respectively. If, however, the birth and clutter intensities are known a priori to be $b_k(x)$ and $\lambda_k(z)$, then the predictions $\hat{b}_k(x)$ and $\hat{\lambda}_k(z)$ can be replaced by $b_k(x)$ and $\lambda_k(z)$. This is the basic strategy taken by the PHD filter.

The use of a posteriori methods makes good sense in many applications. For example, they can help regularize parameter estimates. These methods can also incorporate information not included a priori in the Bayes filter. For example, Jazwinski [54] uses an a posteriori method to derive the Schmidt-Kalman filter for bias compensation. These methods may improve performance, i.e., if the a priori birth and clutter intensities are more accurate or stable than their on-line estimated counterparts, the PHD filter may provide better tracking performance.

Given these substitutions, the augmented space is no longer needed and can be eliminated. This requires some care. If the recursion is simply restricted to $\mathcal{S}$ and no other changes are made, the filter will not be able to discard targets and the target count may balloon out of control. To balance the target birth process, the PHD filter uses a death probability before propagating the multitarget intensity $f_{k-1|k-1}(x)$. This probability was intentionally omitted from the intensity filter because transition into ϕ is target death, and it is redundant to have two death models.

The death process is a Bernoulli thinning process applied to the PPP at time t_{k-1} before targets transition and are possibly detected. Let $d_{k-1}(x)$ denote the probability that a target at time t_{k-1} dies before transitioning to time t_k. The surviving target point process is a PPP and its intensity is $(1 - d_{k-1}(x)) \, f_{k-1|k-1}(x)$. Adding Bernoulli death and restricting the recursion to $\mathcal{S}$ reduces the intensity filter to the PHD filter.

6.2.2 Marked Multisensor Intensity Filter (MMIF)

The intensity filter assumes targets have the same motion model, and that the sensor measurement likelihood function is the same for all targets and data. Such assumptions are idealized at best. An alternative approach is to develop a parameterized intensity filter that accommodates heterogeneous target motion models and measurement pdfs by using target-specific parameterizations. The notion of target-specific parameterizations in the context of PPP target modeling seems inevitably to lead to the idea of modeling individual targets as a PPP, and then using superposition to obtain the aggregate PPP target model. Parameter estimation using the EM method is natural to superposition problems, as shown in Chapter 3. The marked multisensor intensity filter (MMIF) is one instance of such an approach.

The MMIF builds on the basic idea that a target at state x is "marked" with a measurement z. If the target is modeled as a PPP, then the joint measurement-target vector (z, x) is a PPP on the Cartesian product of the measurement and target spaces. This is an intuitively reasonable result, but the details needed to see that it is true are postponed to Section 8.1. The MMIF uses a linear Gaussian target motion and measurement model for each target and superposes them against a background clutter model. Since the Gaussian components correspond to different targets, they need not have the same motion model. Similarly, different sensor measurement models are possible. Superposition therefore leads to an affine Gaussian sum intensity function on the joint measurement-target space. The details of the EM method and the final MMIF recursion are given in Appendix E.

The MMIF adheres to the "at most one measurement per target rule" but only *in the mean*, or on average. It does this by reinterpreting the single target pdf as a PPP intensity function, and by interpreting measurements as the target marks. The expected number of targets that the PPP on the joint measurement-target space produces is one.

Another feature of the MMIF is that the EM weights depend on the Kalman filter innovations. The weights in other Gaussian sum filters often involve scaled multiples of the measurement variances, resulting in filters that are somewhat akin to "nearest neighbor" tracking filters.

The limitation of MMIF and other parameterized sum approaches is the requirement to use a fixed number of terms in the sum. This strongly affects its ability to model the number of targets. In practice, various devices can compensate for this limitation, but they are not intrinsic to the filter.

6.3 Implementation

Simply put, targets correspond to the local peaks of the intensity function and the areas of uncertainty correspond to the contours, or isopleths, of the intensity. Very often in practice, isopleths are approximated by ellipsoids in target state space corresponding to error covariance matrices. Methods for locating the local peak concentrations of intensity and finding appropriate covariance matrices to measure the width of the peaks are discussed in this section.

Implementation issues for intensity filters therefore concern two issues. Firstly, it is necessary to develop a computationally viable representation of the information updated intensity function of the filter. Two basic representations are proposed, one based on particles and the other on Gaussian sums. Secondly, postprocessing procedures are applied to the intensity function representation to extract the number of detected targets, together with their estimated states and corresponding error covariance matrices. Analogous versions of both issues arise in classical single target Bayesian filters.

The fact remains, however, that a proper statistical interpretation of target point estimates and their putative error covariances is lacking for intensity filters. The concern may be dismissed in practice because they are intuitively meaningful and closely resemble their single target Bayesian analogs. The concern is nonetheless worrisome and merits further study.

6.3.1 Particle Methods

The most common and by far the easiest implementation of nonlinear filters is by particle, or sequential Monte Carlo (SMC), methods. In such methods the posterior pdf is represented nonparametrically by a set of particles in target state space, together with a set of associated weights, and estimated target count. Typically these weights are uniform, so the spatial distribution of particles represents the variability of the posterior density. An excellent discussion of SMC methods for Bayesian single target tracking applications is found in the first four chapters of [104].

Published particle methods for the general intensity filter are limited to date to the PHD filter. Extensions to the intensity filter are not reported here. An early and well described particle methodology (as well as an interesting example for tracking on roads) for PHD filters is given in [111]. Particle methods and their convergence properties for the PHD filter are discussed in detail in a series of papers by Vo et al. [137]. Interested readers are urged to consult them for specifics.

Tracking in a surveillance region $\mathcal{R}$ using SMC methods starts with an initial set of particles and weights at time t_{k-1} together with the estimated number of targets in $\mathcal{R}$:

$$\left\{ \left(x_{k-1|k-1}(\ell), w_{k-1|k-1}(\ell) \right) : \ell = 1, L_{SMC} \right\} \text{ and } N_{k-1|k-1},$$

Table 6.2 PHD filter on the state space $\mathcal{S}$

INPUTS:

Data: $z_{1:m} = \{z_1, \ldots, z_m\} \subset \mathcal{T}$ at time t_k

Target death probability function: $d_{k-1}(x) = \Pr\left[\text{target death at state } x \in \mathcal{S} \text{ at time } t_{k-1}\right]$

Target birth probability function: $b_k(x) = \Pr\left[\text{target birth at state } x \in \mathcal{S} \text{ at time } t_k\right]$

Probability of target detection: $P_k^D(x)$ at time t_k

Measurement clutter intensity function: $\lambda_k(z)$ at time t_k

OUTPUT: $f_{k|k}(x) = \text{PHDFilter}\left[f_{k-1|k-1}(x),\ d_{k-1}(x),\ b_k(x),\ \lambda_k(z),\ P_k^D(x),\ z_{1:m}\right]$

- Predicted target intensity : For $x \in \mathcal{S}$,
    - Surviving targets:
        $$S_k(x) = (1 - d_{k-1}(x))\ f_{k-1|k-1}(y)$$
    - Propagated targets:
        $$S_k(x) \leftarrow \int_{\mathcal{S}} \Psi_{k-1}(x \mid y)\, S_k(y)\, dy$$
    - Predicted target intensity:
        $$f_{k|k-1}(x) = b_k(x) + S_k(x)$$
- Predicted measurement intensity:
    IF $m = 0$, THEN $f_{k|k}(x) = \left(1 - P_k^D(x)\right) f_{k|k-1}(x)$ STOP
    FOR $j = 1 : m$,
    - Intensity contributions from predicted target intensity:
        $$\hat{v}_k(z_j) = \int_{\mathcal{S}} p_k(z_j \mid x)\, P_k^D(x)\, f_{k|k-1}(x)\, dx$$
    - Predicted measurement intensity:
        $$\lambda_{k|k-1}(z_j) = \lambda_k(z_j) + \hat{v}_k(z_j)$$
    END FOR
- Information updated target intensity: For $x \in \mathcal{S}$,
$$f_{k|k}(x) = \left(1 - P_k^D(x) + \sum_{j=1}^{m} \frac{p_k(z_j \mid x)\, P_k^D(x)}{\lambda_{k|k-1}(z_j)}\right) f_{k|k-1}(x)$$

- END

---

where $w_{k-1|k-1}(\ell) = 1/L_{SMC}$ for all ℓ. For PHD filters the particle method proceeds in several steps that mimic the procedure outlined in Table 6.2:

- *Prediction.* In the sequential importance resampling (SIR) method, prediction involves thinning a given set of particles with the survival probability $1 - d_k(x)$ and then stochastically transforming the survivors into another particle set using the target motion model $\Psi_{k-1}(y \mid x)$ with weights adjusted accordingly. Additional new particles and associated weights are generated to model new target initializations.
- *Updating.* The particle weights are multiplicatively (Bayesian) updated using the measurement likelihood function $p_k(z \mid x)$ and the probability of detection $P_k^D(x)$. The factors are of the form

$$\left(1 - P_k^D(x) + \sum_{j=1}^{m} \frac{p_k(z_k(j) \mid x)\, P_k^D(x)}{\lambda_{k|k-1}(z_j)}\right), \tag{6.20}$$

where $z_k(1), \ldots, z_k(m)$ are the measurements at time t_k. The updated particle weights are nonuniform.

- *Normalization.* Compute the scale factor, call it $N_{k|k}$, of the sum of the updated particle weights. Divide all the particle weights by $N_{k|k}$ to normalize the weights.
- *Resampling.* Particles are resampled by choosing i.i.d. samples from the discrete pdf defined by the normalized weights. Resampling restores the particle weights to uniformity.

If the resampling step is omitted, the SMC method leads to particle weight distributions that rapidly concentrate on a small handful of particles and therefore poorly represents the posterior intensity. There are many ways to do the resampling in practice.

By computing $N_{k|k}$ before resampling, it is easy to see that

$$N_{k|k} \approx \int_{\mathcal{R}} \lambda_{k|k}(x)\, \mathrm{d}x$$
$$= E\left[\text{Number of targets in } \mathcal{R}\right]. \tag{6.21}$$

The estimated number of targets in any given subset $\mathcal{R}_0 \subset \mathcal{R}$ is

$$N_{k|k}(\mathcal{R}_0) = \left(\frac{\text{Number of particles in } \mathcal{R}_0}{L_{SMC}}\right) N_{k|k}. \tag{6.22}$$

The estimator is poor for sets $\mathcal{R}_0$ that are only a small fraction of the total volume of $\mathcal{R}$.

The primary limitations of particle approaches in many applications are due to the so-called Curse of Dimensionality[1]: the number of particles needed to represent the intensity function grows exponentially as the dimension of the state space increases. Most applications to date seem to be limited to four or five dimensions. The curse is so wicked that Moore's Law (the doubling of computational capability every 18 months) by itself will do little to increase the effective dimensional limit over a human lifetime. Moore's Law and improved methods together will undoubtedly increase the number of dimensions for which particle filters are practical, but it remains to be seen if general filters of dimension much larger than say six can be treated directly.

6.3.2 Mean Shift Algorithm

As is clear from the earlier discussion of PET, it is intuitively reasonable to think of the multitarget intensity filter as a sequential image processing method. In this

---

[1] The name was first used in 1961 by Richard E. Bellman [9]. The name is apt in very many problems; however, some modern methods in machine learning actually exploit high dimensional embeddings.

interpretation an image comprises the "gray scales" of a set of multidimensional voxels in the target state space $\mathbb{R}^{n_x}$. Such an interpretation enables a host of image processing techniques to be applied to the estimated intensity function.

One technique that lends itself immediately to extracting target point estimates from a particle set approach is the "mean-shift" algorithm. This algorithm, based on ideas first proposed in [38], is widely used in computer vision applications such as image segmentation and tracking. The mean-shift algorithm is an EM algorithm for Gaussian kernel density estimators and a generalized EM algorithm for non-Gaussian kernels [12]. The Gaussian kernel is computationally very efficient in the mean-shift method.

Denote the non-ϕ particle set representing the PPP intensity at time t_k by

$$\left\{ x_{k|k}(\ell) \, : \, \ell = 1, \ldots, L_{SMC} \right\} .$$

The intensity function is modeled as a scalar multiple of the kernel estimator

$$\lambda_k(x) \; = \; I_k \sum_{\ell=1}^{L_{SMC}} \mathcal{N}(x \, ; \, x_{k|k}(\ell), \, \Sigma_{\mathrm{ker}}) , \tag{6.23}$$

where $\mathcal{N}(x \, ; \, x_{k|k}(\ell), \, \Sigma_{\mathrm{ker}})$ is the kernel. The covariance matrix Σ_{ker} is specified, not estimated. Intuitively, the larger Σ_{ker}, the fewer the number of local maxima in the intensity (6.23), and conversely. The scale factor $I_k > 0$ is estimated by the particle filter and is taken as known here.

The form (6.23) has no parameters to estimate, so extend it by defining

$$\lambda_k(x \, ; \, \mu) \; = \; I_k \sum_{\ell=1}^{L_{SMC}} \mathcal{N}\left(x \, ; \, x_{k|k}(\ell) - \mu, \, \Sigma_{\mathrm{ker}}\right) , \tag{6.24}$$

where μ is an unknown rigid translation of the intensity (6.23). It is not hard to see that the ML estimate of μ is a local maximum of the kernel estimate, that is, a point estimate for a target. The vector μ is estimated from data using the EM method. The clever part is using an artificial data set with only one point in it, namely, the origin.

Let $r = 0, 1, \ldots$ denote the EM iteration index, and let $\mu^{(0)}$ be a specified initial value for the mean. The auxiliary function is given by (3.20) with $m = 1$, $x_1 = 0, L = L_{SMC}, \theta_\ell = \mu$, and

$$\lambda_\ell(x \, ; \, \mu) \; \equiv \; I_k \, \mathcal{N}\left(x \, ; \, x_{k|k}(\ell) - \mu, \, \Sigma_{\mathrm{ker}}\right) .$$

The bounded surveillance region $\mathcal{R}$ is taken to be $\mathbb{R}^{n_x}$. Define the weights

$$w_\ell\left(\mu^{(r)}\right) = \frac{\mathcal{N}\left(0;\, x_{k|k}(\ell) - \mu^{(r)},\, \Sigma_{\mathrm{ker}}\right)}{\sum_{\ell'=1}^{L_{SMC}} \mathcal{N}\left(0;\, x_{k|k}(\ell') - \mu^{(r)},\, \Sigma_{\mathrm{ker}}\right)}$$

$$= \frac{\mathcal{N}\left(x_{k|k}(\ell);\, \mu^{(r)},\, \Sigma_{\mathrm{ker}}\right)}{\sum_{\ell'=1}^{L_{SMC}} \mathcal{N}\left(x_{k|k}(\ell');\, \mu^{(r)},\, \Sigma_{\mathrm{ker}}\right)}. \tag{6.25}$$

The auxiliary function in the present case requires no sum over j as done in (3.20), so

$$Q(\mu;\, \mu^{(r)}) = -I_k + \sum_{\ell=1}^{L_{SMC}} w_\ell\left(\mu^{(r)}\right) \log\left\{I_k\, \mathcal{N}\left(0;\, x_{k|k}(\ell) - \mu,\, \Sigma_{\mathrm{ker}}\right)\right\}. \tag{6.26}$$

The EM update of μ is found by taking the appropriate gradient, yielding

$$\mu^{(r+1)} = \frac{\sum_{\ell=1}^{L_{SMC}} w_\ell\left(\mu^{(r)}\right) x_{k|k}(j)}{\sum_{\ell=1}^{L_{SMC}} w_\ell\left(\mu^{(r)}\right)}. \tag{6.27}$$

Substituting (6.25) and canceling the common factor gives the classical mean-shift iteration:

$$\mu^{(r+1)} = \frac{\sum_{\ell=1}^{L_{SMC}} \mathcal{N}\left(x_{k|k}(\ell);\, \mu^{(r)},\, \Sigma_{\mathrm{ker}}\right) x_{k|k}(\ell)}{\sum_{\ell=1}^{L_{SMC}} \mathcal{N}\left(x_{k|k}(\ell);\, \mu^{(r)},\, \Sigma_{\mathrm{ker}}\right)}. \tag{6.28}$$

The update of the mean is a convex combination of the particle set. Convergence to a local maximum $\mu_k^{(r)} \to \hat{x}_{k|k}$ is guaranteed as $r \to \infty$.

Different initializations are needed for different targets, so the mean shift algorithm needs a preliminary clustering method to initialize it, as well as to determine the number of peaks in the data correspond to targets. Also, the size of the kernel depends somewhat on the number of particles and may need to be adjusted to smooth the intensity surface appropriately.

6.3.3 Multimode Algorithms

Identifiability remains a problem with the mean shift algorithm, that is, there is no identification of the point estimate to a target except through the way the starting point of the iteration is chosen. This may cause problems when targets are in close proximity.

One way to try to resolve the problem is to use the particles themselves as points to feed into another tracking algorithm. This method exploits serial structure in the filter estimates and may disambiguate closely spaced targets. However, particles are serially correlated and do not satisfy the conditional independence assumptions of measurement data, so the resulting track estimates may be biased.

6.3.4 Covariance Matrices

6.3.4.1 Matrix CRBs

The true error covariance matrix of the ML point estimate computed by the mean shift method is not available; however, the CRB can be evaluated using (4.21). This CRB is appropriate if the available data are reasonably modeled as realizations of a PPP with intensity (6.24). If the intensity (6.24) function is a high fidelity model of the intensity, the FIM of μ is

$$J(\mu) \,=\, I_k\, \Sigma_{\text{ker}}^{-1}\, \Sigma(\mu)\, \Sigma_{\text{ker}}^{-1}\,, \tag{6.29}$$

where the matrix $\Sigma(\mu)$ is

$$\Sigma(\mu) \,=\, \sum_{\ell=1}^{L} \sum_{\ell'=1}^{L} \int_{\mathbb{R}^{n_x}} w(x,\, \mu\,;\, \ell,\, \ell')\, \big(x - x_{k|k}(\ell) + \mu\big) \big(x - x_{k|k}(\ell') + \mu\big)^T \, \mathrm{d}x \tag{6.30}$$

and the weighting function is

$$w(x,\, \mu\,;\, \ell,\, \ell') \,=\, \frac{\mathcal{N}\big(x\,;\, x_{k|k}(\ell) - \mu,\, \Sigma_{\text{ker}}\big)\, \mathcal{N}\big(x\,;\, x_{k|k}(\ell') - \mu,\, \Sigma_{\text{ker}}\big)}{\sum_{\ell''=1}^{L_{SMC}} \mathcal{N}\big(x\,;\, x_{k|k}(\ell'') - \mu,\, \Sigma_{\text{ker}}\big)}\,. \tag{6.31}$$

The CRB is $J^{-1}(\mu)$ evaluated at the true value of μ. Because the integral is over all of $\mathbb{R}^{n_x}$, a change of variables shows that the information matrix $J(\mu)$ is independent of the true value of μ. This means that the FIM is not target specific.

A local bound is desired, since the mean shift algorithm converges to a local peak of the intensity. By restricting the intensity model to a specified bounded gate $G \subset \mathbb{R}^{n_x}$, the integral in (6.30) is similarly restricted. The matrix $\Sigma(\mu)$ is thus a function of G. The gated CRB is local to the gate, i.e., it is a function of the target within the gate.

6.3.4.2 OIM: The Surrogate CRB

The OIM is the Hessian matrix of the negative loglikelihood evaluated at the MAP point estimate; it is often used as a surrogate for Fisher information when the likelihood function is complicated. Its inverse is the surrogate CRB of the estimate $\hat{x}_{k|k}$. The construction of OIMs for the mean-shift algorithm is implicit in [12].

The loglikelihood function of μ using the intensity function (6.23) and the one data point $x_1 = 0$ is

$$\log p(\mu) \,=\, -I_k\, L_{SMC} + \log\left[\sum_{\ell=1}^{L_{SMC}} \mathcal{N}(x_{k|k}(\ell)\,;\, \mu,\, \Sigma_{\text{ker}})\right].$$

Direct calculation gives the general expression

$$\nabla_\mu \left[\nabla_\mu \log p(\mu)\right]^T = \Sigma_{\text{ker}}^{-1} + \frac{1}{\kappa^2}\left[\sum_{\ell=1}^{L_{SMC}} \mathcal{N}(\mu\,;\,x_{k|k}(\ell),\,\Sigma_{\text{ker}})\,\Sigma_{\text{ker}}^{-1}\left(\mu - x_{k|k}(\ell)\right)\right]$$

$$\times \left[\sum_{\ell=1}^{L_{SMC}} \mathcal{N}(\mu\,;\,x_{k|k}(\ell),\,\Sigma_{\text{ker}})\,\Sigma_{\text{ker}}^{-1}\left(\mu - x_{k|k}(\ell)\right)\right]^T$$

$$-\frac{1}{\kappa}\Sigma_{\text{ker}}^{-1}\left[\sum_{\ell=1}^{L_{SMC}} \mathcal{N}(\mu;x_{k|k}(\ell),\Sigma_{\text{ker}})\left(\mu-x_{k|k}(\ell)\right)\left(\mu-x_{k|k}(\ell)\right)^T\right]\Sigma_{\text{ker}}^{-1}\,,$$

where the normalizing constant is

$$\kappa = \sum_{\ell=1}^{L_{SMC}} \mathcal{N}(\mu\,;\,x_{k|k}(\ell),\,\Sigma_{\text{ker}})\,.$$

The observed information matrix is evaluated at the MAP estimate $\mu = \hat{x}_{k|k}$. The middle term is proportional to $\left[\nabla_\mu\, p(\mu)\right]\left[\nabla_\mu\, p(\mu)\right]^T$ and so is zero any stationary point of $p(\mu)$, e.g., at the MAP estimate $\mu = \hat{x}_{k|k}$. The OIM is therefore

$$OIM\left(\hat{x}_{k|k}\right) = \Sigma_{\text{ker}}^{-1}$$

$$-\Sigma_{\text{ker}}^{-1}\left[\frac{\sum_{\ell=1}^{L_{SMC}} \mathcal{N}(\hat{x}_{k|k};x_{k|k}(\ell),\Sigma_{\text{ker}})\left(\hat{x}_{k|k}-x_{k|k}(\ell)\right)\left(\hat{x}_{k|k}-x_{k|k}(\ell)\right)^T}{\sum_{\ell=1}^{L_{SMC}} \mathcal{N}(\hat{x}_{k|k};x_{k|k}(\ell)!\Sigma_{\text{ker}})}\right]\Sigma_{\text{ker}}^{-1}\,,$$

$$(6.32)$$

The CRB surrogate is $OIM^{-1}\left(\hat{x}_{k|k}\right)$. The inverse exists because the OIM is positive definite at $\hat{x}_{k|k}$. This matrix is, in turn, a surrogate for the error covariance matrix.

The OIM for $\hat{x}_{k|k}$ can be computed efficiently in conjunction with any EM method (see [69] for a general discussion). As noted in Section 4.7, the statistical interpretation of the OIM is unresolved in statistical circles. Its utility should be carefully investigated in applications.

6.3.5 Gaussian Sum Methods

An alternative to representing the PPP intensity using particle filters is to use a Gaussian sum instead. One advantage of this method is that the clustering and track extraction issues are somewhat simpler, that is, target state estimates and covariance matrices are extracted from the means and variances of the Gaussian sum instead of a myriad of particles. Gaussian sum methods are also potentially more computationally practical than particle methods for very large numbers of targets.

The Gaussian sum approach is especially attractive when constant survival and detection functions are assumed. This assumption means that the PPP thinning functions are independent of state. In this case, assuming also linear Gaussian target motion and measurement models, the prediction and information update steps are closed form.

Gaussian sum implementations of the PHD filter are carefully discussed by Vo and his colleagues [139]. An unnormalized Gaussian sum is used to approximate the intensity function. These methods are important because they have the potential to be useful in higher dimensions than particle filters. Nonetheless, despite some comments to the contrary, Gaussian sum intensity filters do not escape the curse of target state space dimensionality.

Gaussian sum methods for intensity estimation comprise several steps:

- *Prediction.* The target intensity at time t_{k-1} is a Gaussian sum, to which is added a target birth process that is modeled by a Gaussian sum. The prediction equation for every component in the Gaussian sum is identical to a Kalman filter prediction equation.
- *Component Update.* For each point measurement, the predicted Gaussian components are updated using the usual Kalman update equations. The update therefore increases the number of terms in the Gaussian sum if there is more than one measurement. This step has two parts. In the first, the means and covariance matrices are evaluated. In the second, the coefficients of the Gaussian sum are updated by a multiplicative procedure.
- *Merging and Pruning.* The components of the Gaussian sum are merged and pruned to obtain a "nominal" number of terms. Various reasonable strategies are available for such purposes, as detailed in [139]. This step is the analog of resampling in the particle method.

Some form of pruning is necessary to keep the size of the Gaussian sum bounded over time, so the last—and most heuristic—step cannot be omitted.

Left out of this discussion are details that relate the weights of the Gaussian components to the estimated target count. These details can be found in [139]. For nonlinear target motion and measurement models, [139] proposes both the extended and the unscented Kalman filters. Vo and his colleagues also present Gaussian sum implementations of the CPHD filter in [138, 139].

6.3.6 Regularization

Intensity filters are in the same class of stochastic inverse problems as image reconstruction in emission tomography—the sequence $t_0, t_1, \ldots, t_k$ of intensity filter estimates $f_{k|k}(x)$ is essentially a movie (in dimension n_x) in target state space. As discussed in Section 5.7, such problems suffer from serious noise and numerical artifacts. The high dimensionality of the PPP parameter, i.e., the number of voxels of the intensity function, makes regularization a priority in all applications. Regularization for intensity filters is a relatively new subject. Methods such as cardinalization are inherently regularizing.

Grenander's method of sieves used in Section 5.7.1 for regularizing PET adapts to the intensity filter, but requires some additional structure. The sieve kernel $k_0(x \mid u)$ is a pdf on $\mathcal{S}^+$, so that

$$\int_{\mathcal{S}^+} k_0(x \mid u)\, dx = 1 \qquad (6.33)$$

for all points u in the discrete-continuous space $\mathcal{U}^+ = \mathcal{U} \cup \phi$. As before, the choice of kernel and space $\mathcal{U}^+$ is very flexible. The multitarget intensity at every time t_k, $k = 0, 1, \ldots$, is restricted to the collection of functions of the form

$$f_{k|k}(x) = \int_{\mathcal{U}^+} k_0(x \mid u)\, \zeta_{k|k}(u)\, du \qquad \text{for some } \zeta_{k|k}(u) > 0. \qquad (6.34)$$

The kernel k_0 can be a function of time t_k if desired. The restriction (6.34) is also imposed on the predicted target intensity:

$$f_{k|k-1}(x) = \int_{\mathcal{U}^+} k_0(x \mid u)\, \zeta_{k|k-1}(u)\, du \qquad \text{for some } \zeta_{k|k-1}(u) > 0. \qquad (6.35)$$

Substituting (6.35) into the predicted measurement intensity (6.3) gives

$$\lambda_{k|k-1}(z) = \int_{\mathcal{U}^+} \widetilde{p}_k(z \mid u)\, \zeta_{k|k-1}(u)\, du, \qquad (6.36)$$

where

$$\widetilde{p}_k(z \mid u) = \int_{\mathcal{S}^+} p_k(z \mid x)\, k_0(x \mid u)\, dx. \qquad (6.37)$$

is the regularized measurement likelihood function.

An intensity filter is used to update $\zeta_{k|k}(u)$. This filter employs a transitic function $\Phi_{k-1}(\cdot \mid \cdot)$ to provide a dynamic connection between the current int $\zeta_{k-1|k-1}(u)$ and predicted intensity $\zeta_{k|k-1}(u)$. The function $\Phi_{k-1}(\cdot \mid \cdot)$ is for all u and v in $\mathcal{U}^+$ and is, in principle, any reasonable function, but linked to the target motion model $\Psi_{k-1}(\cdot \mid \cdot)$.

One way to define $\Phi_{k-1}(\cdot \mid \cdot)$ requires defining an additio (6.34) into the predicted detected target intensity to obtain

$$f_{k|k-1}(x) = \int_{\mathcal{S}^+} f_{k-1|k-1}(y)\,\Psi_{k-1}(x\,|\,y)\,dy$$

$$= \int_{\mathcal{S}^+} \left(\int_{\mathcal{U}^+} k_0(y\,|\,u)\,\zeta_{k-1|k-1}(u)\,du \right) \Psi_{k-1}(x\,|\,y)\,dy$$

$$= \int_{\mathcal{U}^+} \widetilde{\Psi}_{k-1}(x\,|\,u)\,\zeta_{k-1|k-1}(u)\,du\,,$$

where

$$\widetilde{\Psi}_{k-1}(x\,|\,u) = \int_{\mathcal{S}^+} \Psi_{k-1}(x\,|\,y)\,k_0(y\,|\,u)\,dy\,. \tag{6.38}$$

Define a Bayesian kernel $k_1(v\,|\,x)$ so that

$$\int_{\mathcal{U}^+} k_1(v\,|\,x)\,dv = 1 \tag{6.39}$$

for all points $x \in \mathcal{S}^+$. Like the sieve kernel $k_0(\cdot)$, the Bayesian kernel $k_1(v\,|\,x)$ is very flexible. It is easily verified that the function

$$\Phi_{k-1}(v\,|\,u) = \int_{\mathcal{S}^+} k_1(v\,|\,x)\,\widetilde{\Psi}_{k-1}(x\,|\,u)\,dx \tag{6.40}$$

is a valid transition function for all $k_0(\cdot)$ and $k_1(\cdot)$.

Given the intensity $\widehat{\zeta}_{k-1|k-1}(u)$ from time t_{k-1}, the predicted intensity $\widehat{\zeta}_{k|k-1}(u)$ at time t_k is defined by

$$\widehat{\zeta}_{k|k-1}(u) = \int_{\mathcal{U}^+} \Phi_{k-1}(u\,|\,v)\,\widehat{\zeta}_{k-1|k-1}(v)\,dv\,.$$

The information updated intensity $\widehat{\zeta}_{k|k}(u)$ is evaluated via the intensity filter using the regularized measurement pdf (6.37) and the predicted measurement intensity (6.36). The regularized target state intensity at time t_k is the integral

$$f_{k|k}(x) = \int_{\mathcal{U}^+} k(x\,|\,u)\,\widehat{\zeta}_{k|k}(u)\,du\,. \tag{6.41}$$

The regularized intensity $f_{k|k}(x)$ depends on the sieve and Bayesian kernels $k_0(\cdot)$ and $k_1(\cdot)$.

The question of how best to define the $k_0(\cdot)$ and $k_1(\cdot)$ kernels depends on the application. It is common practice to define kernels using Gaussian pdfs. As mentioned in Section 5.7, the sieve kernel is a kind of measurement smoothing kernel. ...) $< \dim(\mathcal{S})$, the Bayesian kernel disguises observability issues, that is, ...s $x \in \mathcal{S}$ map with the same probability to a given point $u \in \mathcal{U}$. This ...chanism for target state space smoothing.

6.4 Estimated Target Count

Estimating the number of targets present in the data is a difficult hypothesis testing problem. In a track before detect (TBD) approach, it is a track management function that is integrated with the tracking algorithm. Intensity filters seem to offer an alternative way to integrate, or fuse, the multitarget track management and state estimation functions.

6.4.1 Sources of Error

Accurate knowledge of the target detection probability function $P_k^D(x)$ is crucial to correctly estimating target count. An incorrect value of $P_k^D(x)$ is a source of systematic error. For example, if the filter uses the value $P_k^D(x) = 0.5$ but in fact *all* targets *always* show up in the measured data, the estimated mean target count will be high by a factor of two. This example is somewhat extreme but it makes the point that correctly setting the detection probability is an important task for the track management function. The task involves executive knowledge about changing sensor performance characteristics, as well as executive decisions external to the tracking algorithm about the number of targets actually present—decisions that feedback to validate estimates of $P_k^D(x)$. Henceforth, the probability of detection function $P_k^D(x)$ is assumed accurate.

There are other possible sources of error in target count estimates. Birth-death processes can be difficult to tune in practice, regardless of whether they are modeled implicitly as transitions into and out of a state ϕ in the intensity filter, or explicitly as in the PHD filter. If in an effort to detect new targets early and hold track on them as long as possible, births are too spontaneous and deaths are too infrequent, the target count will be too high on average. Conversely, it will be too low with delayed initiation and early termination. Critically damped designs, however that concept is properly defined in this context, would seem desirable in practice. In any event, tuning is a function of the track management system.

Under the PPP model, the estimated expected number of targets in a given region, $\mathcal{A}$, is the integral over $\mathcal{A}$ of the estimated multitarget intensity. Because the number is Poisson distributed, the variance of the estimated number of targets is equal to the mean number. This large variance is an unhappy fact of life. For example, if 10 targets are present, the standard deviation on the estimated number is $\sqrt{10} \approx 3$.

It is therefore foolhardy in practice to assume that the estimated of the number of targets is the number that are *actually* present. Variance reduction in the target count estimate is a high priority from the track management point of view for both intensity and PHD filters.

6.4.2 Variance Reduction

The multisensor intensity filter discussed in Section 6.5 reduces the variance by averaging the sensor-level intensity functions over the number of sensors that contribute to the filter. Consequently, if the individual sensors estimate target count correctly, so does the multisensor intensity filter.

Moreover, and just as importantly, the variance of the target count estimate of the multisensor intensity filter is reduced by a factor of M compared to that of a single sensor, where M is the number of sensors, assuming for simplicity that the sensor variances are identical.

This important variance reduction property is analogous to estimators in other applications. An especially prominent and well known example is power spectral estimation of wideband stationary time series. For such signals the output bins of the DFT of a non-overlapped blocks of sampled data are distributed with a mean level equal to the signal power in the bin, and the variance equal to the mean. This property of the periodogram is well known, as is the idea of time averaging the periodogram, i.e., the non-overlapped DFT outputs, to reduce the variance of spectral estimates.[2] The Wiener-Khinchin theorem justifies averaging the short term Fourier transforms of nonoverlapped data records as a way to estimate the power spectrum with reduced variance. In practice, the number of DFT records averaged is often about 25.

The multisensor intensity filter is low computational complexity, and applicable to distributed heterogeneous sensor networks. It is thus practical and widely useful. Speculating now for the sheer fun of it, if the number of data records in a power spectral average carries over to multisensor multitarget tracking problems, then the multisensor intensity filter achieves satisfactory performance for many practical purposes with about 25 sensors.

6.5 Multiple Sensor Intensity Filters

To motivate the discussion, consider the SPECT imaging application of Section 5.4. In SPECT, a single gamma camera is moved to several view angles and a snapshot is taken of light observed emanating from gamma photon absorption events. The EM recursion given by (5.69) is the superposition of the intensity functions estimated by each of the camera view angles. Intuitively, different snapshots cannot contain data from the same absorption event, so the natural way to fuse the multiple camera images into one image is to add them, after first weighting by the fraction of radioisotope that is potentially visible in each. The theoretical justification is that, since the number of decays is unknown and Poisson distributed, the estimates of the spatial distribution of the radioisotope obtained from different view angles are independent, *not* conditionally independent, so the intensity functions (images) are superposed.

The general multisensor multitarget filtering problem is not concerned with radioisotope decays, but rather with physical entities (aircraft, ships, etc.) that persist over long periods of time—target physics are very different from the physics

---

[2] Averaging trades off variance reduction and spectral resolution. It was first proposed by M. S. Bartlett [6] in 1948.

of radioisotopes and gamma photons. Nonetheless, for reasons discussed at the beginning of this chapter, a PPP multitarget model is used for single sensor multitarget tracking. It is assumed, very reasonably, that the same PPP target model holds regardless of how many sensors are employed to detect and track targets. The analogy with SPECT is now clear: each sensor in the multisensor filtering problem is analogous to a camera view angle in SPECT, and the sensor-level data are analogous to the camera snapshot data.

The PPP multitarget model has immediate consequences. The most important is that conditionally independent sensors are actually independent due to the independence property of PPPs discussed in Section 2.9. In other words, sensors provide independent estimates of the multitarget intensity function. The multisensor intensity filter averages the sensor-level intensities [124]. Averaging gives the ML estimate of target count when multiple independent Poisson distributed estimates of a fixed intensity are available. (See (6.53) below.) The intuitive reason for averaging intensity functions and not simply adding them as in SPECT is that targets are persistent, unlike radioisotope decays—adding the intensities will result in "over counting" targets. Further discussion and an outline of the derivation of the multisensor filter are given in Section 6.5.1.

The possibility of regularization at the multisensor level is not considered explicitly. Although perhaps obvious, the multisensor intensity filter is fully compatible with sensor-level regularization methods.

Let the number of sensors be $M \geq 1$. It is assumed that the target detection probability functions, $P_k^D(x ; \ell)$, $\ell = 1, \ldots, M$, are specified for each sensor. The sensor-specific state space coverage is defined by

$$C_k(\ell) = \left\{ x \in \mathcal{S} : \quad P_k^D(x ; \ell) > 0 \right\} . \tag{6.42}$$

In homogeneous problems the sensor coverages are identical, i.e., $C_k(\ell) \equiv C_k$ for all ℓ. Heterogeneous problems are those that are not homogeneous.

Two sensors with the same coverage need not have the same, or even closely related, probability of detection functions. As time passes, homogeneous problems may turn into heterogeneous ones, and vice versa. In practice, it is probably desirable to set a small threshold to avoid issues with very small probabilities of detection. Homogeneous and heterogeneous problems are discussed separately.

6.5.1 Identical Coverage Sensors

For $\ell = 1, \ldots, M$, let the measurement space of sensor ℓ be $\mathcal{Z}(\ell)$. Denote the measurement pdf by $p_k(z \mid x ; \ell)$, where $z(\ell) \in \mathcal{Z}(\ell)$ is a point measurement. The predicted measurement intensity is

$$\lambda_{k|k-1}(z ; \ell) = \int_{\mathcal{S}^+} p_k(z \mid x ; \ell) \, P_k^D(x ; \ell) \, f_{k|k-1}^{\text{Fused}}(x) \, dx , \quad z \in \mathcal{Z}(\ell), \tag{6.43}$$

where $f_{k|k-1}^{\text{Fused}}(x)$ is the predicted target intensity based on the fused estimate $f_{k-1|k-1}^{\text{Fused}}(x)$ at time t_{k-1}. The measured data from sensor ℓ is

$$\xi_k(\ell) = (m_k(\ell), \{z_k(1\,;\,\ell), \ldots, z_k(m_k(\ell)\,;\,\ell)\})\,. \tag{6.44}$$

The sensor-level intensity filter is

$$f_{k|k}(x\,;\,\ell) = L_k(\xi_k(\ell)\,|\,x\,;\,\ell)\,f_{k|k-1}^{\text{Fused}}(x)\,, \tag{6.45}$$

where the Bayesian information update factor is

$$L_k(\xi_k(\ell)\,|\,x\,;\,\ell) = 1 - P_k^D(x\,;\,\ell) + \sum_{j=1}^{m_k(\ell)} \frac{p_k(z_k(j\,;\,\ell)\,|\,x\,;\,\ell)\,P_k^D(x\,;\,\ell)}{\lambda_{k|k-1}(z_k(j\,;\,\ell)\,;\,\ell)}\,. \tag{6.46}$$

The multisensor intensity filter is the average:

$$\begin{aligned}
f_{k|k}^{\text{Fused}}(x) &= \frac{1}{M} \sum_{\ell=1}^{M} f_{k|k}(x\,;\,\ell) \\
&= \frac{1}{M} \left(\sum_{\ell=1}^{M} L_k(\xi_k(\ell)\,|\,x\,;\,\ell) \right) f_{k|k-1}^{\text{Fused}}(x)\,. \tag{6.47}
\end{aligned}$$

If the sensor-level intensity filters are maintained by particles, and the number of particles is the same for all sensors, the multisensor averaging filter is implemented merely by pooling all the particles (and randomly downsampling to the desired particle size, if desired).

Multisensor fusion methods sometimes rank sensors by some relative quality measure. This is unnecessary for the multisensor intensity filter. The reason is that sensor quality, as measured by the probability of detection functions $P_k^D(x\,;\,\ell)$ and the sensor measurement pdfs $p_k(z\,|\,x\,;\,\ell)$, is automatically included in (6.47).

The multisensor intensity filter estimates the number of targets as

$$\begin{aligned}
N_{k|k}^{\text{Fused}} &= \int_{\mathcal{S}} f_{k|k}^{\text{Fused}}(x)\,\mathrm{d}x \\
&= \frac{1}{M} \sum_{\ell=1}^{M} \int_{\mathcal{S}} f_{k|k}(x\,;\,\ell)\,\mathrm{d}x \\
&= \frac{1}{M} \sum_{\ell=1}^{M} N_{k|k}(\ell)\,, \tag{6.48}
\end{aligned}$$

where $N_{k|k}(\ell)$ is the number of targets estimated by sensor ℓ. Taking the expectation of both sides gives

$$E\left[N_{k|k}^{\text{Fused}}\right] = \frac{1}{M}\sum_{\ell=1}^{M} E\left[N_{k|k}(\ell)\right] . \qquad (6.49)$$

If the individual sensors are unbiased on average, or in the mean, then $E[N_{k|k}(\ell)] = N$ for all ℓ, where N is the true number of targets present. Consequently, the multi-sensor intensity filter is also unbiased.

The estimate $N_{k|k}(\ell)$ is Poisson distributed, and the variance of a Poisson distribution is equal to its mean, so

$$\text{Var}[N_{k|k}(\ell)] = N , \qquad \ell = 1, \ldots, M .$$

The variance of the average in (6.48) is the average of the variances, since the terms in the sum are independent. Thus,

$$\begin{aligned}
\text{Var}\left[N_{k|k}^{\text{Fused}}\right] &= \frac{1}{M^2}\sum_{\ell=1}^{M} \text{Var}\left[N_{k|k}(\ell)\right] \\
&= \frac{N}{M} . \qquad (6.50)
\end{aligned}$$

In words, the standard deviation of the estimated target count in the multisensor intensity filter is smaller than that of individual sensors by a factor of $\sqrt{M}$, where M is the number of fielded sensors. This is an important result for spatially distributed networked sensors.

The averaging multisensor intensity filter is derived by Bayesian methods in [124]. It is repeated here in outline. The Bayesian derivation of the single sensor intensity filter in Appendix D is a good guide to the overall structure of most of the argument.

The key is to exploit the PPP target model on the augmented space $\mathcal{S}^+$. Following the lead of (D.5) in Appendix D, the only PPP realizations with nonzero likelihood have $m_k = \sum_{\ell=1}^{M} m_k(\ell)$ microtargets. The m_k PPP microtargets are paired with the m_k sensor data points, so the overall joint likelihood function is the product of the sensor data likelihoods given the microtarget assignments. This product is then summed over all partitions of the m_k microtargets into parts of size $m_k(1), \ldots, m_k(M)$.

The sum over all partitions is the Bayes posterior pdf on the event space $\mathcal{E}(\mathcal{S}^+)$. It is a very complex sum, but it has important structure. In particular, the single target marginal pdfs are identical, that is, the integrals over all but one microtarget state are all the same. After tedious algebraic manipulation, the single target marginal pdf is seen to be

$$p_X^{\text{Fused}}(x) = \frac{1}{m_k} \sum_{\ell=1}^{M} L_k(\xi_k(\ell) \mid x ; \ell) \, f_{k|k-1}^{\text{Fused}}(x) , \qquad x \in \mathcal{S}^+ . \tag{6.51}$$

The mean field approximation is now invoked as in (D.13). Under this approxima-tion, $f_{k|k}^{\text{Fused}}(x) = c \, p_X^{\text{Fused}}(x)$, where the constant $c > 0$ is estimated. From (6.17) and (6.18), the measurement intensity is

$$\lambda_{k|k}^{\text{Fused}}(z) = c \int_{S^+} p_k(z \mid x) \, p_X^{\text{Fused}}(x) \, \mathrm{d}x , \tag{6.52}$$

so the likelihood function of c given the data sets $\xi_k(\ell)$ is

$$\mathcal{L}(c ; \xi_k(1), \ldots, \xi_k(M)) = \prod_{\ell=1}^{M} \left\{ e^{-\int_{S^+} c \, p_X^{\text{Fused}}(x) \mathrm{d}x} \prod_{j=1}^{m_k(\ell)} \lambda_{k|k}^{\text{Fused}}(z_k(j ; \ell)) \right\}$$
$$\propto e^{-c \, M} \, c^{m_k} . \tag{6.53}$$

Setting the derivative with respect to c to zero and solving gives ML estimate $\hat{c}_{ML} = \frac{m_k}{M}$. The multisensor intensity filter is $\hat{c}_{ML} \, p_X^{\text{Fused}}(x)$. Further purely technical details of the Bayesian derivation provide little additional insight, so they are omitted.

The multiplication of the conditional likelihoods of the sensor data happens at the PPP event level, where the correct associations of sensor data to targets is assumed unknown. The result is that the PPP parameters—the intensity functions—are aver-aged, not multiplied. The multisensor intensity filter therefore cannot reduce the area of uncertainty of the extracted target point estimates. In other words, the mul-tisensor intensity averaging filter cannot improve spatial resolution. Intuitively, the multisensor filter achieves variance reduction in the target count by foregoing spatial resolution of the target point estimates.

6.5.2 Heterogeneous Sensor Coverages

When the probability of detection functions are not identical, the multisensor inten-sity filter description is somewhat more involved. At each target state x the only sensors that are averaged are those whose detection functions are nonzero at x. This leads to a "quilt-like" fused intensity that may have discontinuities at the boundaries of sensor detection coverages.

The Bayesian derivation of (6.47) outlined above assumes that all the microtar-gets of the PPP realizations can be associated to any of the M sensors. If, however, any of these microtargets fall outside the coverage set of a sensor, then the assign-ment is not valid. The way around the problem is to partition the target state space appropriately.

The total coverage set

$$C = \cup_{\ell=1}^{M} C(\ell) \tag{6.54}$$

contains points in target state space that are covered by at least one sensor. Partition C into disjoint, nonoverlapping sets B_ρ that comprise points covered by exactly ρ sensors, $\rho = 1, \ldots, M$. Now partition B_ρ into subsets $B_{\rho,1}, \ldots, B_{\rho,j_\rho}$ that are covered by different combinations of ρ sensors. To simplify notation, denote the sets $\{B_{\rho j}\}$ by $\{A_\omega\}$, $\omega = 1, 2, \ldots, \Omega$. The sets are disjoint and their union is all of C:

$$C = \cup_{\omega=1}^{\Omega} A_\omega, \qquad A_i \cap A_j = \varnothing \ \text{for} \ i \neq j. \tag{6.55}$$

No smaller number of sets satisfies (6.55) and also has the property that each set A_ω in the partition is covered by the same subset of sensors.

The overall multisensor intensity filter operates on the partition $\{A_\omega\}$. The assignment assumptions of the multisensor intensity filter are satisfied in each of the sets A_ω. Thus, the overall multisensor filter is

$$f_{k|k}^{\text{Fused}}(x) = \frac{1}{|A_\omega|} \left(\sum_{\ell \in \mathscr{I}(A_\omega)} L_k(\xi_k(\ell) \,|\, x \,;\, \ell) \right) f_{k|k-1}^{\text{Fused}}(x), \quad x \in A_\omega. \tag{6.56}$$

where $\mathscr{I}(A_\omega)$ are the indices of the sensors that contribute to the coverage of A_ω, and $|A_\omega|$ is the number of sensors that do so.

The multisensor intensity filter is thus a kind of "patchwork" with the pieces being the sets A_ω of the partition. The variance of the multisensor filter is not the same throughout C—the more sensors contribute to the coverage of a set in the partition, the smaller the variance in that set.

A simple way to write the multisensor filter in the general case is

$$f_{k|k}^{\text{Fused}}(x) = \left\{ \frac{\sum_{\ell=1}^{M} w_k(x \,;\, \ell) \, L_k(\xi_k(\ell) \,|\, x \,;\, \ell)}{\sum_{\ell=1}^{M} w_k(x \,;\, \ell)} \right\} f_{k|k-1}^{\text{Fused}}(x), \tag{6.57}$$

where

$$w_k(x \,;\, \ell) = \begin{cases} 1, & \text{if } P_k^D(x \,;\, \ell) > 0, \\ 0, & \text{if } P_k^D(x \,;\, \ell) = 0, \end{cases} \tag{6.58}$$

is the coverage indicator function for sensor ℓ.

6.6 Historical Note

The PHD (Probability Hypothesis Density) filter ([74], [76] and references therein) was developed by Mahler beginning about 1994. Mahler was influenced in this work [74, p. 1154] by an intriguing approach to additive evidence accrual proposed in 1993 by Stein and Winter [121]. The recent emphasis in this line of research is in Mahler's cardinalized PHD (CPHD) filtering [75]. Kastella independently developed a unified approach to multiple target detection and tracking called the joint multitarget probability density (JMPD) method in [58–60]. The JMPD extends his earlier work on event-averaged maximum likelihood estimation. The JMPD can be derived using methods of finite point processes [85]. Alternative and insightful bin-based derivations of both the PHD and CPHD filters were discovered in 2008 by Erdinc, Willett, and Bar Shalom [30, 31].

The intensity filter of Streit and Stone [130] is very similar to the PHD filter, differing from it primarily in its use of an augmented target state space, S^+, instead of birth-death processes to model target initiation and termination. The PHD filter is recovered from the intensity filter by modifying the posterior intensity function. This paper is the source of the Bayesian approach to the intensity filter given in Section 6.1. The other two approaches presented in that section follow the discussion given in [125]. It draws on the connections to PET to gain intuitive insight into the interpretation of the PPP target model. These approaches greatly simplify the mathematical discussion surrounding intensity filters since they build on earlier work in PET imaging.

The multisensor intensity filter was first derived in 2008 by Streit [124] using a Bayesian methodology, followed by the same kind of PPP approximation as is used in the single sensor intensity filter. The general multisensor problem was presented there for both homogeneous and heterogeneous sensor coverage. The theoretical and practical importance of the variance reduction property of the averaging multisensor filter was also discussed in the same paper. The relationship of the multisensor intensity filter to SPECT is new.

Mahler [74] reports a product form for the multisensor PHD filter. It is unclear if the product form estimates target count correctly. The problem arises from the need for each of the sensor-level integrals of intensity as well as the multisensor integral of intensity to estimate the target count. In any event, the multisensor intensity filter and the multisensor PHD filter take quite different forms, and therefore are different filters.

The MMIF tracking filter is new. It was developed by exploring connections between intensity filters and the PMHT (Probabilistic Multiple Hypothesis Tracking) filter, a Gaussian mixture (*not* Gaussian sum) approach to multitarget tracking developed by Streit and Luginbuhl [126–128] that dates to 1993. These connections reveal the PPP underpinnings of PMHT.

Chapter 7
Distributed Sensing

How can I tell what I think till I see what I say?
E.M. Forster, Aspects of the Novel, *1927*

Abstract PPPs make several important, albeit somewhat disjointed, contributions to distributed sensor network detection, tracking, and communication connectivity. The focus in this chapter is on detection and communication since tracking problems are specialized forms of the multisensor intensity filter presented in Section 6.5. Communication path lengths of a randomly distributed sensor field are characterized by distance distributions. Distance distributions are obtained for sensors located at the points of a nonhomogeneous PPP realization. Both sensor-to-target and sensor-to-sensor distances are discussed. Communication diversity, that is, the number of communication paths between sensors in a distributed sensor field is discussed as a threshold phenomenon a geometric random graph and related to the abrupt transition phenomenon of such graphs. Detection coverage is discussed for both stationary and drifting sensor fields using Boolean models. These problems relate to classical problems in stochastic geometry and geometric probability. The connection between stereology and distributed sensor fields is presented as a final topic.

Keywords Distributed sensor detection · Distance distributions · Sensors to target · Between sensors · Slivnyak's Theorem · Geometric random graph · Expected vertex degree · Communication diversity · Erdos-Renyi random graph · Stationary sensor fields · Drifting sensor fields · Poisson distribution approximation for k-coverage · Anisotropy · Stereology

Two central problems of distributed sensor fields are sensor communication and target detection. Both problems are discussed in this chapter. Sensor communication problems deal with connectivity issues. Two aspects of connectivity are discussed, one local and the other global. Local issues are addressed by the distribution of distances from a target to the nearest sensor in the field. This distance relates to target detection capability of the fielded sensors. A different but closely related notion is the distribution of distances between sensors in the field. This distance relates to the capability of two sensors to communicate with each other. These distributions are obtained for sensors that are located at the points of a nonhomogeneous PPP. The distinction between these distributions is highlighted by Slivnyak's Theorem.

R.L. Streit, *Poisson Point Processes*, DOI 10.1007/978-1-4419-6923-1_7,

Communication issues for the sensor field as a whole are not addressed directly by distance distributions. Global connectivity issues are discussed in terms of the statistical properties of the sensor connectivity graph. This work relates concepts of communication channel diversity to that of k-connectivity of a geometric random graph. Communication diversity is one example of a threshold phenomenon, that is, of a property of a random graph that abruptly takes hold as some appropriate scale is increased. The recognition that threshold phenomena are ubiquitous in geometric random graphs is exciting new research.

Target detection by a fielded multisensor array also has local and global aspects. The local issue deals with the ability of a single sensor to detect to target. The target to nearest sensor distance distributions address this issue. The probability that a fielded sensor array will detect a target is a global, or field-level, detection issue. Field level detection capability is modeled as a coverage problem. Homogeneous sensor fields with isotropic detection, that is, fields of identical sensors with the same detection range regardless of sensor location, are the easiest to analyze. Extensions to fields with directional sensors with uniformly distributed orientations are also straightforward. Further extensions to anisotropic problems with preferred directional orientations are also tractable.

Detection problems bring ideas from stochastic geometry into play naturally. There are six million stories in stochastic geometry. Only a few are told here. Further details of the beautiful connections to integral geometry are banished (cruelly) to the references. (Fortunately, [108] is a delight to read.)

A word of caution is in order. Sensor fields are often deployed in a systematic manner that, while it may have random aspects, is not amenable to exact or approximate PPP modeling. In such cases, the distance distributions and detection probabilities of PPP models are low fidelity, or even inappropriate. Model limitations can be mitigated in some applications by using the extensions and variations of PPPs given in Chapter 8. For example, the points in realizations of the Matérn hard core point process are not closer together than a specified distance, h. This gives a more efficient, i.e., nonoverlapping, sensor spatial distribution than does a PPP; however, they are also harder to analyze because the hard core constraint means that the points are not i.i.d. conditioned on their number. In other words, the minimum separation constraint *reduces* the spatial variability of the points. A different example is that of a cluster process. These processes increase the spatial variability of the points. Distance distributions for cluster point processes are given in [140]; see also [8].

7.1 Distance Distributions

In distributed sensor detection applications, the target is detected by a field of sensors whose locations are modeled as the points of a PPP. Together with propagation conditions and sensor characteristics, the distances from the target to the sensors govern the probability of target detection P_d by one or more sensors in the field. Defined this way, P_d is an ensemble probability evaluated over all possible sensor fields distributed according to the PPP intensity—it is not the detection probability of a specific realization of the sensor field. Consequently, this P_d relates to detection capability of a system and its deployment model.

There are two kinds of distance distributions. One is "point to event," where the point is the target and the event comprises the sensor locations, that is, the points of the PPP realization. The distances of interest are those from a specified point to the nearest sensor, the next nearest sensor, etc. These distributions are of long standing interest in diverse applications. They are treated in the first section below.

The other distance distributions are "event to event." In this case, the distances of interest are the distances between the closest pair of sensors, the second closest pair, etc. This problem is substantially different from the first, except when the PPP is homogeneous. The nature of this difference is clarified by Slivnyak's Theorem [113]. This result, which dates only to 1962, deals with the expectation of a random sum that is pertinent to this problem. It is discussed in Section 7.1.2.

7.1.1 From Sensors To Target

For the moment, suppose the target is located at the origin. Let $D_1 \leq D_2 \leq \cdots$ denote the distances arranged in increasing order from the origin to the points of a PPP with intensity $\lambda(s)$ on $\mathcal{S} \subset \mathbb{R}^m$. The pdf of these distances are easily computed. The approach here is based on the essentially geometric method of [132]. More accessible references are [16, 43, 52]. It is assumed that $\int_{\mathcal{S}} \lambda(s)\, ds = \infty$, so there are infinitely many points in the realization.

The distances D_n are random variables, since they depend on the realization of the PPP. Let r_n be a realization of D_n, and let $r_0 = 0$. Let

$$S(r) \;=\; S \cap \{x \,:\, \|x\| \,<\, r\}. \tag{7.1}$$

Thus, for $a < b$, $S(b) - S(a)$ is the shell centered at the origin of $\mathbb{R}^m$ with inner and outer radii a and b, respectively, intersected with S. Let N_A denote the number of points of the PPP in $\mathcal{A} \subset \mathcal{S}$.

The event $0 < r_1 < r_2 < \cdots < r_n$ comprises several events: no points are in $S(r_1) - S(0)$, one point is in $S(r_1 + \Delta r_1) - S(r_1)$, no points are in $S(r_2) - S(r_1 + \Delta r_1)$, one point is in $S(r_2 + \Delta r_2) - S(r_2)$, etc. These shells are nested and not overlapped, so their probabilities are independent. Hence, setting $r_0 + \Delta r_0 \equiv 0$,

$$\Pr[0 < r_1 < r_2 < \cdots < r_n]$$

$$= \prod_{j=1}^{n} \Pr[N_{S(r_j) - S(r_{j-1} + \Delta r_{j-1})} = 0]\, \Pr[N_{S(r_j + \Delta r_j) - S(r_j)} = 1]. \tag{7.2}$$

Let

$$\mu(r) \;=\; \int_{S(r)} \lambda(s)\, ds \,.$$

Then

$$\Pr[N_{S(r_j) - S(r_{j-1} + \Delta r_{j-1})} = 0] = e^{-\mu(r_j) + \mu(r_{j-1} + \Delta r_{j-1})} \tag{7.3}$$

and

$$\Pr[N_{S(r_j + \Delta r_j) - S(r_j)} = 1] = e^{-\mu(r_j + \Delta r_j) + \mu(r_j)} \left(\mu(r_j + \Delta r_j) - \mu(r_j) \right) .$$

By the Mean Value Theorem of Calculus, there is a point $\tilde{r}_j \in [r_j, r_j + \Delta r_j]$ such that

$$\mu(r_j + \Delta r_j) - \mu(r_{j-1}) = \mu'(\tilde{r}_j) \, \Delta r_j \, ,$$

where $\mu'(r)$ is the derivative of $\mu(r)$ with respect to r. Then,

$$\Pr[N_{S(r_j + \Delta r_j) - S(r_j)} = 1] = e^{-\mu(r_j + \Delta r_j) + \mu(r_j)} \mu'(\tilde{r}_j) \, \Delta r_j \, . \tag{7.4}$$

Substituting (7.3) and (7.4) into (7.2) gives

$$\Pr[0 < r_1 < r_2 < \cdots < r_n] = e^{-\mu(r_n + \Delta r_n)} \prod_{j=1}^{n} \mu'(\tilde{r}_j) \, \Delta r_j \, .$$

Dividing by $\Delta r_1 \cdots \Delta r_n$ and taking the limits as $\Delta r_j \to 0$ gives the pdf of the ordered event $0 < r_1 < r_2 < \cdots < r_n$:

$$p(r_1, \ldots, r_n) = e^{-\mu(r_n)} \prod_{j=1}^{n} \mu'(r_j). \tag{7.5}$$

Integrating over $r_1, \ldots, r_{n-1}$ gives $p_{D_n}(r_n)$, the pdf of D_n. The required integrals are consistent with the event order $0 < r_1 < r_2 < \cdots < r_n$, so that

$$p_{D_n}(r_n) = \int_0^{r_n} \cdots \int_0^{r_2} p(r_1, \ldots, r_n) \, dr_1 \cdots dr_{n-1} \, .$$

The result is

$$p_{D_n}(r) = \frac{\mu^{n-1}(r)}{(n-1)!} e^{-\mu(r)} \frac{d}{dr} \mu(r) \, . \tag{7.6}$$

The densities (7.6) hold for distances to the origin.

Distances to a target located at the non-origin point, say a, are obtained by centering the spheres at a. Let

$$S(r \, ; a) = \{x \, : \, \|x - a\| < r\} , \tag{7.7}$$

and set

$$\mu(r \, ; a) = \int_{S(r \, ; a)} \lambda(s) \, ds . \tag{7.8}$$

Proceeding as before leads to densities (7.6) that depend parametrically on the point a for nonhomogeneous PPPs. This dependence can be quite dramatic. Consider, e.g., a strongly peaked unimodal intensity. The distance distributions for a point a far removed from the mode of the PPP intensity will have large means compared to the means of the distance distributions obtained from a point b that is near the mode of the intensity.

Example 7.1 Homogeneous PPPs. For homogeneous PPPs the densities (7.6) are independent of the choice of a. In this case,

$$\mu(r) \equiv \lambda \, c_m \, r^m , \tag{7.9}$$

where λ is the homogeneous intensity and

$$c_m = \frac{\pi^{m/2}}{\Gamma\left(\frac{m}{2} + 1\right)} \tag{7.10}$$

is the volume of the unit radius sphere in $\mathbb{R}^m$. From (7.6),

$$p_{D_n}(r) = \frac{m}{(n-1)!} \, (\lambda \, c_m)^n \, r^{mn-1} \, e^{-\lambda \, c_m \, r^m} , \tag{7.11}$$

The j-th moment of (7.11) is

$$\begin{aligned}
E\left[D_n^j\right] &= \int_0^\infty r^j \, p_{D_n}(r) \, dr \\
&= \frac{\Gamma\left(n + \frac{j}{m}\right)}{(n-1)!} \, (\lambda \, c_m)^{-\frac{j}{m}} .
\end{aligned} \tag{7.12}$$

Example 7.2 Distances in the plane. In the plane, the pdf of the distance from a target to the nearest sensor is

$$p_{D_1}(r) = 2 \lambda \pi \, r \, e^{-\lambda \pi r^2} .$$

Its mean is $E[D_1] = 1/\left(2\sqrt{\lambda}\right)$, and its variance is

$$\mathrm{Var}[D_1] = E\left[D_1^2\right] - E[D_1]^2 = \frac{4 - \pi}{4\,\lambda\,\pi}\,.$$

If $\lambda = \pi/4$, the mean distance of a fixed target to the nearest sensor is $\pi^{-1/2} \approx 0.564$, and its standard deviation is $\sqrt{4 - \pi}/\pi \approx 0.295$. Thus, a back of the envelop calculation suggests that if the sensor detection range is $r_0 = 0.5$, the field-level detection probability is

$$\int_0^{r_0} p_{D_1}(r)\,\mathrm{d}r = 1 - \exp\left(-\lambda\,\pi\,r_0^2\right) \approx 0.460\,.$$

This kind of calculation is reasonable for sensors with idealized "cookie cutter" detection regions.

Example 7.3 Distances to level curves (or "shells"). The derivation of (7.6) holds for more general functions also: Let $h(x)$, $x \in \mathbb{R}^m$, be a continuous function such that

- $\min_x h(x) = h(\mathcal{O}) = 0$, where $\mathcal{O}$ is the origin in $\mathbb{R}^m$,
- the sets $\tilde{S}(r) \equiv \{x : h(x) \leq r\}$ are nested, and
- $\tilde{S}(r) \to S$ as $r \to \infty$.

Let

$$\tilde{\mu}(r) = \int_{\tilde{S}(r)} \lambda(s)\,\mathrm{d}s\,.$$

Then, replacing μ by $\tilde{\mu}$ in (7.6) gives the pdf of the n-th ordered value of $\{h(\cdot)\}$ evaluated at the points of the PPP realizations. Examples include:

- $h_\Gamma(x) = \left(x^T \, \Gamma\, x\right)^{1/2}$, where Γ is a positive definite matrix, and
- $h_p(x) = \left(\sum_{i=1}^m |x_i|^p\right)^{1/p}$, $p > 0$.

For $0 < p < 1$, the function $h_p(x)$ is not a generalization of the concept of distance since it is not convex. The functions $h_p(x)$ are of great interest in compressive sensing.

Example 7.4 Connection to Extreme Value Distributions. For $n = 1$, the nearest neighbor distribution in $\mathbb{R}^m$ is the integral of the pdf of the distance D_1:

$$F(x) = \int_0^x p_{D_1}(r)\,\mathrm{d}r\,, \qquad x > 0,$$

$$= \int_0^x m\,\lambda\,c_m\,r^{m-1}\,e^{-\lambda\,c_m\,r^m}\,\mathrm{d}x$$

$$= 1 - e^{-\lambda\,c_m\,r^m}\,. \tag{7.13}$$

The rate of decay of this function increases very rapidly as the dimension m increases, yet another symptom of the Curse of Dimensionality.

The function $F(x)$ is the famous Weibull distribution. It is one of the three stable laws of extreme value statistics that are allowed by the so-called Trinity Theorem. The other two stable laws are named for Fréchet and Gumbel. The Trinity Theorem holds whenever the cumulative distribution function of the underlying sample distribution is continuous and has an inverse.

The nearest neighbor is the minimum of a large number of realizations of i.i.d. random variables, so the Trinity Theorem must hold for nonhomogeneous intensities. The limiting distribution for nonhomogeneous PPPs is currently unavailable, but it is likely that it is Weibull with an intensity equal to the local intensity $\lambda(a)$ at the point a from which nearest neighbor distances are centered. Higher order correction terms would need to be determined by asymptotic methods.

7.1.2 Between Sensors

The distances between a sensor and all the other sensors in a given realization is a conceptually more difficult problem. For one thing there is no legitimate concept of a reference sensor since different realizations comprise different points.

One way to approach the problem is to average over the sensors in a given realization, and then to seek the expectation of this sum over all PPP realizations. For example, the distance between a point in a realization and its nearest neighbor can be averaged over all the points of the realization. Such sums are random sums, but of a more specialized kind than are used in Campbell's Theorem. Evaluating the expectation of these random sums is the goal of Slivnyak's Theorem.

7.1.2.1 Slivnyak's Theorem

Let $\xi = (n, \{x_1, \ldots, x_n\})$ denote a realization of a point process with outcomes in the event space $\mathcal{E}(\mathcal{S})$. Let f denote a real-valued function with two arguments, the first in the space $\mathcal{S}$ and the second in the event space

$$\mathbb{E}(\mathcal{S}) = \{\varnothing\} \cup_{n=1}^{\infty} \left\{ \{x_1, \ldots, x_n\} \: : \: x_j \in \mathcal{S}, \; j = 1, \ldots, n \right\} . \qquad (7.14)$$

The event space $\mathbb{E}(\mathcal{S})$ is identical to the event space $\mathcal{E}(\mathcal{S})$ with the integer counts removed. An example of such a function is the nearest neighbor distance from an arbitrary point $x \in \mathcal{S}$ to the points in ξ :

$$f_{NN}(x, \{x_1, \ldots, x_n\}) = \min_{1 \leq i \leq n} \|x_i - x\| , \qquad (7.15)$$

with $f_{NN}(x, \{\varnothing\}) = 0$. The pdf of f_{NN} is given in the previous section when ξ is a realization of a PPP.

For distributed sensors a more interesting example is the nearest neighbor distance from an arbitrary vertex x_j in ξ to the other points in the same realization:

$$
\begin{aligned}
f_{VV}\left(x_j, \{x_1, \ldots, x_n\} \setminus x_j\right) &\equiv f_{VV}\left(x_j, \{x_1, \ldots, x_{j-1}, x_{j+1}, \ldots, x_n\}\right) \\
&= \min_{\substack{1 \le i \le n \\ i \neq j}} \|x_i - x_j\| .
\end{aligned}
\tag{7.16}
$$

The sum of the nearest neighbor distances of the vertices in the PPP realization ξ is

$$
\mathcal{F}(\xi) = \sum_{j=1}^{n} f_{VV}\left(x_j, \{x_1, \ldots, x_n\} \setminus x_j\right) .
\tag{7.17}
$$

For $n = 0$, $\mathcal{F}(\xi) \equiv 0$. Slivnyak's Theorem is concerned with the expected value of random sums of the form (7.17) for general functions f.

Denote the intensity of the PPP by $\lambda(x)$, $x \in \mathcal{S} \subset \mathbb{R}^{n_x}$. By definition (2.26), the expectation of $\mathcal{F}(\xi)$ for a general function f, of which f_{NN} and f_{VV} are examples, is

$$
\begin{aligned}
E\left[\mathcal{F}\right] &= E\left[\sum_{j=1}^{n} f\left(x_j, \{x_1, \ldots, x_n\} \setminus x_j\right) \right] \\
&= \sum_{n=1}^{\infty} \frac{e^{-\int_\mathcal{S} \lambda(x)\mathrm{d}x}}{n!} \sum_{j=1}^{n} \int_\mathcal{S} \cdots \int_\mathcal{S} f\left(x_j, \{x_1, \ldots, x_n\} \setminus x_j\right) \\
&\qquad \times \prod_{j=1}^{n} \lambda(x_j)\, \mathrm{d}x_1 \cdots \mathrm{d}x_n \\
&= \sum_{n=1}^{\infty} \frac{e^{-\int_\mathcal{S} \lambda(x)\mathrm{d}x}}{(n-1)!} \int_\mathcal{S} \cdots \int_\mathcal{S} f\left(x_n, \{x_1, \ldots, x_{n-1}\}\right) \prod_{j=1}^{n} \lambda(x_j)\, \mathrm{d}x_1 \cdots \mathrm{d}x_n .
\end{aligned}
$$

The second step moves the integrals of the expectation inside the sum over j. The last step follows from recognizing that the n-fold integrals over $\mathcal{S}$ are identical. Now, for every n, the integral over x_n is unchanged by replacing x_n with the dummy variable, x. Therefore,

$$
\begin{aligned}
E\left[\mathcal{F}\right] &= \int_\mathcal{S} \left\{ \sum_{n=1}^{\infty} \frac{e^{-\int_\mathcal{S} \lambda(x)\mathrm{d}x}}{(n-1)!} \int_\mathcal{S} \cdots \int_\mathcal{S} f\left(x, \{x_1, \ldots, x_{n-1}\}\right) \right. \\
&\qquad\left. \times \prod_{j=1}^{n-1} \lambda(x_j)\, \mathrm{d}x_1 \cdots \mathrm{d}x_{n-1} \right\} \lambda(x)\, \mathrm{d}x \\
&= \int_\mathcal{S} E\left[f\left(x, \{x_1, \ldots, x_n\}\right)\right] \lambda(x)\, \mathrm{d}x ,
\end{aligned}
$$

where the last step is merely shifting the index $n \leftarrow n - 1$ so that the infinite sum starts at $n = 0$. This gives Slivnyak's Theorem:

$$E\left[\sum_{j=1}^{n} f\left(x_j, \{x_1, \ldots, x_n\} \setminus x_j\right)\right] = \int_{S} E\left[f\left(x, \{x_1, \ldots, x_n\}\right)\right] \lambda(x)\,\mathrm{d}x.$$

(7.18)

The result relates two different kinds of averages. One is how a point in a realization relates to other points in the very same realization. The other is how a given point relates to the points of an arbitrary PPP realization. The latter average can be easier to evaluate analytically.

7.1.2.2   Geometric Random Graphs

A graph G comprises a finite number n of vertices V and the edges E that link them. In general the vertices are points in an arbitrary space. In distributed sensor applications, the vertices correspond to sensors, and the (undirected) edges of G represent communication links between sensor pairs. Typically, the vertices are the point locations of the sensors and lie in a bounded set $\mathcal{R} \subset \mathbb{R}^{n_x}$, where $n_x = 2$.

The number n of vertices in an Erdös-Rényi random graph is fixed, but the edges are drawn uniformly at random between vertex pairs. More carefully, in a realization of the random Erdös-Rényi graph, $n(n - 1)/2$ i.i.d. Bernoulli trials with success probability p are performed. Each trial corresponds to a possible edge: edges are drawn only between vertices with a successful Bernoulli trial.

A geometric random graph, or nearest neighbor graph, is quite different from Erdös-Rényi. The vertices of G are the locations of a realization of a point process, so the number of sensors is a random variable. Edges are drawn between pairs of vertices that are within a specified distance $r \geq 0$ of each other. Denote this graph by G_r. For a given realization of the vertices, G_r is a subgraph of G_s if $r \leq s$.

The vertex degree $d_G(v)$ of a vertex $v \in G$ is the number of edges with an endpoint at v. For a specified realization of a geometric random graph, the vertex degree is a monotone increasing step function with minimum zero and maximum value n. The average behavior over all possible geometric random graphs generated by a PPP with intensity $\lambda(x)$ is obtained from Slivnyak's Theorem. To see this, let

$$f_r\left(x, \{x_1, \ldots, x_n\}\right) = N\left[\{i : \|x - x_i\| \leq r\}\right],$$

(7.19)

where $N[\cdot]$ is a set function that counts the number of points in its argument. In terms of the function f_r, the degree of vertex x_j is

$$f_r\left(x_j, \{x_1, \ldots, x_n\} \setminus x_j\right),$$

(7.20)

so the aggregate vertex degree, summed over all vertices in a given realization of the graph G_r, is the random sum

$$\sum_{j=1}^{n} f_r \left(x_j, \{x_1, \ldots, x_n\} \setminus x_j\right) . \tag{7.21}$$

Let $\overline{d}_{\Sigma(r)}$ denote the mean total vertex degree.

The expected value on the right hand side of (7.18) is straightforward for the function (7.19). For $x \in \mathcal{R}$, by a fundamental property of PPPs,

$$E\left[f_r\left(x, \{x_1, \ldots, x_n\}\right)\right] = \int_{S(r\,:\,x)} \lambda(y)\,dy$$
$$= \int_{S(r)} \lambda(y + x)\,dy\,,$$

where $S(r)$ is given by (7.1) and $S(r\,;\,x)$ by (7.7). Hence, from (7.18), the expected total vertex degree is

$$\overline{d}_{\Sigma(r)} = \int_{\mathcal{S}} \left(\int_{S(r)} \lambda(y + x)\,dy\right) \lambda(x)\,dx . \tag{7.22}$$

Interchanging the order of integration and dividing by 2 gives

$$\overline{d}_{\Sigma(r)} = \int_{S(r)} \left(\int_{\mathcal{S}} \lambda(y + x)\,\lambda(x)\,dx\right) dy . \tag{7.23}$$

The mean vertex degree is the ratio of $\overline{d}_{\Sigma(r)}$ and the expected number of vertices. This ratio, written in a form that emphasizes convexity, is

$$\overline{d}(r) = \int_{S(r)} \left(\frac{\int_{\mathcal{S}} \lambda(y + x)\,\lambda(x)\,dx}{\int_{\mathcal{S}} \lambda(x)\,dx}\right) dy . \tag{7.24}$$

The average vertex degree $\overline{d}(r)$ is an increasing function of r. It is strictly increasing if $\lambda(x) > 0$.

Another example is the nearest neighbor distance function f_{NN} given by (7.15). In this case the expected value on the left hand side of (7.18) is the expected value of the sum of the nearest vertex distances. The right hand side is evaluated for non-homogeneous intensities after first computing the mean in the manner described just before Example 7.2. This can be done in specific cases.

The mean in the homogeneous case is easy because the required expectation is given in Example 7.2 as $1/\left(2\sqrt{\lambda}\right)$. Thus, the right hand side of (7.18) for f_{NN} is $(|\mathcal{S}|\lambda)/\left(2\sqrt{\lambda}\right)$, where $|\mathcal{S}|$ is the area of $\mathcal{S}$. Dividing both sides by the expected number of points (vertices) in $\mathcal{S}$ gives the average nearest vertex distance:

$$\overline{D_1^V} = \frac{(|\mathcal{S}|\lambda)\,/\left(2\sqrt{\lambda}\right)}{E[\text{Number of vertices}]}$$

$$= \frac{1}{2\sqrt{\lambda}}\,. \tag{7.25}$$

The average nearest vertex distance is identical to the mean nearest neighbor distance for homogeneous PPPs, a result that accords well with intuition.

7.2 Communication Diversity

The question considered in this section is "How many communication paths are there between pairs of sensors in the network?" The question is answered *with high probability*, a phrase that is commonplace in the theory of random graphs. As is implicitly clear, the answer provided uses only the i.i.d. property of PPPs.

Several standard definitions from graph theory are useful. Let G denote a given (nonrandom) graph.

- The minimum degree of any vertex in G is denoted by $\delta(G)$, that is,

$$\delta(G) = \min_{v \in G} d_G(v)\,. \tag{7.26}$$

- A path in G from the vertex $a \in G$ to the vertex $b \in G$ is a sequence of edges that connect a and b.
- Two paths in G are independent if any vertex common to both paths is an end point of both paths.
- The graph G is k-connected if there are at least k pairwise independent paths connecting every pair of vertices in G.
- The maximum value of k for which G is k-connected is the connectivity, $\kappa(G)$, of G.

Let the number of vertices of G_r be n, and let their locations be i.i.d. independent samples on a bounded set $\mathcal{R} \subset \mathbb{R}^{n_x}$. Penrose [94] shows that, for all $k \geq 1$,

$$\lim_{n \to \infty} \Pr\left[\min_r \{\kappa(G_r) \geq k\} = \min_r \{\delta(G_r) \geq k\}\right] = 1. \tag{7.27}$$

This result holds if $n_x \geq 2$ and $\mathcal{R}$ is the n_x-dimensional unit hypercube. A small modification of (7.27) is needed for $n_x = 1$.

If the vertices of G_r are realizations of a homogeneous PPP on $\mathcal{R}$, the conditions for (7.27) hold as the intensity goes to infinity. It is tempting to think that some modified version of (7.27) also holds for Matérn hard core processes whose intensity is given by (8.9), but this is currently unknown.

Despite the limitations of (7.27), the result is important in sensor network applications. Paraphrasing [94], for sufficiently many sensors in $\mathcal{R}$, with high probability, if the sensor communication graph G_r is "grown" by slowly increasing r from some small initial value, then edges are added consecutively, and the increasingly complex graph becomes k-connected for the *very same* value of r at which the minimum vertex degree becomes k. If this value of r, denoted r_{thresh}, is less than the transmission range r_{tran}, then the network has at least k independent communication paths between any pair of sensors. The numerical value of r_{thresh} is determined by simulation.

In practical applications, if $r_{\text{thresh}} \ll r_{\text{tran}}$, it may be possible to reduce the number of sensors n in the network to reduce cost, or to reduce the sensor transmit power to extend the lifetime of the fielded sensors. A discussion of these and related topics in the context of multi-objective Pareto optimization of sensor networks is given in [13].

The result (7.27) is related to a well known result for (nongeometric) Erdös-Rényi random graphs [10, Section 7.2]. The graph in which every pair of the n vertices are connected by an edge is the complete graph, K_n, with $\binom{n}{2}$ edges. Suppose, when drawing K_n, that edges are drawn one at a time. The $\binom{n}{2}!$ possible ways to draw K_n in this manner correspond to the $\binom{n}{2}!$ orderings of the edges. If the orderings are equally likely, then, with high probability, the graph becomes k-connected at the same time the minimum vertex degree becomes k. The difference between this result and (7.27) is that the edges of G_r are not drawn uniformly at random.

7.3 Detection Coverage

Coverage refers to set containment. For example, in the classic papers of Robbins [105, 106], a germ-grain Boolean model is used. In this model the germs are the centers x_j of n grains, or discs, $\mathscr{D}(x_j)$ of radius $r > 0$. The germs are distributed uniformly in the plane so that every grain has nonempty intersection with a given rectangle $\mathcal{R}$. The coverage of $\mathcal{R}$ is the set

$$C = \left(\cup_{j=1}^{n} \mathscr{D}(x_j) \right) \cap \mathcal{R} .$$

Let $|\mathcal{A}|$ denote the area of the set $\mathcal{A}$. The expected coverage is $E[\,|C|\,]$, where the expectation is over the disc/grain centers. The probability that any given point x in $\mathcal{R}$ is covered by (contained in) C is defined by

$$\Pr[x \in C] = \frac{E[\,|C|\,]}{|\mathcal{R}|} . \tag{7.28}$$

The connection of this problem to detection with distributed sensors is clear—the grains are the detection areas of a sensor and the spatial distribution of the germs is the manner in which sensors are deployed in the application. If the germs are distributed in a well defined random manner, the probability of target detection is the probability that any given point is covered by at least one grain.

Connections between coverage and distributed detection of a moving target with fixed (i.e., motionless) sensors are presented in Section 7.3.1 using isotropic coverage models. Isotropic models are independent of direction of target motion.

Anisotropic models are needed in two situations. One is when the target moves, and the other is when the sensor field experiences a "bulk" drift. Anisotropic problems are discussed in Section 7.3.2 under the assumption that motion is linear and that the direction of motion is randomly distributed with known pdf.

Recent work on distributed detection using coverage models are discussed in [67, 88, 107, 142]. They discuss extensions to sensors with nonconvex detection regions and a heterogeneous mix of sensors.

The results presented in this section are intended only to reveal the flavor of the kind of results that are possible. The number of points is assumed known and is not a random variable. The points are i.i.d. with uniform pdf on the surveillance set $\mathcal{R}$. Extending these results to homogeneous PPPs in which the number of points is Poisson distributed is straightforward, but not discussed here. Extensions to nonhomogeneous pdfs seems not at all straightforward.

7.3.1 Stationary Sensor Fields

Defining probability as the ratio of areas as is done in (7.28) is a natural procedure. Ratios of other quantities of the same dimension, such as a ratio of lengths and volumes, is also natural. Extended use is now made of this fact.

A line ℓ in the plane is uniquely parameterized by two parameters, say (ρ, θ), so the set of lines $\mathcal{L}$ is identified with the set of points:

$$\mathcal{L} \quad \Leftrightarrow \quad G \equiv \{(\rho, \theta) : \quad \ell(\rho, \theta) \text{ is a line in } \mathcal{L}\} \subset \mathbb{R}^2. \tag{7.29}$$

The coordinate system is chosen so that the double integral over the point set G on the right hand side of (7.29) does not change when the lines in $\mathcal{L}$ are subjected to a rigid motion in the plane.[1] It can be verified that the natural coordinate system with this invariance property is the so-called normal form of the line:

$$x \cos\theta + y \sin\theta = \rho, \tag{7.30}$$

where $\rho \geq 0$ is the distance from the origin to the foot of the perpendicular to the line ℓ, and θ, $0 \leq \theta < 2\pi$, is the angle measured counterclockwise from the positive x-axis to the perpendicular line. The angle θ is not limited to $[0, \pi)$ because the perpendicular is a line segment with one endpoint always at the origin. See Fig. 7.1. The double integral over the point set G in (7.29) has units of length.

---

[1] The importance of invariance was not recognized in early discussions of the concept of a random line. Bertrand's paradox (1888) involves three different answers to a question about the length of a random chord of a circle. The history is reviewed in [61, Chapter 1].

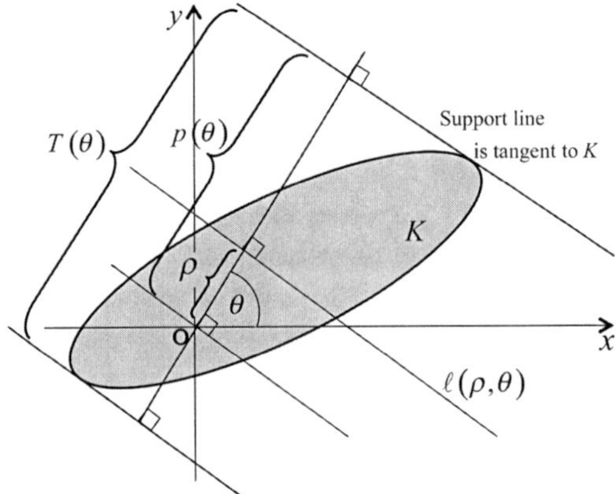

Fig. 7.1 Depiction of coordinate system and the line $\ell(\rho,\ \theta)$. The support function $p(\theta)$ and support line for a convex set K is also shown. The origin $\mathscr{O}$ is any specified point interior to K. The line $\ell(\rho,\ \theta)$ intersects K whenever $0 \le \rho \le p(\theta)$. The thickness $T(\theta)$ is defined later in (7.45)

Let K be a bounded convex set in the plane, and take as the origin any point interior to K. The double integral over all lines that intersect K is very simple:

$$\iint_{G\cap K \neq \varnothing} \mathrm{d}\rho\, \mathrm{d}\theta = L\,, \tag{7.31}$$

where L is the perimeter of K. To see this, note that

$$\iint_{G\cap K \neq \varnothing} \mathrm{d}\rho\, \mathrm{d}\theta = \int_0^{2\pi} \int_0^{p(\theta)} \mathrm{d}\rho\, \mathrm{d}\theta = \int_0^{2\pi} p(\theta)\, \mathrm{d}\theta\,, \tag{7.32}$$

where $p(\theta)$ is the distance from the origin at angle θ to the tangent line to K. The function $p(\theta)$ is called the support function of K, and the tangent line is called a support line. The support function and line are depicted in Fig. 7.1. It can be shown that $p(\theta) + p''(\theta) > 0$ and that the infinitesimal of arclength of K is

$$\mathrm{d}s = \big(p(\theta) + p''(\theta)\big)\, \mathrm{d}\theta\,. \tag{7.33}$$

Thus,

$$L = \int_0^L \mathrm{d}s = \int_0^{2\pi} \big(p(\theta) + p''(\theta)\big)\, \mathrm{d}\theta = \int_0^{2\pi} p(\theta)\, \mathrm{d}\theta\,, \tag{7.34}$$

where the integral over $p''(\theta)$ is zero because $p(\theta)$ is periodic. Comparing (7.32) and (7.34) gives (7.31).

Let K_1 denote a convex subset of K. Then

$$\iint_{G\cap K_1 \neq \varnothing} \mathrm{d}\rho\,\mathrm{d}\theta = L_1, \tag{7.35}$$

where L_1 is the perimeter of K_1. Therefore, with the same notion of probability as used in (7.28), the probability that a random line ℓ intersects K_1 given that it intersects K is

$$\Pr\left[\ell \cap K_1 \neq \varnothing \mid \ell \cap K \neq \varnothing\right] = \frac{|K_1|}{|K|}. \tag{7.36}$$

Further details are found in [108, p. 30].

This remarkably simple and elegant result (due to Crofton, 1885) translates immediately into the distributed sensor detection application: if a surveillance region K is a disc of radius R and the detection region of a sensor is a disc K_r of radius r, the probability that the sensor has the opportunity to detect a random, but constant course, target traversing K is r/R. The possibly more intuitive result, namely the ratio of areas r^2/R^2, is thus overly pessimistic.

Continuing this example, the mean time that the target is within the surveillance region K of the sensor field is the duration of the target detection opportunity. This "dwell" time is proportional (depending on target velocity) to the average chord length, that is, the average length of the line segment that lies in the convex set K. The average chord length is given by another of Crofton's Theorems:

$$E[\text{chord length}] = \frac{\pi\,|K|}{L}, \tag{7.37}$$

where L is the perimeter of K. To see this, let the line $\ell(\rho, \theta)$ intersect K. Denote by $\sigma(\rho, \theta)$ the length of the line segment, or chord, subtended by K. The mean chord length of lines that intersect K is the ratio

$$E[\text{chord length}] = \frac{\iint_{G\cap K \neq \varnothing} \sigma(\rho, \theta)\,\mathrm{d}\rho\,\mathrm{d}\theta}{\iint_{G\cap K \neq \varnothing} \mathrm{d}\rho\,\mathrm{d}\theta}.$$

From (7.31), the denominator is L. The numerator is

$$\int_0^{2\pi} \int_0^{p(\theta)} \sigma(\rho, \theta)\,\mathrm{d}\rho\,\mathrm{d}\theta.$$

For any angle θ, the infinitesimal $\sigma(\rho, \theta)\,\mathrm{d}\rho$ is an element of area for K. One part of K is covered by area elements with angle θ, and the other part is covered by elements with angle $\theta + \pi$. The area elements are not overlapped, so integrating over all area elements gives the numerator as

$$\int_0^\pi \left(\int_0^{p(\theta)} \sigma(\rho, \theta)\, \mathrm{d}\rho + \int_0^{p(\theta + \pi)} \sigma(\rho, \theta)\, \mathrm{d}\rho \right) \mathrm{d}\theta = \int_0^\pi |K|\, \mathrm{d}\theta = \pi\, |K|.$$
(7.38)

The establishes (7.37).

If the sensor surveillance region K is not convex, that is, if K is any closed and bounded subset of $\mathbb{R}^2$, then

$$E[\text{chord length}] = \frac{\pi\, |K|}{\int_K \mathrm{d}p\, \mathrm{d}\theta},$$
(7.39)

In the general case, chord length is replaced by the sum of the subtended line segments [2].

A field of n sensors with *congruent* detection regions, denoted $K_1, \ldots, K_n$, are dropped at random so that for every j the region K_j intersects the surveillance region K_0. The orientation of the sensors is also random, an assumption that matters only when the sensor detection capability is *not* circular.

Let I_k be the area of $\mathcal{R}$ that is covered by exactly k sensors. Let F_0 and L_0 denote the area and perimeter of K_0, respectively. Similarly, let F and L denote the area and perimeter of the sets K_j. The expected value of the area of I_k is

$$E[|I_k|] = \binom{n}{k} \frac{(2\pi F)^k\, (2\pi F_0 + L_0 L)^{n-k}\, F_0}{(2\pi(F + F_0) + L_0 L)^n}.$$
(7.40)

It is no accident that this expression resembles the binomial theorem; see [108, pp. 98–99]. To see this result it is necessary to introduce the concept of mixed areas (also called *quermasse* integrals) that were first studied by Minkowski (c. 1903). The product $L_0 L$ in (7.40) is a mixed area. This endeavor is left to the references.

Dividing (7.40) by the area of the surveillance region gives the probability that a target is detected by precisely k sensors:

$$\Pr[k\text{-coverage}] = \frac{E[|I_k|]}{F_0}$$
$$= \binom{n}{k} \frac{(2\pi F)^k\, (2\pi F_0 + L_0 L)^{n-k}}{(2\pi(F + F_0) + L_0 L)^n}.$$
(7.41)

For $k = 0$ the result gives the probability that a target is not detected by even one sensor. The set I_0 is called the vacancy, and it is of much interest in modern coverage theory [44].

Example 7.5 Poisson Approximation to k-Coverage. The Poisson distribution approximates the probability that a random point is detected by exactly k sensors. To see this, let the surveillance region K_0 expand to the entire plane in such a way that the ratio of the number of sensors n and the area of K_0 satisfy

$$\frac{n}{F_0} = \lambda \quad \Leftrightarrow \quad \frac{F}{F_0} = \frac{\lambda F}{n} .$$

Regardless of the shape of K_0, the ratio of its perimeter to its area $L_0/F_0 \to 0$. Manipulating (7.41) and taking the limit as $n \to \infty$ gives

$$\Pr[\,k\text{-coverage}\,] = \frac{(\lambda F)^k}{k!} e^{-\lambda F} . \tag{7.42}$$

To see this, a variation of the Poisson approximation to the binomial distribution is used. Let

$$\frac{2\pi F}{2\pi F + 2\pi F + L_0 L} = \frac{\kappa \lambda F}{n} , \tag{7.43}$$

where

$$\kappa = \left(\frac{\lambda F}{n} + 1 + \frac{L_0 L}{2\pi F_0} \right)^{-1} .$$

Clearly, $\kappa \to 1$ as $n \to \infty$. Now, substituting (7.43) into (7.41) and manipulating the algebra gives the identity

$$\Pr[\,k\text{-coverage}\,] = \frac{(\kappa \lambda F)^k}{k!} \frac{n!}{(n-k)!\,n^k} \left(1 - \frac{\kappa \lambda F}{n} \right)^{-k} \left(1 - \frac{\kappa \lambda F}{n} \right)^{n} .$$

In the limit as $n \to \infty$, the middle terms both go to one, and the last term goes to the exponential. This establishes (7.42).

7.3.2 Drifting Fields and Anisotropy

Anisotropic geometric distributions arise in several applications. Anisotropic coverage arises when the entire sensor field experiences a bulk motion, or drift. An equivalent problem arises when targets move through a fixed sensor field with a preferred direction of motion. If a sensor has a circular detection region with detection range R_D and drifts for a known time period $\Delta t = t - t_0$ with known constant velocity v, its detection coverage is the "pill-shaped" region depicted in Fig. 7.2. The overall length of the pill depends on the speed and duration of drift, and the sensor detection range. The preferred orientation of the pill-shaped areas corresponds to the direction of sensor drift or to target motion, as the case may be. The total field-level detection coverage is the union of the pill-shaped regions.

Figure 7.3a,b depict two different fixed drift angle configurations for 10 sensors. They show that the field-level coverage changes as a function of drift angle. Figure 7.3c depicts 10 sensors with drift angles that vary independently about a

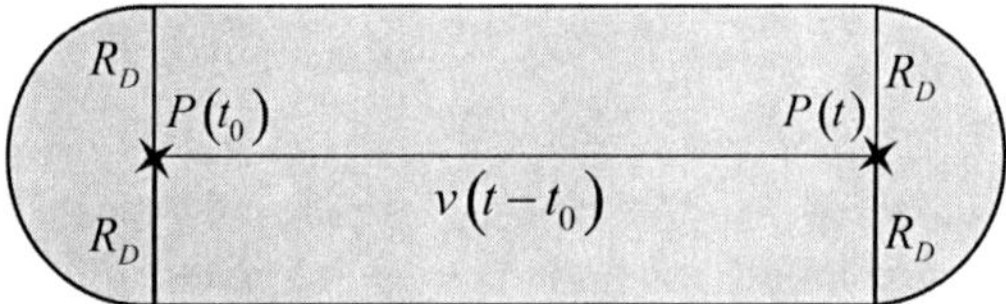

Fig. 7.2 Depiction of the pill-shaped detection region of a sensor with a circular detection region with detection range R_D drifting for a known time period $\Delta t = t - t_0$ and known constant velocity v. The point $P(t_0)$ is the initial sensor location, and $P(t)$ is its location at time t

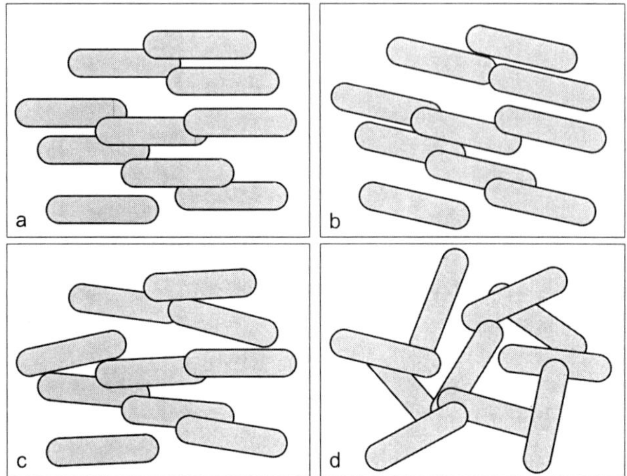

Fig. 7.3 Depiction of the varying coverage of 10 drifting sensors pill-shaped detection regions. The length of the rectangular section of the pill is proportional to drift speed and duration of drift; the width is the detection range of the sensor. (**a**) All sensors drift to the east. (**b**) If, instead, all sensors drift to the southeast, the total coverage and sensor overlap changes. (**c**) All 10 drift to the east on average, but each with a slightly different angle. (**d**) Random orientations as required by the coverage theory of Section 7.3.1

mean drift direction. As depicted in Fig. 7.3d, pills scattered randomly in angle and location do not exhibit a preferred orientation. The coverage results of the previous section are only valid for the situation depicted in Fig. 7.3d. Therefore, they are either inappropriate or at best an approximation if the anisotropy is small, that is, if the pill-shaped regions approximate a circle because $v \approx 0$.

A different kind of example of the need for anisotropy arises in barrier problems. In these problems the convex surveillance set K is typically long and narrow, and long traverses are unrealistic. These infrequent, but long, traverses make the average chord length longer than may be reasonable in some applications. For example, the average chord length for a rectangular barrier region K of width ϵ and length $D \gg \epsilon$ is, by Crofton's Theorem (7.37),

$$E[\text{chord length}] = \frac{\pi \epsilon D}{2\epsilon + 2D} \approx \frac{\pi \epsilon}{2} \approx 1.57\,\epsilon\,. \tag{7.44}$$

Long transit lengths, while infrequent, nonetheless make the average transit length significantly larger than the shortest possible transit ϵ. If target trajectories are more likely to be roughly perpendicular to the long side of K, the isotropic model is of low fidelity and needs modification. Given a pdf of the angle θ of transit, an expression for the expected transit length can be developed using the methods discussed below.

Anisotropic problems are studied by Dufour [25] in $\mathbb{R}^\kappa$, $\kappa = 2, 3$. Much of this work seems to be unpublished, but accessible versions of some of the results are given in [120, pp. 55–74] and also [108, p. 104]. The former discusses anisotropy generally and gives an interesting application of anisotropy to motor vehicle traffic flow analysis. Most if not all of Dufour's results are analytically explicit in the sense that they are amenable to numerical calculation. Of the many results in [25], two are discussed here to illustrate how anisotropy enters the problem.

The distribution $dF(\theta)$ of the orientation angle is assumed known. This notation accommodates diverse angle distributions, including the case when there is only one possible orientation. If F is differentiable, then $F'(\theta)$ is the pdf of orientation angle. Let $K_\perp(\theta)$ denote the orthogonal projection of a bounded set $K \subset \mathbb{R}^2$ onto a line through the origin with angle θ. The set $K_\perp(\theta)$ is an interval on the line if K is connected (i.e., if K is not the union of two or more disjoint sets). The thickness of K is defined by

$$T(\theta) = \int_{K_\perp(\theta)} d\rho\,. \tag{7.45}$$

For convex sets, $T(\theta) = p(\theta) + p(\theta + \pi)$, as is illustrated in Fig. 7.1. The expected thickness of K is the average thickness over all angles:

$$E[T] = \int_0^\pi T(\theta)\,dF(\theta)\,. \tag{7.46}$$

The mean thickness is an important quantity in several kinds of intersection problems.

A line segment of length R is characterized by its midpoint and orientation. The midpoint is uniformly distributed over an area that contains the bounded convex set K_0, and the orientation of the line is distributed as $dF(\theta)$. Such a line is an anisotropic random line segment. Let K_1 be a convex subset of K_0. The probability that a random line segment intersects K_1 given that it intersects K_0 is

$$\Pr\left[\text{line segment intersects } K_1 \mid \text{line segment intersects } K_0\right]$$
$$= \frac{|K_1| + R\,E[T_1]}{|K_0| + R\,E[T_0]}\,, \tag{7.47}$$

where $E[T_i]$ is the mean thickness of K_i, $i = 0, 1$. As the line segment length $R \to \infty$, the probability simplifies to

$$\Pr\left[\text{line segment intersects } K_1 \mid \text{line segment intersects } K_0\right] = \frac{E\left[T_1\right]}{E\left[T_0\right]}.$$

These probabilities can be computed analytically for specific sets K_0 and K_1.

A very different problem involves determining the mean area of the intersection of n random convex sets K_i and a specified convex set K_0. The random sets are anisotropic with distribution $dF(x)$. The ratio of the mean area to the total area of K_0 is interpreted as the probability that a target presents n sensor detection opportunities to a drifting sensor field. The expected area of the intersection $\cap_{i=0}^{n} K_i$ is the product of n terms involving the areas of K_i and the expected mixed area of K_i with K_0 [25, Theorem 14]. Roughly speaking, the expected mixed area is defined in a manner analogous to the expected thickness. The mean perimeter of the intersection can also be computed. The expression in this case is somewhat more complicated than the expected area, but it also involves the expected mixed areas. The expected area and perimeter are both amenable to numerical computation.

7.4 Stereology

Stereology is most often defined as the study of three dimensional objects by means of plane cross-sections through the object. A 3-D object in $\mathbb{R}^3$ can be studied by 2-D and 1-D sections, or by combination of them. More generally, it is the study of the structure of objects in $\mathbb{R}^m$ that are known only via measurements of lower dimensional linear sections. Stereology is of great practical importance in many fields ranging from geology (e.g., core samples) to biology (e.g., microscope slide smears, tissue sections, needle biopsies).

Tomography might be seen as a specialized form of stereology because it involves (nondestructive) reconstruction of 3-D properties from thin 2-D sections. However, tomography computes as many plane sections as are needed to image an object fully, and is therefore typically data rich compared to stereology. Only data starved tomographic applications (e.g., acoustic tomography of the ocean volume) might be considered comparable to the problems of stereology.

Distributed multiple sensor problems are not currently thought of as problems of stereology. The field of view (FOV) of any given sensor in a distributed sensor field is a limited cross-section of the medium in which the sensors reside. Interpreting this limited data to understand and estimate statistical properties of the medium as a whole is a stereological interpretation of the problem. Admittedly, the connection seems tenuous at best, but the potential of stereology to contribute to understanding the problems involved is real enough to justify inclusion of a brief mention of the topic here.

Problems in stereology are difficult, but sometimes have surprising and pleasing solutions. An excellent example of this is Delesse's principle from mineralogy, and it dates to 1848. For concreteness, suppose a rock sample is sliced by a saw and the cross-section polished. The cross-section is examined for the presence of a specified mineral. The ratio of the area of the cross-section that intersected the mineral to the total area of the cross-section is called the area fraction, denoted A_A. Similarly, the ratio of the total volume of the mineral to the total volume of the rock sample is the volume fraction, denoted by V_V. Assuming the rock sample is a representative specimen from a much larger homogeneous volume, Delesse's principle says that the expected volume fraction equals the expected area fraction:

$$A_A = V_V.$$

The expectation is over all possible two-dimensional slices through the rock. The area fraction is an unbiased estimator of the volume fraction. The variance of the estimate is much harder to evaluate. Delesse's principle is very practical since the area fractions of several rock samples are much easier to measure then volume fraction.

Rosiwal extended the result in 1898 to line fractions. Draw an equispaced grid of parallel lines on the polished rock surface. Measure the length of the line segments that overlay the mineral of interest, and divide by the total length of the line segments. This ratio is the line fraction L_L. Rosiwal's principle says that

$$L_L = A_A = V_V ,$$

provided, as before, the rock is a sample from a larger homogeneous volume. Line fractions are even easier to measure than area fractions.

Similar applications in microscopy are complicated by the simple fact that even the sharpest knife may not cut a cell, but simply push it aside, unless the tissue is frozen first. Freezing is not always an option. There are other practical issues as well.

These methods do *not* provide estimates of the number of objects in a volume. In the biological application, for example, the number of cells cut by a two-dimensional tissue slice does not in general indicate the number of cells in the tissue volume.

Modern stereology is a highly interdisciplinary field with many connections to stochastic geometry. An excellent modern book on the subject is [3], which also gives references to the original papers of Delesse and Rosiwal. The relationships between stereology and integral geometry, convexity, and Crofton-style formulae are discussed in [2] and the references cited therein. An informative overview is also given in [44, Section 1.9].

Example 7.6 Connections. An old puzzle related to Delesse's principle goes as follows: a region in the plane is such that every line through the origin intersects it in a line segment of length 2. Is the region a unit radius circle centered at the origin? The answer is no, and the cardioid $r = 1 - \cos\theta$ is a nice counter-example. If, however, the region is centrally symmetric, the answer is yes. Extended to three dimensions, the problem is: an object in $\mathbb{R}^3$ is such that the area of its intersection with every plane through the origin is π. Is the object the unit sphere centered at the origin? The answer [32] in this case is the same as in the plane, but the solution is also much deeper mathematically. Counter-examples are found by using spherical harmonics. This problem is a special case of the Funk-Hecke theorem [29, 109] for finding the spherical harmonic expansion of a function knowing only its integrals on all the great circles. While seeking the analog of this theorem in the plane (given integrals of a function on all straight lines), Radon found in 1917 what is now called the Radon transform, which is recognized as the mathematical basis of tomography.

Part III
Beyond the Poisson Point Process

Chapter 8
A Profusion of Point Processes

The time has come, the Walrus said,
To talk of many things:
Of shoes–and ships–and sealing-wax–
Lewis Carroll, The Walrus and The Carpenter
in Through the Looking-Glass and
What Alice Found There, *1872*

Abstract Generalizations of PPPs are useful in a large variety of applications. A few of the better known of these point processes are presented here, with emphasis on the processes themselves, not the applications. Marked processes are relatively simple extensions of PPPs that model auxiliary phenomena related to the point distribution. Other processes model the point-to-point correlation that may exist between the otherwise random occurrences of points. These processes include hard core processes and cluster processes. Cox processes are briefly reviewed, along with two stochastic processes, namely, Markov modulated Poisson processes and filtered processes. Gibbs (or Markov) point processes are not straightforward generalizations of PPPs. Generating realizations of Gibbs processes is typically done using MCMC methods.

Keywords Marked process · Marking Theorem · Hard core process · Matern process · Cluster processes · Poisson cluster process · Neyman-Scott cluster process · Cox process · Bartlett's Theorem · Markov modulated Poisson process (MMPP) · Gibbs point process

This chapter discusses only a few of the bewilderingly large variety of point processes that are not PPPs and are also useful in applications. The MCMC revolution, as it has been called, will undoubtedly increase the variety and number of non-PPPs that find successful application in real world problems.

Many useful point processes are built upon a PPP foundation, which assists in their simulation and aids in their theoretical development. Marked PPPs are discussed first. Despite appearances, marked PPPs turn out in the end to be equivalent to PPPs.

Hard core processes are discussed next. They have less spatial variability than PPPs because points are separated by a specified minimum distance. Loosely

R.L. Streit, *Poisson Point Processes*, DOI 10.1007/978-1-4419-6923-1_8,
© Springer Science+Business Media, LLC 2010

speaking, the points are "mutually repelled" from each other. The Matérn hard core processes use dependent thinning to enforce point separation.

Cluster processes are presented next. They have greater spatial variability than PPPs because, as the name suggests, points tend to gather closely together—points are in a loose sense "mutually attractive." Poisson and Neyman-Scott cluster processes are discussed. A special case of the latter is the Matérn cluster process, which uses a resampling procedure to encourage point proximity.

The Cox, or doubly stochastic, processes are discussed next. These are complex processes that push the concept of the ensemble to include the intensity function. In other words, the intensity function is itself a random variable. It is shown that a Cox process whose intensity function is a random sum is a Neyman-Scott process.

Many useful point processes are not directly related to PPPs. One of these—the Gibbs, or Markov, point process—is discussed briefly here.

8.1 Marked Processes

Marked PPPs model problems in which the points are accompanied by a "mark." The mark is commonly called a "feature vector" in applications. The mark can be discrete, continuous, or discrete-continuous. One of the oldest examples is in forestry, where a point represents the location of a tree, and its mark comprises one or more of the following kinds of information: its species and health (discrete/categorical), circumference at a fixed height above the ground (continuous), or both. Other examples are easily conceived.

Realizations of a marked PPP are in principle almost as easy to generate as those of a PPP:

- Use the two-step procedure of Section 2.3 to generate a realization

$$\xi = (m, \{x_1, \ldots, x_m\})$$

  of a PPP with intensity $\lambda(s)$ on the space $\mathcal{S}$. Given m, the points x_j are i.i.d. samples of the random variable X whose pdf is $p_X(x) = \lambda(x)/\int_{\mathcal{S}} \lambda(s)\,ds$.
- The mark is a random variable U on a mark space $\mathcal{U}$ with conditional pdf $p_{U|X}(u \mid x)$. Given the PPP realization ξ, generate the marks

$$\{u_1, \ldots, u_m\} \subset \mathcal{U}$$

  as independent realizations of $p_{U|X}(\cdot \mid x_j)$, $j = 1, \ldots, m$.
- Pair the points with their corresponding marks:

$$\xi' \equiv (m, \{(x_1, u_1), \ldots, (x_m, u_m)\}) . \tag{8.1}$$

The realization of the marked PPP is ξ'.

The mark space $\mathcal{U}$ can be very general, but it is typically either a discrete set or a subset of the Euclidean space $\mathbb{R}^\kappa$. An example is the MMIF of Section 6.2.2 and Appendix E. In this application, the targets are the points of a PPP, the measurements are the marks, and the measurement likelihood function is the conditional mark pdf.

In the simplest case, the marks are independent of ξ, so that

$$p_{U \mid X} (u \mid x_j) \equiv p_U (u) .$$

The marks in this case are independent of the locations of the points in ξ as well as the number m. This kind of marked PPP is called a compound PPP by Snyder, and its theory is well developed in [118, Chapter 3] and [119, Chapter 4].

Several compound PPPs are used earlier in the book, but without comment. The "Coloring Theorem" of Example 2.7 in Chapter 2 is an example of a marked PPP. The marks are the colors assigned to the points. In Chapter 3, marks are introduced as part of the complete data; that is, the marks are the missing data of the EM method. A specific example is the complete data (3.12) for estimating superposed PPPs, which is a realization of a compound PPP in which the mark space is $\mathcal{U} = \{1, \ldots, L\}$.

Yet another example, one that could have been discussed as a marked PPP but was not, is the intensity filter of Chapter 6 when the target PPP is split during the prediction step into the detected and undetected target PPPs. In this case, the detection process is equivalent to a marking procedure with marks $\mathcal{U} = \{0, 1\}$, where zero/one denotes target non-detection/detection.

8.1.1 Product Space and Marking Theorem

Marked PPPs are intuitive models with a rich structure that enables them to model phenomena in diverse applications. The mark structure might make them seem fundamentally different from ordinary PPPs, but this is not so. Marked PPPs are equivalent to PPPs on the Cartesian product of the space $\mathcal{S}$ on which the PPP Ξ is defined and the mark space $\mathcal{U}$ with a joint intensity function μ given by

$$\mu(x, u) = p_{U \mid X}(u \mid x) \, \lambda(x) . \tag{8.2}$$

This result is important as well as insightful. The similarity of (8.2) to conditional Bayesian factorization is self evident.

First observe that the realization ξ' of (8.1) is an element of the PPP event space $\mathcal{E}(\mathcal{S} \times \mathcal{U})$, so it has the form of a realization of a PPP on $\mathcal{S} \times \mathcal{U}$. Since the two-step procedure is the definition of a PPP, it is only necessary to verify that the intensity function of the point process with realizations ξ' takes the form (8.2). In light of the result of Section 2.6.1, it is enough to show that for functions $f(x, u)$ defined on the Cartesian product $\mathcal{S} \times \mathcal{U}$ the characteristic function of the random sum

$$F(\xi') \equiv F(m, (x_1, u_1), \ldots, (x_m, u_m))$$

$$= \sum_{j=1}^{m} f(x_j, u_j)$$

is in the form given by Campbell's Theorem, namely,

$$E\left[e^{-F}\right] = \exp\left\{ \int_{\mathcal{S}} \int_{\mathcal{U}} \left(e^{-f(x,u)} - 1\right) p_{U|X}(u \mid x)\, \lambda(x)\, dx\, du \right\}$$

$$= \exp\left\{ \int_{\mathcal{S} \times \mathcal{U}} \left(e^{-f(x,u)} - 1\right) \mu(x, u)\, dx\, du \right\}. \tag{8.3}$$

The random variables $f(X_j, U_j)$ are independent given m, so the expectation of $F(\xi')$ with respect to the marks, conditioned on the points $x_1, \ldots, x_m$, is

$$E\left[e^{-F} \mid x_1, \ldots, x_m\right] = E_{U_1 \cdots U_m \mid X_1 \cdots X_m}\left[e^{-\sum_{j=1}^{m} f(x_j, U_j)}\right]$$

$$= E_{U_1 \cdots U_m \mid X_1 \cdots X_m}\left[\prod_{j=1}^{m} e^{-f(x_j, U_j)}\right]$$

$$= \prod_{j=1}^{m} E_{U_j \mid X_j}\left[e^{-f(x_j, U_j)}\right]$$

$$= \prod_{j=1}^{m} \int_{\mathcal{U}} e^{-f(x_j, u_j)}\, p_{U|X}(u_j \mid x_j)\, du_j$$

$$= e^{-\sum_{j=1}^{m} g(x_j)}, \tag{8.4}$$

where, for any $x \in \mathcal{S}$,

$$g(x) = -\log \int_{\mathcal{U}} e^{-f(x,u)}\, p_{U|X}(u \mid x)\, du. \tag{8.5}$$

Applying Campbell's Theorem gives the expectation of the expression (8.4) with respect to the PPP ξ with intensity function $\lambda(x)$:

$$E\left[e^{-F}\right] = E_{\Xi}\left[e^{-\sum_{j=1}^{M} g(X_j)}\right]$$

$$= \exp\left\{ \int_{\mathcal{S}} \left(e^{-g(x)} - 1\right) \lambda(x)\, dx \right\}.$$

Substituting (8.5) gives

$$E\left[e^{-F}\right] = \exp\left\{\int_{\mathcal{S}}\left(\int_{\mathcal{U}} e^{-f(x,\,u)}\, p_{U|X}(u\,|\,x)\,du - 1\right)\lambda(x)\,dx\right\}$$

$$= \exp\left\{\int_{\mathcal{S}}\left(\int_{\mathcal{U}}\left(e^{-f(x,\,u)} - 1\right) p_{U|X}(u\,|\,x)\,du\right)\lambda(x)\,dx\right\}.$$

The last expression is equivalent to (8.3).

This result applies with only minor changes when the mark space is discrete.

8.1.2 Filtered Processes

A filtered Poisson process is the output of a function that is similar to (2.30). Let $h(x,\,y;\,u)$ be a real valued function defined for all $x,\,y \in \mathcal{S}$ and $u \in \mathcal{U}$. Given the realization $\xi' \equiv (m,\,\{(x_1,\,u_1),\,\ldots,\,(x_m,\,u_m)\})$ of the marked PPP, define the random sum

$$F(y) = \begin{cases} 0, & \text{if } m = 0, \\ \sum_{j=1}^{m} h(y,\,x_j;\,u_j), & \text{if } m \geq 1, \end{cases} \tag{8.6}$$

where the function $h(y,\,x;\,u)$ is the response at y to a point at x with mark u. If the marked PPP is a compound PPP, then (8.6) is called a filtered Poisson process.

Filtered Poisson processes are often discussed as stochastic processes when they are defined in the time domain, so that $\mathcal{S} \equiv \mathbb{R}^1$ and h is the impulse response function of some specified linear system. If $h(y,\,x;\,u) = 0$ for $y < x$, then h is said to be causal. The simplest example in this case is shot noise, for which $h(y,\,x;\,u) \equiv h(y,\,x)$. Shot noise is an model for white noise. Shot noise is extensively studied in [101] and in [18, Chapter 7]. These discussions are classical applications of PPPs to signal processing.

Interesting topics such as the central limit theorem for filtered Poisson processes and Poisson driven Markov processes are discussed in [118, Chapter 4] and [119, Chapter 5]. These excellent discussions are widely available, so the topic is not pursued further here.

8.1.3 FIM for Unbiased Estimators

A marked PPP is equivalent to a PPP in the Cartesian product space $\mathcal{S} \times \mathcal{U}$, so the FIM (4.21) for unbiased estimators applies unchanged to general marked PPPs. The data comprise a realization of the PPP with intensity function $\lambda(s\,;\,\theta)$ together with the corresponding marks.

For compound PPPs with continuous marks in the space $\mathcal{U}$ with pdf

$$p_U(u) \equiv p_U(u\,;\,\theta),$$

the FIM for unbiased estimators of θ is

$$F(\theta) = \int_{\mathcal{U}} \int_{\mathcal{S}} \frac{1}{\lambda(s\,;\,\theta)\,p_U(u;\,\theta)} \tag{8.7}$$

$$\left[\nabla_\theta\left\{\lambda(s\,;\,\theta)\,p_U(u\,;\,\theta)\right\}\right]\left[\nabla_\theta\left\{\lambda(s\,;\,\theta)\,p_U(u\,;\,\theta)\right\}\right]^T \, ds\, du\,.$$

For discrete marks in the space $\mathcal{U} \equiv \{u_1,\, u_2,\, \ldots\}$, the FIM is

$$F(\theta) = \sum_{k=1}^{\infty} \int_{\mathcal{S}} \frac{1}{\lambda(s\,;\,\theta)\,p_U(u_k\,;\,\theta)} \tag{8.8}$$

$$\left[\nabla_\theta\left\{\lambda(s\,;\,\theta)\,p_U(u_k\,;\,\theta)\right\}\right]\left[\nabla_\theta\left\{\lambda(s\,;\,\theta)\,p_U(u_k;\,\theta)\right\}\right]^T \, ds\,.$$

Further discussion of this case is also given in [119, p. 213].

8.2 Hard Core Processes

A hard core process is a point process in which no two points in a realization are closer together than a specified minimum distance, say $2h$. The points of such a process are the centers of non-overlapping m-dimensional spheres with radius h. The (solid) spheres are the "hard cores" of the process. The points of a hard core process are not independent, since positioning one point constrains the placement of the others. Hard core processes often differ significantly in the spatial dependence of the points, with the details depending on the application.

Hard core models correspond to hard-packing problems and are important in physics, material science, and coding theory. A highly readable account by Diaconis [24, Section 4] sketches the interesting connections between high density close packing problems and the very important problem (in physics) of solid-liquid-gas phase transitions. The references therein are a good entrée to the subject. The larger purpose of [24] is to review MCMC (Markov chain Monte Carlo) methods, but one of his examples generates realizations of a high density hard core process using a Metropolis algorithm.

Hard core processes are important in less physics-based kinds of applications too. For example, the sensors in spatially distributed fields are often deployed to avoid overlapping detection coverage.

Matérn [77] gives two procedures for generating realizations a hard core process on $\mathbb{R}^m$ by thinning a homogeneous PPP [77, 78]. Both yield homogeneous point processes, but the second method accommodates higher packing density than the first. These methods are given in the next two examples.

Example 8.1 Matérn Method II. Let $\xi = (\ell, \{x_1,\, \ldots,\, x_\ell\})$ denote a realization of a homogeneous PPP with intensity λ_0. The points $\{x_j\}$ of ξ are marked by samples $\{u_j\}$ of a uniform random variable on $[0,\, 1]$. The marks are independent of each other and of the location of the points x_j. The point x_j with mark u_j is retained if the sphere of radius h centered at x_j contains no points with marks smaller than u_j. Points are deleted from the realization only after *all* points are determined to be either thinned or retained. This kind of thinning is *not* Bernoulli independent thinning.

The intensity function of the resulting Matérn hard core process is

$$\lambda_{\text{Matérn(II)}} = \frac{1}{c_m \, h^m} \left(1 - e^{-\lambda_0 \, c_m \, h^m} \right), \qquad (8.9)$$

where c_m is the volume of the unit radius sphere in $\mathbb{R}^m$ (see (7.9)). To see this, follow the method of [123]: note that the process that deletes points x with mark $u < t$ is a thinned PPP whose intensity is $\lambda_0 t$. Hence, $r(t) = \exp\left(-\lambda_0 \, c_m \, h^m \, t\right)$ is the probability that the thinned PPP has no points in the sphere of radius t. Equivalently, $r(t)$ is the probability that a point at x with mark t is retained. Hence,

$$p = \int_0^1 r(t) \, dt = \frac{1 - e^{-\lambda_0 \, c_m \, h^m}}{\lambda_0 \, c_m \, h^m}$$

is the probability that the point x is retained. The intensity $\lambda_{\text{Matérn(II)}}$ is the product of p and the PPP intensity λ_0.

The intensity of the Matérn process increases with increasing initial intensity λ_0. From (8.9), the limiting intensity is

$$\lambda_{\text{MaxMatérn(II)}} = \lim_{\lambda_0 \to \infty} \lambda_{\text{Matérn}} = \frac{1}{c_m \, h^m} .$$

The limiting intensity is one point per unit sphere. For $m = 2$ and $h = 1$,

$$\lambda_{\text{MaxMatérn(II)}} = \frac{1}{\pi} \approx 0.318 .$$

For comparison, regular hexagonal packing gives the maximum possible intensity of one point per hexagon (inscribed in the unit circle), or $\lambda_{\text{Hex}} = 2\sqrt{3}/9 \approx 0.385$ with $h = 1$.

Example 8.2 Matérn Method I. This method starts with the same realization ξ as in the previous example, but more aggressively thins the points of the PPP. Here, pairs of points in the realization ξ that are separated by a distance $2h$ or less from each other are removed from ξ. Pairs of points are removed only after first identifying all pairs to be removed. The intensity function of the resulting hard core process is

$$\lambda_{\text{Matérn(I)}} = \lambda_0 \, e^{-\lambda_0 \, c_m \, h^m}, \qquad (8.10)$$

To see this it is only necessary to see that the probability that a given point is retained is $e^{-\lambda_0 \, c_m \, h^m}$, and to multiply this probability by λ_0.

Variable size hard core models are defined by specifying a mark that denotes, say, the radius of the hard core. Soft core models are defined in [123, p. 163]. Hard core processes are very difficult to analyze theoretically. In most applications, numerical modeling and simulation is often necessary in practice. As pointed out by Diaconis [24], MCMC methods play an important role in the modern developments of these problems.

8.3 Cluster Processes

A cluster process Ξ_C is a finite point process generated from realizations of a parent process and a family of daughter processes. The parent process is denoted by Ξ, and the daughter processes are denoted by $\Xi(x)$ because there is one process for each point $x \in \mathcal{S}$. These point processes are finite point processes and none, in general, are PPPs. The set of points in a realization of the daughter process is called a cluster. It is assumed that the clusters corresponding to realizations of the daughter processes $\Xi(x)$ and $\Xi(y)$ are distinct with probability one if $x \neq y$.

Let $\xi = (n, \{x_1, \ldots, x_n\})$ denote a realization of the parent process Ξ. Every point x_j in ξ is *replaced* by a realization, denoted by $\xi(j)$, of the daughter process $\Xi(x_j)$. The union of the clusters

$$\xi_C = \cup_{j=1}^{n} \xi(j)$$

is a realization of the cluster process Ξ_C. The parent points x_j are not in the realization ξ_C. (In some applications, parent points are retained [103].) Cluster processes are very general, and it is desirable to specialize them further.

8.3.1 Poisson Cluster Processes

Poisson cluster processes are cluster processes in which the parent process is a nonhomogeneous PPP. The family of daughter processes remain general finite point processes.

The Neyman-Scott process is a Poisson cluster process in which the daughter processes take a special form; realizations are generated via the following procedure. The first step yields a realization $\xi = (n, \{x_1, \ldots, x_n\})$ of the parent PPP Ξ whose intensity function is $\lambda(x)$, $x \in \mathcal{S}$. The second step draws i.i.d. samples k_j, $j = 1, \ldots, n$, from a discrete random variable K on the nonnegative integers with specified probabilities

$$p_K(k) = \Pr[K = k] \equiv p_k, \quad k = 0, 1, 2, \ldots . \tag{8.11}$$

The discrete variable K is not in general Poisson distributed, and it is independent of the parent process Ξ. The parent point x_j produces k_j daughters. The output

$$(n, \{(x_1, k_1), \ldots, (x_n, k_n)\})$$

is equivalent to that of a marked PPP. Let $h(x)$, $x \in \mathcal{S}$, be a specified pdf. The final step draws k_j i.i.d. samples from the shifted pdf $h(x - x_j)$. Denote these samples by x_{ij}, $i = 1, \ldots, k_j$. The i-th child of x_j is x_{ij}. Let $k = \sum_{j=1}^{n} k_j$. The realization of the Neyman-Scott cluster process is

$$\xi_C = \left(k, \{x_j + x_{ij} : i = 1, \ldots, k_j \text{ and } j = 1, \ldots, n\}\right) \in \mathcal{E}(\mathcal{S}). \quad (8.12)$$

The defining parameters of the Neyman-Scott cluster process are the PPP parent intensity function $\lambda(x)$, the distribution of the number K, and the daughter pdf $h(x)$. The probability generating functional of the general Neyman-Scott process is given in the next section.

A special case of the Neyman–Scott process is the Matérn cluster process. In this case, the clusters are homogeneous PPPs with intensity function λ_0 on the sphere of radius R. The discrete pdf K in the simulation is in this case the Poisson distribution with parameter $c_\kappa \lambda_0$, where c_κ is the volume of the unit sphere in $\mathbb{R}^\kappa$. The points x_{ij} of the simulation are i.i.d. and uniformly distributed in the sphere of radius R. The defining parameters of the Matérn cluster process are the PPP intensity function $\lambda(x)$, the homogeneous PPP intensity λ_0, and the radius R.

8.3.2 Neyman-Scott Processes

The characteristic functional of the general Neyman-Scott process Ξ on the Euclidean space $\mathcal{S} = \mathbb{R}^\kappa$ is evaluated by a straightforward calculation. Let Ξ_{parent} denote the parent PPP, and let $\lambda(x)$, $x \in \mathcal{S}$, denote its intensity function. Let

$$\Xi_{\text{parent}} = (N, \mathcal{X}|N),$$

where N is the random number of points and

$$\mathcal{X}|N = \{X_1, \ldots, X_N\}$$

are the conditionally i.i.d. points of Ξ_{parent}. Denote the number of daughter points generated by the parent point X_j by the discrete random variable K_j. By definition of the Neyman-Scott process, the variables $K_1, \ldots, K_N$ are i.i.d. with pdf given by (8.11). Let $\eta(t)$ denote the probability generating function of the discrete random variable K:

$$\eta(t) = \sum_{k=0}^{\infty} p_K(k) \, t^k. \quad (8.13)$$

The daughter points of the parent point X_j are the random variables

$$X_{1j}, \ldots, X_{K_j,j} \, .$$

By definition of the Neyman-Scott process, they are i.i.d. with conditional pdf

$$p_{X_{ij}|X_j}\left(x \mid X_j = x_j\right) = h(x - x_j), \quad i = 1, \ldots, K_j, \tag{8.14}$$

where the cluster pdf $h(x)$ is given. The characteristic functional of Ξ evaluated for a given function f is, by definition,

$$G_\Xi(f) = E_\Xi\left[\prod_{j=1}^{N}\prod_{i=1}^{K_j} f(X_{ij})\right].$$

The random variables X_j of the parent points do not appear in the product because they are not points of the output process. The expectation is evaluated in the nested form:

$$G_\Xi(f) = E_{\Xi_{\text{parent}}}\left[\prod_{j=1}^{N} E_{K_j|X_j}\left[E_{X_{1j}X_{2j}\cdots X_{K_j,j}|K_jX_j}\left[\prod_{i=1}^{K_j} f(X_{ij})\right]\right]\right]. \tag{8.15}$$

Because the daughter points are conditionally i.i.d., the innermost expectation factors into the product of expectations:

$$E_{X_{1j}X_{2j}\cdots X_{K_j,j}|K_jX_j}\left[\prod_{i=1}^{K_j} f(X_{ij})\right] = \prod_{i=1}^{K_j} E_{X_{ij}|X_j}\left[f(X_{ij})\right]$$

$$= \prod_{i=1}^{K_j}\int_S f(x)h(x - X_j)\,dx = \prod_{i=1}^{K_j}\int_S f(x + X_j)h(x)\,dx$$

$$= \left(\int_S f(x + X_j)h(x)\,dx\right)^{K_j}. \tag{8.16}$$

The expectation over $K_j|X_j$ is expressed in terms of the probability generating function $\eta(\cdot)$ as

$$E_{K_j|X_j}[\,\cdot\,] = \sum_{k_j=0}^{\infty}\left(\int_S f(x + X_j)h(x)\,dx\right)^{k_j} p_{k_j}$$

$$= \eta\left(\int_S f(x + X_j)h(x)\,dx\right). \tag{8.17}$$

The outermost expectation over Ξ_{parent} is simply

$$G_\Xi(f) = E_{\Xi_{\text{parent}}} \left[\prod_{j=1}^{N} \eta \left(\int_S f(x + X_j) h(x)\,dx \right) \right].$$

Using the generating functional of the parent PPP (see Eqn. (2.53)) gives the characteristic functional of the Neyman-Scott cluster process as

$$G_\Xi(f) = \exp \left\{ \int_S \left[\eta \left(\int_S f(x + s) h(x)\,dx \right) - 1 \right] \lambda(s)\,ds \right\}. \tag{8.18}$$

This expression is used in the next section.

8.4 Cox (Doubly Stochastic) Processes

A Cox process is a PPP in which the intensity function is randomly selected from a well defined space of possible intensities, say Λ. Thus, realizations of a Cox process are the result of sampling first from the space Λ to obtain the intensity function $\lambda(x)$, and then finding a realization of the PPP with intensity $\lambda(x)$ via the two step procedure of Section 2.3. The idea is that the intensity space Λ characterizes possible environments in which a PPP might be a good model—provided the right intensity is used.

To set ideas, consider a homogeneous PPP whose intensity λ is selected so that the mean number of points $\mu = \lambda \mathcal{R}$ in the window $\mathcal{R}$ is selected from an exponential pdf:

$$p_M(\mu) = \frac{1}{\mu_0} \exp \left(-\frac{\mu}{\mu_0} \right), \tag{8.19}$$

where $\mu_0 > 0$ is a specified (Bayesian) parameter. The pdf of the number, n, of points in the bounded window $\mathcal{R}$ is then

$$\begin{aligned}
p_N(n) &= \int_0^{\infty} p_{N|M}(n \mid \mu)\, p_M(\mu)\,d\mu \\
&= \int_0^{\infty} \left(\frac{1}{n!} e^{-\mu} \mu^n \right) \left(\frac{1}{\mu_0} e^{-\mu/\mu_0} \right) d\mu \\
&= \frac{\mu_0^n}{(1 + \mu_0)^{n+1}}.
\end{aligned} \tag{8.20}$$

The one dimensional version of this example is used in [83] to model neural spike trains. Carrying the general notion over to PPPs gives the Cox process.

A natural way to specify the random intensity function of a Cox process is to use realizations of a nonhomogeneous PPP. One class of Cox process is equivalent to a Neyman-Scott cluster process, as is seen in the first subsection below. The subsections following it discuss two stochastic processes that appear in various applications.

8.4.1 Equivalent Neyman-Scott Process

An interesting example of a Cox process is one in which the intensity function on $\mathcal{S} \subset \mathbb{R}^\kappa$ is the random sum:

$$\Lambda(x ; N, \mathcal{X} \mid N) = \mu \sum_{j=1}^{N} h\left(x - X_j\right), \tag{8.21}$$

where $\mu > 0$ is a known scale constant, $h(x)$ is a specified pdf on $\mathcal{S}$, and where the points $\mathcal{X} = \{X_1, \ldots, X_N\}$ are the random points of a realization of a PPP with intensity function $\lambda(x)$ on $\mathcal{S}$. Bartlett (1964) showed that this Cox process is a Neyman-Scott cluster process with a Poisson distributed number of daughter points, i.e., the random variable K in (8.11) is Poisson distributed.

To see Bartlett's result, evaluate the characteristic functional of the Cox process Ξ. Let $\{U_1, \ldots, U_N\}$ denote N points of a realization of Ξ. The required expectation is equivalent to the nested expectations

$$G_\Xi(f) = E_\Lambda \left[E_{\Xi \mid \Lambda} \left[\prod_{j=1}^{N} f(U_j) \right] \right].$$

The inner expectation is the characteristic functional of the PPP with intensity $\Lambda(x ; N, \mathcal{X} \mid N)$; explicitly,

$$E_{\Xi \mid \Lambda}[\cdot] = \exp\left\{ \int_\mathcal{S} (f(u) - 1)\, \Lambda(x ; N, \mathcal{X} \mid N)\, \mathrm{d}u \right\}$$

$$= \exp\left\{ \int_\mathcal{S} (f(u) - 1)\, \mu \sum_{j=1}^{N} h\left(u - X_j\right) \mathrm{d}u \right\}$$

$$= \prod_{j=1}^{N} \Psi(X_j),$$

where

$$\Psi(X_j) = \exp\left\{ \mu \left(\int_\mathcal{S} f(u + X_j)\, h(u)\, \mathrm{d}u - 1 \right) \right\}.$$

Since the points X_j are the points of a PPP with intensity $\lambda(x)$, the characteristic functional of Ξ is

$$
\begin{aligned}
G_\Xi(f) &= E_\Lambda\left[\prod_{j=1}^{N} \Psi(X_j)\right] \\
&= \exp\left\{\int_S [\Psi(x) - 1]\lambda(x)\,\mathrm{d}x\right\}.
\end{aligned}
\tag{8.22}
$$

Comparing this to the Neyman-Scott characteristic functional of f shows that

$$
\eta(s) = e^{\mu(s-1)},
\tag{8.23}
$$

where

$$
s = \int_S f(u + x)h(u)\,\mathrm{d}u.
\tag{8.24}
$$

The right hand side of (8.23) is the characteristic function of the discrete Poisson distribution with mean μ. Therefore, the Cox process has a Poisson distributed number of points that are distributed around parent points with pdf $h(u)$, $u \in S$. Since probability generating functionals characterize finite orderly point processes ([16, p. 625]), the Cox process with random intensity function given by (8.21) is a Neyman-Scott process with a Poisson distributed number of points. (General conditions under which a Poisson cluster process is a Cox process are not known. Further details are given in [16, pp. 663–664].)

8.4.2 Intensity Function as Solution of an SDE

Cox processes are useful in applications in which the parameter vector of the intensity function is the solution of a stochastic differential equation (SDE). An example is the intensity function (4.38) when the mean μ_t of the Gaussian component is time dependent and satisfies the Ito diffusion equation [90, p. 104]

$$
\mathrm{d}\mu_t = b(\mu_t)\,\mathrm{d}t + \sigma(\mu_t)\,\mathrm{d}B_t, \quad t \geq 0, \quad \mu_0 \equiv a_0,
\tag{8.25}
$$

where B_t is Brownian motion of appropriate dimension, and the functions b and σ satisfy certain regularity conditions. PPPs in time and space are also very interesting in tracking applications. See [119, Chapter 7], as well as Snyder's charmingly titled paper [117] and the extended paper [36]. These topics lead to interesting tracking filters, but they are outside the scope of this book

Temporal (one dimensional) Cox processes are a special case of self-exciting point processes [119, p. 348]. Self-exciting point processes are in turn closely related to renewal processes (see [119, Chapters 6 and 7]). These processes are well discussed elsewhere.

8.4.3 Markov Modulated Poisson Processes

A special kind of Cox process is the Markov modulated Poisson process (MMPP). It is defined in conjunction with a continuous time Markov chain $\mathcal{A}$ on a countable number of states. Denote the states by $C \equiv \{c(1), c(2), \ldots\}$, the transition probability function by

$$A_{c(i),c(j)}(s, t) \equiv \Pr\left[\mathcal{A}(t) = c(j)\,\middle|\,\mathcal{A}(s) = c(i)\right] \quad \text{for all } s \leq t, \qquad (8.26)$$

and the initial state probability mass function by $\pi(\cdot)$ on C. The initial time is $t_0 \equiv t(0)$. Homogeneous PPPs are defined for each state $c(j)$ with intensity $\lambda_{c(j)}$. (Nonhomogeneous PPPs can also be used.)

A realization of a MMPP on the interval $[t(0), T]$ is obtained from a two stage procedure. The first stage generates a realization of $\mathcal{A}$ on the time interval $[t(0), T]$ (see [100]). The initial state $c(t(0))$ at time $t(0)$ is a realization of $\pi(\cdot)$. Including $t(0)$, the switching times and states of the Markov chain are $\{t(0), t(1), \ldots, t(\ell)\}$ and $\{c(t(0)), c(t(1)), c(t(2)), \ldots, c(t(\ell))\}$, respectively. Let $t(\ell + 1) = T$. The second stage generates a realization of the PPP with intensity $\lambda_{c(t(j))}$ on the time interval $[t(j), t(j + 1))$, $j = 0, 1, \ldots, \ell$. The concatenation of the realizations of these Markov switched PPPs is a realization of the MMPP.

MMPPs are stochastic processes often used to model bursty phenomena in various applications, especially in telecommunications. Other applications include load modeling that changes abruptly depending on the outcome of certain events, say, an alternating hypothesis SPRT (sequential probability ratio test) that controls arrival rates in a queue. The superposition of MMPPs is an MMPP, a fact that facilitates applications. The transition function of the superposed MMPP in terms of the component MMPPs can be found in [95].

8.5 Gibbs Point Processes

The Gibbs distribution is defined on sets, or configurations, of n points in $\mathbb{R}^m$, where n is given. All that is needed to make the distribution into the Gibbs point process is method for generating n randomly. Physicists refer to the case with random n as the *grand canonical ensemble*. Another name for Gibbs processes is Markov point processes. Many choices for the distribution of n are possible.

For current purposes, the Gibbs pdf is written

$$p(x_1, \ldots, x_n \mid n, \lambda) = \frac{1}{Z_n\, n!} \exp\left(-E(x_1, \ldots, x_n)\right), \qquad (8.27)$$

where $E(\cdot)$ is a specified "energy" function of the points, or configuration, and Z_n is the "configurational" partition function that normalizes (8.27) so that it is a pdf. If $E(x_1, \ldots, x_n) = \lambda n$, the conditional density of the points is that of a homogeneous PPP, although the number of points need not be Poisson distributed.

The standard choice for the energy function is a sum over "interaction potentials." The potentials (i.e., the summands) may involve on any number of particles, depending on the application. Common choices are one and two particle potentials, nearest neighbor potentials, and clique potentials. Further discussion of the many interesting properties of Gibbs point processes is outside the scope of this paper. A good starting point for further reading is [123, Chapter 5.5].

Chapter 9
The Cutting Room Floor

Finished products are for decadent minds.
Isaac Asimov, Second Foundation, 1953

Abstract Several topics of interest not discussed elsewhere in this book are mentioned here.

The main properties of PPPs that seem (to the author) insightful and useful in applications are reviewed in this book. They do not appear to be gathered into one place elsewhere. The applications to medical imaging and tomography, multiple target tracking, and distributed sensor detection are intended to provide insight into methods and models of PPPs and related point processes, and to serve as an entry point to exciting new applications of PPPs that are of active research interest.

Many interesting topics are naturally omitted from the book. For reasons already mentioned in the introductory Chapter 1, nearly all discussion of one dimensional point processes is omitted. Other omissions are detailed below in a chapter by chapter review. A brief section on possible directions for further work follows.

Keywords History of PPPs · PET · SPECT · Posterior Cramer-Rao bound (PCRB) · Hammersley-Clifford-Robbins Bound · PCRB for multitarget tracking · Monotone functions of random graphs · Threshold property · Palm intensity · Papangelou intensity · Coupling from the past

9.1 Further Topics

Chapter 1 is missing a section on the history of PPPs. The Poisson distribution dates to the 1840s, but Poisson point processes seem not to have really begun until probability was formulated on a measure theoretic foundation in the 1930s. The name "point process" apparently was first coined by Palm in his 1943 paper [92]. The history of point processes and PPPs is quite short when compared to mainstream mathematical topics. Appreciating that much of the work on PPPs is relatively recent may encourage readers to seek new applications and methods. Omitting the history may not be egregious, but it is unfortunate.

R.L. Streit, *Poisson Point Processes*, DOI 10.1007/978-1-4419-6923-1_9,
© Springer Science+Business Media, LLC 2010

Chapter 2 on the basics of PPPs omits many topics simply because they are not used in the applications presented later in the book. Many of these topics have diverse applications in one dimensional processes, and they are also interesting in themselves. The connections with Poisson random measures is not mentioned. More generally, the connections between stochastic processes and point processes are only briefly mentioned. Markov point processes are omitted. This is especially unfortunate because it means that physically important point processes with spatial point to point correlations or interactions are treated only via example. The Matérn hard core processes of Section 8.2 are excellent examples of Markov point processes.

Chapter 3 on estimation is heavily weighted toward superposition because of the needs of applications. Superposition leads rather naturally to an emphasis on the method of EM. The dependence of the convergence rate of the EM method on the OIM is not discussed. Since the convergence rate of EM based algorithms is ultimately only linear, other algorithms merit consideration. Some methods use the OIM to accelerate EM algorithm convergence; others use hybrid techniques that switch from EM to other more rapidly convergent algorithms. MCMC methods are also omitted from the discussion.

Chapter 4 discussion of the CRB is fairly complete for the purposes of PPPs. The Bayesian CRB, or posterior CRB (PCRB), is not treated even though it is useful in some applications, e.g., tracking. The notion of a lower bound for parameters that are inherently discrete is available, but not discussed here, even though such bounds are potentially useful in applications. The Hammersley-Chapman-Robbins (HCR) bound holds for discrete parameters [14, 45]. It is a special case of the earlier Barankin bound [5, 134]. When the parameter is continuously differentiable, the HCR bound becomes the CRB.

Chapter 5 on PET, SPECT, and transmission tomography only scratches the surface. Absent for the chapter are details about specific technical issues involved in practical applications of the Shepp-Vardi and related algorithms. These details are important to specialists and not without interest to others. Overcoming these issues often involves exploiting the enormous flexibility of the algorithm. Examples that illustrate to current state of the art would entice readers to learn more about these topics, without delving too deeply into the intricacies of the methods. Happily, the CRB is already available for PET and methods are available for computing the CRB for subsets of the full PET image [49].

Chapter 6 on multitarget tracking applications of PPPs is an active area of research. It is reasonable to ask about the CRB of the intensity filter, since the intensity is the defining parameter. In this case, however, the target motion model is a Bayesian prior and the CRBs of Chapter 4 must be extended to include this case. The PCRB for the intensity filter is not available, but it is important for understanding the quality of the intensity filter estimate. Computing it will surely be a complicated affair because the intensity function is the small cell limit of a sequence of step functions in the single target state space. Presumably, after first finding the PCRB for a finite number of cells, taking the small cell limit of a suitably normalized version yields a function on the product space $\mathcal{S} \times \mathcal{S}$.

Chapter 7 on distributed sensor detection does not mention specific telecommunication applications. This is a rich and fertile area that deserves more attention than it is given. The connections between k-connectivity in geometric random graphs and percolation in lattices of two or more dimensions is omitted. The so-called threshold behavior of monotone properties of geometric random graphs is not discussed. The critical probability, p_c, beyond which the probability that a given vertex is part of an infinitely large clump, is an early example of a threshold.

Chapter 8 on point processes that are not PPPs is too brief. Concepts from PPPs that generalize to finite point processes are not explored fully. Boolean models are not discussed, although they are mentioned in Chapter 1. Palm and Papangelou intensities are not discussed. Interesting processes are not mentioned, e.g., Strauss hard core point processes and their simulation via MCMC. Estimation methods for non-PPPs are an especially important topic, but they too are omitted. In particular, Ripley's K-function and the pair-correlation function are not discussed.

Chapter 9 is mercifully short. In Italian, it is *a misura d'uomo*—just the right size—for a work of nonfiction.

Appendices. An appendix on measure theory would be helpful in bridging the gap between the measure theory free methods chosen for this book and the measure theoretic methods of much of the literature. Such an appendix would make clear that measure theory is not an arcane mathematical abstraction of no benefit in applications, but rather the opposite: It explains why the "test sets" that appear occasionally in the book with little explanation are important; it provides a natural model for certain kinds of generalized functions such as the Dirac delta function; and it clarifies important material presented in Chapter 2, e.g., the proof that a PPP after Bernoulli thinning is still a PPP.

An appendix on the many uses of PPPs in classical physics would be worthwhile. The brief mention of Olber's paradox scratches the surface, but it is an example of a common use of PPPs as a model of spatially distributed ambient physical phenomena. For instance, in underwater acoustics, active sonar reverberation processes from the ocean surface, bottom, and volume are modeled as PPPs and then superposed [33]. Another example is [1], where the heavy tails of the K-distribution are fit to active sonar reverberation data. The K-distribution arises from a negative binomial distribution on the number of homogeneously distributed and coherently summed scatterers, and then letting the mean number go to infinity [53]. Except for the use of the negative binomial distribution, the point process generated in this manner seems akin to the binomial point process (BPP) mentioned in Section 1.3.

9.2 Possible Trends

Some argue that theoretical analysis of point processes is of little value in real world applications because only extensive high fidelity simulations can give quantitative understanding of the many variables of practical interest. This argument is easily refuted in applications such as PET imaging, but more difficult to refute fully in applications such as distributed sensing that use point processes to gain insight rather than as an exact model.

The debate between theory and simulation will undoubtedly continue for years, but it is a healthy debate reminiscent of the debate between theoretical and experimental physicists. Practical problems will undoubtedly challenge theory, and the explanatory power of theoretical methods will grow. Because of its extraordinary flexibility, MCMC methods will become a progressively more important tool as demands on model sophistication and fidelity increase. The idea of perfect simulation via the "coupling from the past" (CFTP) technique of Propp and Wilson [96] will undoubtedly enrich applications. These methods will blur the distinction between theory and simulation.

Geometrically distributed sensors and social networks (e.g., the Internet) are both graphs whose vertices (points, sensors, agents, etc.) are connected by edges. Edges represent connectivity that arises from geometric proximity, or from some non-physical social communication link. The empirical evidence shows that the vertex connectivity of geometric random graphs is very different from that of social network graphs—complex social networks typically have a power law distribution on connectivity, whereas geometric random graphs do not. Detecting subgraphs that are proximate both geometrically and socially is an important problem. Adding the element of time makes the problems dynamic and even more realistic. Integral geometry and its methods, especially Boolean models, may eventually be seen as a key component of a mathematical foundation of detection in these kinds of problems. It will be interesting to see what manner of contribution MCMC methods make to the subject.

Appendix A
Expectation-Maximization (EM) Method

Expectation-Maximization (EM) is a *method* for deriving algorithms to maximize likelihood functions—using it with different likelihood functions results in different ML (or MAP, as the case may be) estimators. It is, on first encounter, a mysterious method with strong appeal to those who love to see elaborate mathematical constructions telescope into simple forms.

EM is reviewed in this appendix. The distinctions between the named random variables—incomplete, missing, and complete—are discussed, together with the auxiliary function called Q, and the E- and M-steps. It is shown that increasing the auxiliary function increases the likelihood function. This means that all EM based algorithms monotonically increase the data likelihood function $p_Z(z\,;\,\theta)$. Iterative maximization is a strictly numerical analysis method that interprets the auxiliary function Q as a two parameter family of curves whose envelop is the data likelihood function.

EM works amazingly well in some applications and is seemingly useless in others. It is often a hand-in-glove fit for likelihood functions that involve sums (integrals), for in these cases the natural missing data are the summation indices (variables of integration) and they simplify the likelihood function to a product of the summands (integrands). Such products are often much simpler to manipulate. However, other missing data are possible, and different choices of missing data can lead to different EM convergence rates.

A.1 Formulation

The observed data z are called the *incomplete data*. As is seen shortly, the name makes sense in the context of the EM method. The measurements z are a realization of the random variable Z, and Z takes values in the space $\mathcal{Z}$. The pdf of the data z is specified parametrically by $p_Z(z\,;\,\theta)$, where $\theta \in \Theta$ is an unknown parameter vector. Here, Θ is the set of all valid parameter values.

The likelihood function of θ is the pdf $p_Z(z\,;\,\theta)$ thought of as a function of θ for a given z. It is assumed that the likelihood of the data for any $\theta \in \Theta$ is finite. It is also assumed that the likelihood function of θ is uniformly bounded above, that is, $p_Z(z\,;\,\theta) \leq B < \infty$ for all $\theta \in \Theta$. The latter assumption is important.

R.L. Streit, *Poisson Point Processes*, DOI 10.1007//978-1-4419-6923-1,

The maximum likelihood estimate of θ is

$$\hat{\theta}_{ML} = \arg\max_{\theta \in \Theta} \; p_Z(z\,;\,\theta)\,.$$

For pdfs differentiable with respect to θ, the natural way to compute $\hat{\theta}_{ML}$ is solve the so-called necessary conditions

$$\nabla_\theta \, p_Z(z\,;\,\theta) = 0$$

using, say, a safeguarded Newton-Raphson algorithm. This approach can encounter difficulties in some nonlinear problems, especially when the dimension of θ is large.

The EM method is an alternative way to compute $\hat{\theta}_{ML}$. Let K be a random variable whose realizations k occur in the set $\mathcal{K}$, that is, $k \in \mathcal{K}$. For ease of exposition, the discussion in this appendix assumes that K is a continuous variable. If K is discrete, the discussion below is modified merely by replacing integrals over $\mathcal{K}$ by sums over $\mathcal{K}$.

Now, K is called *missing data* in the sense of EM if

$$p_Z(z\,;\,\theta) = \int_{\mathcal{K}} p_{ZK}(z,\,k\,;\,\theta)\,\mathrm{d}k\,, \tag{A.1}$$

where $p_{ZK}(z,\,k\,;\,\theta)$ is the joint pdf of Z and K. The pair of variables $(Z,\,K)$ are called the *complete data*. In words, (A.1) says that the pdf of the data Z is the marginal of the complete data pdf over the missing data K. The user of the EM method is at liberty to choose K and the joint pdf $p_{ZK}(\cdot)$ as fancy or need dictates, as long as the (A.1) holds. This is crucial. In many applications, however, the choice of missing data is very natural.

The pdf of K conditioned on Z is, by the definition of conditioning,

$$p_{K|X}(k\,|\,x\,;\,\theta) = \frac{p_{ZK}(z,\,k\,;\,\theta)}{p_Z(z\,;\,\theta)} = \frac{p_{ZK}(z,\,k\,;\,\theta)}{\int_{\mathcal{K}} p_{ZK}(z,\,k\,;\,\theta)\,\mathrm{d}k}\,. \tag{A.2}$$

Let $n \geq 0$ denote the EM iteration index. Let $\theta^{(0)} \in \Theta$ be an initial (valid) value of the parameter.

A.1.1 E-step

For $n \geq 1$, the EM auxiliary function is defined by the conditional expectation:

$$\begin{aligned}
Q\left(\theta\,;\,\theta^{(n-1)}\right) &= E_{K|X;\theta^{(n-1)}}\left[\log p_{ZK}(z,\,k\,;\,\theta)\right] \\
&\equiv \int_{\mathcal{K}} (\log p_{ZK}(z,\,k\,;\,\theta))\, p_{K|Z}\left(k\,|\,z\,;\,\theta^{(n-1)}\right)\mathrm{d}k\,. \tag{A.3}
\end{aligned}$$

The sole purpose of the E-step is to evaluate $Q\left(\theta ; \theta^{(n-1)}\right)$. In practice, this is very often a purely symbolic step, that is, the expectation (A.3) is manipulated so that it takes a simple analytic form. When manipulation does not yield an analytic form, it is necessary to replace the expectation with a suitably chosen discrete sum (defined, e.g., via Monte Carlo integration).

A.1.2 M-step

The EM updated parameter is

$$\theta^{(n)} = \arg\max_{\theta \in \Theta} Q\left(\theta ; \theta^{(n-1)}\right) . \tag{A.4}$$

The maximum in (A.4) can be computed by any available method. It is sometimes thought that it is necessary to solve the maximization step explicitly, but this is emphatically not so—a strictly numerical method is quite acceptable.

It is unnecessary to solve for the maximum in (A.4). Instead it is only necessary to find a value $\theta^{(n)}$ such that

$$Q\left(\theta^{(n)} ; \theta^{(n-1)}\right) > Q\left(\theta^{(n-1)} ; \theta^{(n-1)}\right) . \tag{A.5}$$

If the update $\theta^{(n)}$ is chosen in this way, the method is called the generalized EM (GEM) method. The GEM method is very useful in many problems. It is the starting point of other closely related EM-based methods. A prominent example is the SAGE (Space Alternating Generalized EM) algorithm [35].

A.1.3 Convergence

Under mild conditions, convergence is *guaranteed* to a critical point of the likelihood function, that is, $\theta^{(n)} \to \theta^*$ such that

$$\left[\nabla_\theta \, p_Z(z \,|\, \theta)\right]_{\theta = \theta^*} = 0 .$$

Experience shows that in practice, θ^* is almost certainly a local maximum and not a saddle point, so that $\theta^* = \hat{\theta}_{ML}$.

One of the mild conditions that cannot be overlooked is the assumption that the likelihood function of θ is uniformly bounded above. If it is not, then the EM iteration will converge only if the initial value of θ is by chance in the domain of attraction of a point of local maximum likelihood. Otherwise, it will diverge, meaning that the likelihood function of the iterates will grow unbounded. Unboundedness would not be an issue were it not for the fact that the only valid values of θ have finite likelihoods, so the iterates—if they converge to a point—converge to an invalid parameter value, that is, to a point not in Θ. Estimation of heteroscedastic Gaussian sums is notorious for this behavior (due to covariance matrix collapse). See Section 3.4 for further discussion.

The details of the EM convergence proof are given in too many places to repeat all the details here. The book [80] gives a full and careful treatment of convergence, as well as examples where convergence fails in various ways.

For present purposes, it is enough to establish two facts. One is that if the update $\theta^{(n)}$ satisfies (A.5), then it also increases the likelihood function. The other is that a critical point of the auxiliary function is also a critical point of the likelihood function, and conversely.

To see that each EM step monotonically increases the likelihood function, recall that $\log x \leq x - 1$ for all $x > 0$ with equality if and only if $x = 1$. Then, using only the above definitions,

$$
\begin{aligned}
0 < \; & Q\left(\theta ; \theta^{(n-1)}\right) - Q\left(\theta^{(n-1)} ; \theta^{(n-1)}\right) \\
= \; & \int_K \left(\log p_{ZK}(z, k ; \theta)\right) p_{K|Z}\left(k \mid z ; \theta^{(n-1)}\right) dk \\
& - \int_K \left(\log p_{ZK}\left(z, k ; \theta^{(n-1)}\right)\right) p_{K|Z}\left(k \mid z ; \theta^{(n-1)}\right) dk \\
= \; & \int_K \left(\log \left\{\frac{p_{ZK}(z, k ; \theta)}{p_{ZK}\left(z, k ; \theta^{(n-1)}\right)}\right\}\right) p_{K|Z}\left(k \mid z ; \theta^{(n-1)}\right) dk \\
\leq \; & \int_K \left\{\frac{p_{ZK}(z, k ; \theta)}{p_{ZK}\left(z, k ; \theta^{(n-1)}\right)} - 1\right\} \frac{p_{ZK}\left(z, k ; \theta^{(n-1)}\right)}{p_Z(z ; \theta^{(n-1)})} dk \\
= \; & \frac{1}{p_Z\left(z ; \theta^{(n-1)}\right)} \int_K \left\{p_{ZK}(z, k ; \theta) - p_{ZK}\left(z, k ; \theta^{(n-1)}\right)\right\} dk \\
= \; & \frac{1}{p_Z\left(z ; \theta^{(n-1)}\right)} \left\{p_Z(z ; \theta) - p_Z\left(z ; \theta^{(n-1)}\right)\right\} .
\end{aligned}
$$

Clearly, *any* increase in Q will increase the likelihood function.

To see that critical points of the likelihood function are critical points of the auxiliary function, and conversely, simply note that

$$
\begin{aligned}
0 = \; & \nabla_{\theta_0} p_Z(z ; \theta_0) \\
= \; & \int_K \nabla_{\theta_0} p_{ZK}(z, k ; \theta_0) \, dk \\
= \; & \int_K p_{ZK}(z, k ; \theta_0) \, \nabla_{\theta_0} \log p_{ZK}(z, k ; \theta_0) \, dk \\
= \; & p_Z(z ; \theta_0) \int_K \frac{p_{ZK}(z, k ; \theta_0)}{p_Z(z ; \theta_0)} \nabla_{\theta_0} \log p_{ZK}(z, k ; \theta_0) \, dk
\end{aligned}
$$

$$= p_Z\,(z\,;\,\theta_0)\int_{\mathcal{K}} p_{K|Z}\,(k\mid z\,;\,\theta_0)\,\big[\nabla_\theta \log\, p_{ZK}\,(z,\,k\mid\theta)\big]_{\theta\,=\,\theta_0}\, dk$$

$$= p_Z\,(z\,;\,\theta_0)\,[\nabla_\theta\, Q\,(\theta\,;\,\theta_0)]_{\theta\,=\,\theta_0}\;.$$

For further discussion and references to the literature, see [80].

A.2 Iterative Majorization

The widely referenced paper [23] is the synthesis of several earlier discoveries of EM in a statistical setting. However, EM is not essentially statistical in character, but is rather only one member of a larger class of strictly numerical methods called iterative majorization [19–21]. This connection is not often mentioned in the literature, but the insight was noticed almost immediately after the publication of [23].

The observed data likelihood function $p_Z(z\,;\,\theta)$ is the envelop of a two parameter family of functions. This family is defined in the EM method by the auxiliary function, $Q(\theta\,;\,\phi)$. As seen from the results in [23], the data likelihood function majorizes every function in this family. For each specified parameter ϕ, the function $Q(\theta\,;\,\phi)$ is tangent to the $p_Z(z\,;\,\theta)$ at the point ϕ. This situation is depicted in Fig. A.1 for the sequence $\phi = \theta_0,\,\theta_1,\,\theta_2,\,\ldots$. It is now intuitively clear that EM based algorithms monotonically increase the data likelihood function, and that they converge with high probability to a local maximum of the likelihood function $p_Z(z\,;\,\theta)$ that depends on the starting point.

It is very often observed in practice that EM based algorithms make large strides toward the solution in the early iterations, but that progress toward the solution

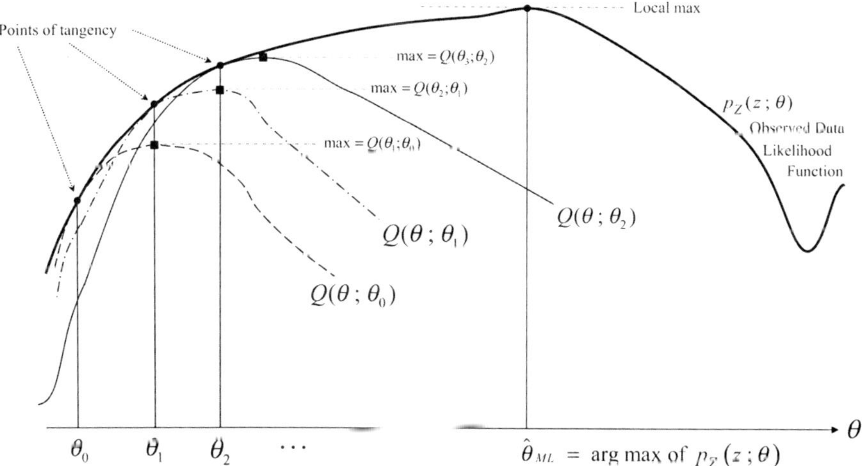

Fig. A.1 Iterative majorization interpretation of the observed (*Incomplete*) data likelihood function as the envelop of the EM auxiliary function, $Q(\theta,\,\phi)$

slows significantly as the iteration progresses. The iterative majorization interpretation of EM provides an intuitive insight into both phenomena. See [80] for further discussion and, in particular, for a proof that the rate of convergence is ultimately only linear. This convergence behavior explains the many efforts proposed in the literature to speed up EM convergence.

A.3 Observed Information

The observed information matrix (OIM) corresponding to the ML estimate $\hat{\theta}_{ML}$ is defined by

$$OIM(\hat{\theta}_{ML}) = -\left[\nabla_\theta \left(\nabla_\theta \, \log \, p_Z(z \, ; \, \theta)\right)^T\right]_{\theta = \hat{\theta}_{ML}}. \qquad (A.6)$$

The matrix $OIM(\theta)$ is an OIM only if θ is an ML estimate. Evaluating the OIM is often a relatively straightforward procedure. Moreover, EM based ML estimation algorithms are easily adapted to compute the OIM as a by-product of the EM iteration at very little additional computational expense. See [69] for details.

The OIM and FIM differ in important ways:

- The FIM computes the expectation of the negative Hessian of the loglikelihood function with respect to the data, while the OIM does not.
- The OIM is evaluated at the ML estimate $\hat{\theta}_{ML}$, while the FIM is evaluated at the true value of the parameter.

The expected value of the OIM is sometimes said to be the FIM, but this is not precisely correct.

The OIM and FIM are alike in that they both evaluate the negative Hessian of the loglikelihood function. Because of this and other similarities between them, the OIM is often used as a surrogate for the FIM when the FIM is unknown or otherwise unavailable. See [26] for further discussion of the OIM, its origins, and potential utility.

Appendix B
Solving Conditional Mean Equations

A conditional mean equation used in intensity estimation is discussed in this appendix. The equation is seen to be monotone, and monotonicity implies that the equation is uniquely solvable.

The conditional mean equation of interest is the vector equation (3.8), which is repeated here:

$$\frac{\int_{\mathcal{R}} s\, \mathscr{N}\,(s\,;\,\mu,\,\Sigma)\,\mathrm{d}s}{\int_{\mathcal{R}} \mathscr{N}\,(s\,;\,\mu,\,\Sigma)\,\mathrm{d}s} = \bar{x}\,, \tag{B.1}$$

where

$$\bar{x} \equiv \frac{1}{m} \sum_{j=1}^{m} x_j \in \mathbb{R}^{n_x}\,.$$

The j-th component of $\bar{x}$ is denoted by $\bar{x}_j$, which should not be confused with the data point $x_j \in \mathbb{R}^{n_x}$. The solution to (B.1) is unique and straightforward to compute numerically for rectangular multidimensional regions

$$\mathcal{R} = [a_1,\, b_1] \times \cdots \times [a_{n_x},\, b_{n_x}]$$

and diagonal covariance matrices $\Sigma = \mathrm{Diag}\left[\sigma_1^2, \ldots, \sigma_{n_x}^2\right]$.

The conditional mean of a univariate Gaussian distributed random variable conditioned on realizations in the interval $[-1,\, 1]$ is defined by

$$M\left[\mu,\, \sigma^2\right] \equiv \frac{\int_{-1}^{1} s\, \mathscr{N}(s\,;\,\mu,\,\sigma^2)\,\mathrm{d}s}{\int_{-1}^{1} \mathscr{N}(s\,;\,\mu,\,\sigma^2)\,\mathrm{d}s}\,, \tag{B.2}$$

where $\mu \in \mathbb{R}$ and $\sigma > 0$. Thus,

$$\lim_{\mu \to \infty} M\left[\mu,\, \sigma^2\right] = 1$$

and

$$\lim_{\mu \to -\infty} M\left[\mu, \sigma^2\right] = -1.$$

The most important fact about the function M is that it is strictly monotone increasing as a function of μ. Consequently, for any number c such that $-1 < c < 1$, and variance σ^2, the solution μ of the general equation

$$M\left[\mu, \sigma^2\right] = c \qquad\qquad (B.3)$$

exists and is unique. To see this, it is only necessary to verify that the derivative $M'\left[\mu, \sigma^2\right] \equiv \frac{\partial}{\partial \mu} M\left[\mu, \sigma^2\right] > 0$ for all μ. The inequality is intuitively obvious from its definiton as a conditional mean. The function $M\left[\mu, \sigma^2\right]$ is plotted for several values of σ in Fig. B.1. Evaluating the inverse function $M^{-1}\left[\mu, \sigma^2\right]$ efficiently is left as an exercise.

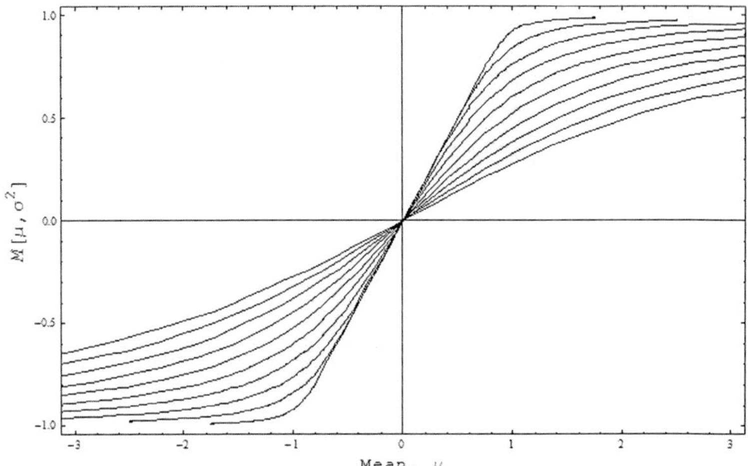

Fig. B.1 Plots of $M[\mu, \sigma^2]$ from $\sigma = 0.1$ to $\sigma = 1$ in steps of 0.1. The monotonicity of $M[\mu, \sigma^2]$ is self evident. The steepness of the transition from -1 to $+1$ increases steadily with decreasing σ

The conditional mean of the j-th component of (B.1) is

$$\frac{\int_{a_1}^{b_1} \cdots \int_{a_{n_x}}^{b_{n_x}} s_j \prod_{i=1}^{n_x} \mathcal{N}\left(s_i ; \mu_i, \sigma_i^2\right) ds_1 \cdots ds_{n_x}}{\int_{a_1}^{b_1} \cdots \int_{a_{n_x}}^{b_{n_x}} \prod_{i=1}^{n_x} \mathcal{N}\left(s_i ; \mu_i, \sigma_i^2\right) ds_1 \cdots ds_{n_x}} = \bar{x}_j . \qquad (B.4)$$

The integrals over all variables except x_j cancel, so (B.4) simplifies to

$$\frac{\int_{a_j}^{b_j} s_j \, \mathcal{N}\left(s_j ; \mu_j, \sigma_j^2\right) \, ds_j}{\int_{a_j}^{b_j} \mathcal{N}\left(s_j ; \mu_j, \sigma_j^2\right) \, ds_j} = \bar{x}_j. \tag{B.5}$$

Substituting

$$s_j = \left(\frac{b_j - a_j}{2}\right) x + \frac{a_j + b_j}{2},$$

into (B.5) and using the function (B.2) gives

$$M\left[\frac{\mu - \frac{a_j + b_j}{2}}{\frac{b_j - a_j}{2}}, \left(\frac{2\sigma}{b_j - a_j}\right)^2\right] + \frac{a_j + b_j}{2} = \bar{x}_j. \tag{B.6}$$

Solving

$$M\left[\tilde{\mu}_j, \left(\frac{2\sigma}{b_j - a_j}\right)^2\right] = \bar{x}_j - \frac{a_j + b_j}{2} \tag{B.7}$$

for $\tilde{\mu}_j$ gives the solution of (B.6) as

$$\mu_j = \left(\frac{b_j - a_j}{2}\right) \tilde{\mu}_j + \frac{a_j + b_j}{2}.$$

The expression (B.4) holds with obvious modifications for more general regions $\mathcal{R}$. However, the multidimensional integral over all the variables except x_j is a function of x_j in general and does not cancel from the ratio as it does in (B.5).

Appendix C
Bayesian Filtering

A brief review of general Bayesian filtering is given in this appendix. The discussion sets the conceptual and notational foundation for Bayesian filtering on PPP event spaces, all without mentioning PPPs until the very end. Gentler presentations that readers may find helpful are widely available (e.g., [4, 54, 104, 122]).

The notation used in this appendix is used in Section 6.1 and also in the alternative derivation of Appendix D.

C.1 General Recursion

Two sequences of random variables are involved: the sequence $\Xi_0, \Xi_1, \ldots, \Xi_k$ models target motion and $\Upsilon_1, \ldots, \Upsilon_k$ models data. These sequences correspond to measurement (or scan) times $t_0, t_1, \ldots, t_k$, where $t_{j-1} < t_j$ for $j = 1, \ldots, k$. The only conditional dependencies between these variables are the traditional ones: the sequence $\{\Xi_j\}$ is Markov, and the conditional variables $\{\Upsilon_j \mid \Xi_j\}$ are independent. The governing equation for Bayesian tracking is the following joint pdf of the track $(x_0, x_1, \ldots, x_k)$ and the measurements $(z_1, \ldots, z_k)$:

$$p_{\Xi_0, \Xi_1, \ldots, \Xi_k, \Upsilon_1, \ldots, \Upsilon_k}(x_0, x_1, \ldots, x_k, z_1, \ldots, z_k)$$

$$= p_{\Xi_0}(x_0) \prod_{j=1}^{k} p_{\Xi_j \mid \Xi_{j-1}}(x_j \mid x_{j-1}) \, p_{\Upsilon_j \mid \Xi_j}(z_j \mid x_j). \tag{C.1}$$

The product form in (C.1) is due to the conditioning.

The Bayesian filter is the posterior pdf of Ξ_k conditioned on all data up to and including time t_k. Conditioning and marginalizing gives

$$p_{\Xi_0, \ldots, \Xi_k \mid \Upsilon_1, \ldots, \Upsilon_k}(x_0, \ldots, x_k \mid z_1, \ldots, z_k)$$

$$= \frac{p_{\Xi_0, \ldots, \Xi_k, \Upsilon_1, \ldots, \Upsilon_k}(x_0, \ldots, x_k, z_1, \ldots, z_k)}{p_{\Upsilon_1, \ldots, \Upsilon_k}(z_1, \ldots, z_k)}$$

$$= \frac{p_{\Xi_0}(x_0) \prod_{j=1}^{k} p_{\Xi_j \mid \Xi_{j-1}}(x_j \mid x_{j-1}) \, p_{\Upsilon_j \mid \Xi_j}(z_j \mid x_j)}{\int_{\mathcal{S}} \cdots \int_{\mathcal{S}} p_{\Xi_0, \ldots, \Xi_k, \Upsilon_1, \ldots, \Upsilon_k}(x_0, \ldots, x_k, z_1, \ldots, z_k) \, \mathrm{d}x_0 \cdots \mathrm{d}x_k}.$$

The posterior pdf is the integral over all states except x_k:

$$p_{\Xi_k | \Upsilon_1, \ldots, \Upsilon_k}(x_k \mid z_1, \ldots, z_k)$$

$$= \int_{\mathcal{S}} \cdots \int_{\mathcal{S}} p_{\Xi_0, \ldots, \Xi_k | \Upsilon_1, \ldots, \Upsilon_k}(x_0, \ldots, x_k \mid z_1, \ldots, z_k) \, \mathrm{d}x_0 \cdots \mathrm{d}x_{k-1}.$$

The multidimensional integral is evaluated recursively.

Let $\Xi_{k-1|k-1}$, $\Xi_{k|k-1}$, and $\Upsilon_{k|k-1}$ denote the random variables Ξ_{k-1}, Ξ_k, and Υ_k conditioned on $\Upsilon_1, \ldots, \Upsilon_{k-1}$, and let $p_{k-1|k-1}(x_{k-1})$, $p_{k|k-1}(x_k)$, and $\pi_{k|k-1}(z_k)$ be their pdfs. Initialize the recursion by setting $p_{0|0}(x_0) = p_{\Xi_0}(x_0)$. The pdf of $\Xi_{k|k}$ is, by Bayes' Theorem,

$$p_{k|k}(x_k) \equiv p_{\Xi_k | \Upsilon_1, \ldots, \Upsilon_k}(x_k \mid z_1, \ldots, z_k)$$

$$= p_{\Upsilon_k | \Xi_k}(z_k \mid x_k) \frac{p_{\Xi_k | \Upsilon_1, \ldots, \Upsilon_{k-1}}(x_k \mid z_1, \ldots, z_{k-1})}{p_{\Upsilon_k | \Upsilon_1, \ldots, \Upsilon_{k-1}}(z_k \mid z_1, \ldots, z_{k-1})}. \tag{C.2}$$

The numerator of (C.2) is the predicted target pdf:

$$p_{k|k-1}(x_k) \equiv p_{\Xi_k | \Upsilon_1, \ldots, \Upsilon_{k-1}}(x_k \mid z_1, \ldots, z_{k-1})$$

$$= \int_{\mathcal{S}} p_{\Xi_k | \Xi_{k-1}}(x_k \mid x_{k-1}) \, p_{\Xi_{k-1} | \Upsilon_1, \ldots, \Upsilon_{k-1}}(x_{k-1} \mid z_1, \ldots, z_{k-1}) \, \mathrm{d}x_{k-1}$$

$$\equiv \int_{\mathcal{S}} p_{\Xi_k | \Xi_{k-1}}(x_k \mid x_{k-1}) \, p_{k-1|k-1}(x_{k-1}) \, \mathrm{d}x_{k-1}. \tag{C.3}$$

The denominator is the pdf of the measurement z_k given that it is generated by the target with pdf $p_{k|k-1}(x_k)$:

$$\pi_{k|k-1}(z_k) \equiv p_{\Upsilon_k | \Upsilon_1, \ldots, \Upsilon_{k-1}}(z_k \mid z_1, \ldots, z_{k-1})$$

$$= \int_{\mathcal{S}} p_{\Upsilon_k | \Xi_k}(z_k \mid x_k) \, p_{\Xi_k | \Upsilon_1, \ldots, \Upsilon_{k-1}}(x_k \mid z_1, \ldots, z_{k-1}) \, \mathrm{d}x_k$$

$$= \int_{\mathcal{S}} p_{\Upsilon_k | \Xi_k}(z_k \mid x_k) \, p_{k|k-1}(x_k) \, \mathrm{d}x_k. \tag{C.4}$$

Substituting (C.3) and (C.4) into (C.2) gives

$$p_{k|k}(x_k) = p_{\Upsilon_k | \Xi_k}(z_k \mid x_k) \frac{p_{k|k-1}(x_k)}{\pi_{k|k-1}(z_k)}. \tag{C.5}$$

The forward recursion is defined by (C.3)–(C.5).

The pdf $\pi_{k|k-1}(z_k)$ is called the partition function in physics and in the machine learning community. In many tracking applications, it is often simply dismissed as a scale factor; however, its form is important in PPP applications to tracking. Since its form is known before the measurement z_k is available, it is called the predicted measurement pdf here.

C.2 Special Case: Kalman Filtering

The classic example of Bayesian filters is the Kalman filter, that is, the single target Bayes-Markov filter with additive Gaussian noise and *no* clutter. In this case Ξ_j represents the state of a single target, the state space is $\mathcal{S} \equiv \mathbb{R}^{n_x}$, $n_x \geq 1$, and Υ_j represents a data point in the event space $\mathcal{T} \equiv \mathbb{R}^{n_z}$, $n_z \geq 1$. The model is usually written in an algebraic form as

$$x_j = F_{j-1}(x_{j-1}) + v_{j-1} \tag{C.6}$$

and

$$z_j = H_j(x_j) + w_j, \tag{C.7}$$

for $j = 1, \ldots, k$, where x_j is a realization of the Markovian state variable Ξ_j, and $\{z_j\}$ is a realization of the conditional variables $\Upsilon_j \mid \Xi_j = x_j$. The process noises v_{j-1} and measurement noises w_j in (C.6) are zero mean, Gaussian and independent with covariance matrices Q_{j-1} and R_j, respectively. The functions $F_{j-1}(\cdot)$ and $H_j(\cdot)$ are known. The pdf of Ξ_0 is $\mathcal{N}(x_0; \bar{x}_0, P_0)$, where $\bar{x}_0$ and P_0 are given. The equivalent system of pdfs is

$$p_{\Xi_0}(x_0) = \mathcal{N}(x_0; \bar{x}_0, P_0) \tag{C.8a}$$
$$p_{\Xi_j \mid \Xi_{j-1}}(x_j \mid x_{j-1}) = \mathcal{N}(x_j; F_{j-1}(x_{J-1}), Q_{j-1}) \tag{C.8b}$$
$$p_{\Upsilon_j \mid \Xi_j}(z_j \mid x_j) = \mathcal{N}(z_j; H_j(x_j), R_j). \tag{C.8c}$$

The joint pdf is found by substituting (C.8a)–(C.8c) into (C.1). The recursion (C.3)–(C.5) gives the posterior pdf on $\mathcal{S} = \mathbb{R}^{n_x}$.

The linear Gaussian Kalman filter assumes that

$$F_{j-1}(x_{j-1}) = F_{j-1}\, x_{j-1} \tag{C.9a}$$
$$H_j(x_j) = H_j\, x_j, \tag{C.9b}$$

where F_{j-1} is an $n_x \times n_x$ matrix called the system matrix, and H_j is $n_z \times n_x$ and is called the measurement matrix. The Kalman filter is, in this instance, a recursion that evaluates the parameters of the posterior pdf

$$p_{k \mid k}(x_k) = \mathcal{N}\left(x_k \mid \hat{x}_{k \mid k}, P_{k \mid k}\right), \tag{C.10}$$

where $\hat{x}_{j \mid k}$ is the point estimate of the target at time t_k and $P_{k \mid k}$ is the associated error covariance matrix. Explicitly, for $j = 0, \ldots, k - 1$,

$$P_{j+1|j} = F_j \, P_{j|j} \, F_j^T + Q_j \tag{C.11a}$$

$$W_{j+1} = P_{j+1|j} \, H_{j+1}^T \left\{ H_{j+1} \, P_{j+1|j} \, H_j^T + R_{j+1} \right\}^{-1} \tag{C.11b}$$

$$P_{j+1|j+1} = P_{j+1|j} - W_{j+1} \, H_{j+1} \, P_{j+1|j} \tag{C.11c}$$

$$\hat{x}_{j+1|j+1} = F_j \, \hat{x}_{j|j} + W_{j+1} \left\{ z_{j+1} - H_{j+1} \, F_j \, \hat{x}_{j|j} \right\} . \tag{C.11d}$$

These equations are not necessarily good for numerical purposes, especially when observability is an issue. In practice, the information form of the Kalman filter is always preferable.

The Kalman filter is often written in terms of measurement innovations when the measurement and target motion models are linear. The predicted target state at time t_j is

$$\hat{x}_{j|j-1} = F_{j-1} \, \hat{x}_{j-1|j-1}, \quad j = 1, \ldots, k, \tag{C.12}$$

and the predicted measurement is

$$\hat{z}_{j|j-1} = H_j \, \hat{x}_{j|j-1}, \quad j = 1, \ldots, k. \tag{C.13}$$

The innovation at time t_j is the difference between the actual measurement and the predicted measurement:

$$v_{j|j-1} = z_j - \hat{z}_{j|j-1}, \quad j = 1, \ldots, k. \tag{C.14}$$

The reason for the name *innovation* is now self-evident. The information updated target state is

$$\hat{x}_{j|j} = \hat{x}_{j|j-1} + W_j \, v_{j|j-1}, \quad j = 1, \ldots, k. \tag{C.15}$$

The update (C.15) is perhaps more intuitive, but it is the same as before.

For completeness, the smoothing (or, lagged) Kalman filter is given here. The posterior pdf is denoted by

$$p_{j|k}(x_j) = \mathcal{N}\left(x_j \,|\, \hat{x}_{j|k}, \, \Sigma_{j|k} \right), \tag{C.16}$$

where $\hat{x}_{j|k}$ is the point estimate of the target at time t_j given all the data $\{z_1, \ldots, z_k\}$ up to and including time t_k, and $\Sigma_{j|k}$ is the associated error covariance matrix. These quantities are computed by the backward recursion: For $j = k-1, \ldots, 0$, the point estimates are

$$\hat{x}_{j|k} = \hat{x}_{j|j} + P_{j|j} \, F_j^T \, P_{j+1|j}^{-1} \left\{ \hat{x}_{j+1} - F_j \, \hat{x}_{j|j} \right\} . \tag{C.17}$$

The innovation form of the filter, if it is desirable to think of the smoothing filter in such terms, is written in terms of the *state* innovations, $\hat{x}_{j+1} - F_j \, \hat{x}_{j|j}$. The corresponding error covariance matrices are

$$\Sigma_{j|k} = P_{j|j} + P_{j|j} \, F_j^T \, P_{j+1|j}^{-1} \left\{ \Sigma_{j+1} - P_{j+1|j} \right\} P_{j+1|j}^{-1} \, F_j \, P_{j|j} \, . \qquad \text{(C.18)}$$

The smoothing recursions (C.17)–(C.18) were first derived in 1965 by Rauch, Tung, and Striebel [98].

C.2.1 Multitarget Tracking

A more exciting example of Bayesian involves multitarget tracking. There are two approaches to modeling the multitarget state. One involves stacking, or concatenating, the states of several targets and proceeding with the Bayesian filter using the stacked state. There are natural symmetries in the joint pdf of this kind of multitarget state that need to be incorporated into the Bayesian filter. Very general measurement likelihood functions can also be used. However, the drawback to this approach is that the target state becomes unmanageably large very quickly. This problem is not easily overcome.

An alternative approach is to model the multitarget state as a PPP random variable Ξ_j. The state space is the PPP event space $\mathcal{S} \equiv \mathcal{E}(\mathbb{R}^{n_x})$, and the random variables Υ_j represent data sets. The conditioning is the same as required for Bayesian filtering, so the posterior pdf is defined on the state space $\mathcal{E}(\mathbb{R}^{n_x})$ and can—in principle only—be evaluated using (C.3)–(C.5). The Bayesian posterior is a finite point process, but it is not a PPP, so it is approximated by one. This approach to multitarget tracking is the subject of Chapter 6. As discussed there, the intensity function of the approximating PPP may be more physically meaningful than the points of the PPP realizations themselves.

Appendix D
Bayesian Derivation of Intensity Filters

The multitarget intensity filter is derived by Bayesian methods in this appendix. The posterior point process is developed first, and then the posterior point process is approximated by a PPP. Finally, the last section discusses the relationship between this method and the "first moment" approximation of the posterior point process.

The steps of the intensity filter are outlined in Fig. 6.1. The PPP interpretations of these steps are thinning, approximating the Bayes update with a PPP, and superposition. The PPP at time t_k is first thinned by detection. The two branches of the thinning are the detected and undetected target PPPs. Both branches are important. Their information updates are different. The undetected target PPP is the lesser branch. Its information update is a PPP. The detected target branch is the main branch, and its information update comprises two key steps. Firstly, the Bayes update of the posterior point process of Ξ_k on $\mathcal{E}(\mathcal{S}^+)$ given data up to and including time t_k is obtained. The posterior is not a PPP, as is seen below from the form of its pdf in (D.10). Secondly, the posterior point process is approximated by a PPP, and a low computational complexity expression for the intensity of the approximating PPP is obtained. The two branches of detection thinning are recombined by superposition to obtain the intensity filter update.

D.1 Posterior Point Process

The random variables $\Xi_{k-1|k-1}$, $\Xi_{k|k-1}$, and $\Upsilon_{k|k-1}$ are defined as in Appendix C. The state space of $\Xi_{k-1|k-1}$ and $\Xi_{k|k-1}$ is $\mathcal{E}(\mathcal{S}^+)$, where $\mathcal{E}(\mathcal{S}^+)$ is a union of sets defined as in (2.1). Similarly, the event space of $\Upsilon_{k|k-1}$ is $\mathcal{E}(\mathcal{T})$, not $\mathcal{T}$.

The process $\Xi_{k-1|k-1}$ is assumed to be a PPP, so it is parameterized by its intensity $f_{k-1|k-1}(s)$, $s \in \mathcal{S}^+$. A realization $\xi_k \in \mathcal{E}(\mathcal{S}^+)$ of $\Xi_{k-1|k-1}$ is transitioned to time t_k via the single target transition function $\Psi_{k-1}(y \mid x)$. Its intensity is, using (2.83),

$$f_{k|k-1}(x) = \int_{\mathcal{S}^+} \Psi_{k-1}(x \mid s) \, f_{k-1|k-1}(s) \, \mathrm{d}s. \tag{D.1}$$

The integral in (D.1) is defined as in (2.97).

The point process $\Xi_{k|k}$ is the sum of detected and undetected target processes, denoted by $\Xi_{k|k}^D$ and $\Xi_{k|k}^U$, respectively. They are obtained from the same realizations of $\Xi_{k|k-1}$, so they would seem to be highly correlated. However, the number of points in the realization is Poisson distributed, so they are actually independent. See Section 2.9.

The undetected target process $\Xi_{k|k}^U$ is the predicted target PPP $\Xi_{k|k-1}$ thinned by $1 - P_k^D(s)$, where $P_k^D(s)$ is the probability of detecting a target at s. Thus $\Xi_{k|k}^U$ is a PPP, and

$$f_{k|k}^U(x) = \left(1 - P_k^D(x)\right) f_{k|k-1}(x) \tag{D.2}$$

is its intensity.

The detected target process $\Xi_{k|k}^D$ is the predicted target PPP $\Xi_{k|k-1}$ that is thinned by $P_k^D(s)$ and subsequently updated by Bayesian filtering. Thinning yields the predicted PPP $\Xi_{k|k-1}^D$, and

$$f_{k|k-1}^D(x) = P_k^D(x) f_{k|k-1}(x) \tag{D.3}$$

is its intensity.

The predicted measurement process $\Upsilon_{k|k-1}$ is obtained from $\Xi_{k|k-1}^D$ via the pdf of a single point measurement $z \in \mathcal{T}$ conditioned on a target located at $s \in \mathcal{S}^+$. The quantity $p_k(z \mid \phi)$ is the likelihood of z if it is a false alarm. See Section 2.12. Thus, $\Upsilon_{k|k-1}$ is a PPP on $\mathcal{T}$ and

$$\lambda_{k|k-1}(z) = \int_{\mathcal{S}^+} p_k(z \mid s) P_k^D(s) f_{k|k-1}(s) \, ds, \tag{D.4}$$

is its intensity.

The measurement set is $\upsilon_k = \{m, \{z_1, \ldots, z_m\}$, where $z_j \in \mathcal{T}$. The conditional pdf of υ_k is defined for arbitrary target realizations $\xi_k = (n, \{x_1, \ldots, x_n\}) \in \mathcal{E}(\mathcal{S}^+)$. All the points x_j of ξ_k, whether they are a true target ($x_j \in \mathbb{R}^{n_x}$) or are clutter ($x_j = \phi$), generate a measurement so that only when $m = n$ is the measurement likelihood non-zero. The correct assignment of point measurements to targets in ξ_k is unknown. All such assignments are equally probable, so the pdf averages over all possible assignments of data to false alarms and targets. Because ϕ is a target state, the measurement pdf is

$$p_{\Upsilon_k|\Xi_k}(\upsilon_k \mid \xi_k) = \begin{cases} \frac{1}{m!} \sum_{\sigma \in \text{Sym}(m)} \prod_{j=1}^m p_k(z_{\sigma(j)} \mid x_j), & m = n \\ 0, & m \neq n, \end{cases} \tag{D.5}$$

where $\text{Sym}(m)$ is the set of all permutations on the integers $\{1, 2, \ldots, m\}$.

The lower branch of (D.5) is a consequence of the "at most one measurement per target" rule together with the augmented target state space $\mathcal{S}^+$. To elaborate, the points in a realization ξ of the *detected* target PPP are targets, some of which have state ϕ. The augmented state space accommodates clutter measurements by using targets in ϕ, so only realizations with $m = n$ points have nonzero probability.

The posterior pdf of $\Xi_{k|k}^D$ on $\mathcal{E}(\mathcal{S}^+)$ is, from (C.5),

$$p_{k|k}(\xi_k) = p_{\Upsilon_k | \Xi_k}(\upsilon_k \mid \xi_k) \frac{p_{k|k-1}(\xi_k)}{\pi_{k|k-1}(\upsilon_k)}. \tag{D.6}$$

The pdf's $p_{k|k-1}(\xi_k)$ and $\pi_{k|k-1}(\upsilon_k)$ of $\Xi_{k|k-1}^D$ and $\Upsilon_{k|k-1}$ are given in terms of their intensity functions using (2.12):

$$p_{k|k-1}(\xi_k) = \frac{1}{m!} \exp\left(-\int_{\mathcal{S}^+} f_{k|k-1}^D(s)\, ds\right) \prod_{j=1}^{m} f_{k|k-1}^D(x_j) \tag{D.7}$$

$$\pi_{k|k-1}(\upsilon_k) = \frac{1}{m!} \exp\left(-\int_{\mathcal{T}} \lambda_{k|k-1}(z)\, dz\right) \prod_{j=1}^{m} \lambda_{k|k-1}(z_j). \tag{D.8}$$

From (D.3) and (D.4),

$$\int_{\mathcal{S}^+} f_{k|k-1}^D(s)\, ds = \int_{\mathcal{T}} \lambda_{k|k-1}(z)\, dz. \tag{D.9}$$

Substituting (D.7), (D.8), and (D.5) into (D.6) and using obvious properties of permutations gives the posterior pdf of $\Xi_{k|k}^D$:

$$p_{k|k}(\xi_k) = \frac{1}{m!} \sum_{\sigma \in \mathrm{Sym}\,(m)} \prod_{j=1}^{m} \frac{p_k(z_{\sigma(j)} \mid x_j)\, P_k^D(x_j)\, f_{k|k-1}(x_j)}{\lambda_{k|k-1}(z_{\sigma(j)})}. \tag{D.10}$$

If ξ_k does not contain exactly m points, then $p_{k|k}(\xi_k) = 0$. Conditioning $\Xi_{k|k}^D$ on m points gives the pdf of the points of the posterior process as

$$p_{k|k}(x_1, \ldots, x_m) = \frac{1}{m!} \sum_{\sigma \in \mathrm{Sym}\,(m)} \prod_{j=1}^{m} \frac{p_k(z_{\sigma(j)} \mid x_j)\, P_k^D(x_j)\, f_{k|k-1}(x_j)}{\lambda_{k|k-1}(z_{\sigma(j)})}. \tag{D.11}$$

The pdf (D.11) holds for $x_j \in \mathcal{S}^+,\ j = 1, \ldots, m$.

D.2 PPP Approximation

The pdf of the posterior point process $\Xi_{k|k}^D$ is clearly not that of a PPP. This causes a problem for the recursion. One way around it is to approximate $\Xi_{k|k}^D$ by a PPP and recursively update the intensity of the PPP approximation.

The pdf $p_{k|k}(x_1, \ldots, x_m) = p_{k|k}(x_{\sigma(1)}, \ldots, x_{\sigma(m)})$ for all $\sigma \in \mathrm{Sym}(m)$; therefore, integrating it over all of arguments except, say, the ℓth argument gives the same result regardless of the choice of ℓ. The form of the "single target marginal" is, using (D.4),

$$
\begin{aligned}
p_{k|k}(x_\ell) &\equiv \int_{\mathcal{S}^+} \cdots \int_{\mathcal{S}^+} p_{k|k}(x_1, \ldots, x_m) \prod_{\substack{i=1 \\ i \neq \ell}}^{m} dx_i \\
&= \frac{1}{m!} \sum_{\sigma \in \mathrm{Sym}\,(m)} \int_{(\mathcal{S}^+)^{m-1}} \prod_{j=1}^{m} \frac{p_k(z_{\sigma(j)}|x_j)\, P_k^D(x_j)\, f_{k|k-1}(x_j)}{\lambda_{k|k-1}(z_{\sigma(j)})} \prod_{\substack{i=1 \\ i \neq \ell}}^{m} dx_i \\
&= \frac{1}{m!} \sum_{r=1}^{m} \sum_{\substack{\sigma \in \mathrm{Sym}\,(m) \\ \text{and } \sigma(\ell)=r}} \frac{p_k(z_{\sigma(\ell)} \mid x_\ell)\, P_k^D(x_\ell)\, f_{k|k-1}(x_\ell)}{\lambda_{k|k-1}(z_{\sigma(j)})} \\
&= \frac{1}{m} \sum_{r=1}^{m} \frac{p_k(z_r \mid x_\ell)\, P_k^D(x_\ell)\, f_{k|k-1}(x_\ell)}{\lambda_{k|k-1}(z_r)} .
\end{aligned} \qquad (D.12)
$$

This identity holds for arbitrary $x_\ell \in \mathcal{S}^+$.

The joint conditional pdf is approximated by the product of its marginal pdf's:

$$
p_{k|k}(x_1, \ldots, x_m) \approx \prod_{j=1}^{m} p_{k|k}(x_j). \qquad (D.13)
$$

The product approximation is called a mean field approximation in the machine learning community [55, pp. 35–36]. Both sides of (D.13) integrate to one.

The marginal pdf is proportional to the intensity of the approximating PPP. Let $f_{k|k}^D(x) = c\, p_{k|k}(x)$ be the intensity. The likelihood function of the unknown constant c is

$$
\mathscr{L}(c \mid \xi_k) = \frac{1}{m!} e^{-\int_{\mathcal{S}^+} c\, p_{k|k}(s)\, ds} \prod_{j=1}^{m} \left(c\, p_{k|k}(x_j) \right) \propto e^{-c}\, c^m .
$$

The maximum likelihood estimate is $\hat{c}_{ML} = m$, so that

$$
f_{k|k}^D(x) = \sum_{r=1}^{m} \frac{p_k(z_r \mid x)\, P_k^D(x)\, f_{k|k-1}(x)}{\lambda_{k|k-1}(z_r)} \qquad (D.14)
$$

is the intensity of the approximating PPP.

D.2.1 Altogether Now

The PPP approximation to the point process $\Xi_{k|k}$ is the sum of the undetected target PPP $\Xi_{k|k}^U$ and the PPP that approximates the detected target process $\Xi_{k|k}^D$. Hence,

$$
\begin{aligned}
f_{k|k}(x) &= f_{k|k}^U(x) + f_{k|k}^D(x), \qquad x \in \mathcal{S}^+ \\
&= \left[1 - P_k^D(x) + \sum_{r=1}^{m} \frac{p_k(z_r \mid x)\, P_k^D(x)}{\lambda_{k|k-1}(z_r)} \right] f_{k|k-1}(x)
\end{aligned}
\tag{D.15}
$$

is the updated intensity of the PPP approximation to $\Xi_{k|k}$.

The intensity filter comprises (D.1), (D.4), and (D.15). The first two equations are more insightful when written in traditional notation. From (D.1),

$$
f_{k|k-1}(x) = \hat{b}_k(x) + \int_{\mathcal{S}} \Psi_{k-1}(x \mid s)\,(1 - d_{k-1}(s))\, f_{k-1|k-1}(s)\, ds, \tag{D.16}
$$

where the predicted target birth intensity is

$$
\hat{b}_k(x) = \Psi_{k-1}(x \mid \phi)\,(1 - d_{k-1}(\phi))\, f_{k-1|k-1}(\phi). \tag{D.17}
$$

Also, from (D.4),

$$
\lambda_{k|k-1}(z) = \hat{\lambda}_k(z) + \int_{\mathcal{S}} p_k(z \mid s)\, P_k^D(s)\, f_{k|k-1}(s)\, ds, \tag{D.18}
$$

where

$$
\hat{\lambda}_k(z) = p_k(z \mid \phi)\, P_k^D(\phi)\, f_{k|k-1}(\phi) \tag{D.19}
$$

is the predicted measurement clutter intensity.

The above derivation of the intensity and PHD filters was first given in [130]. A more intuitive "physical space" approach is given by [30]. An analogous derivation for multisensor multitarget intensity filter is given in [124].

D.3 First Moment Intensity and Janossy Densities

An alternative method due to Mahler [74, 76] is often used in the literature to obtain the intensity function (D.14) for the detected target posterior point process. The process $\Xi_{k|k}^D$ is not a PPP, but it is a special case of the class of finite point processes—its realizations contain exactly m points. The theory of general finite point processes dates to the 1950s. (An excellent reference is [17, Chapter 5].) This theory is now applied to $\Xi_{k|k}^D$.

Let $\xi = (N, \mathcal{X}|N)$ denote a realization of $\Xi_{k|k}^{D}$, where N is the number of points and $\mathcal{X}|N$ is the point set. From [17, Section 5.3], the Janossy probability density of a finite point process is defined by

$$j_n(x_1, \ldots, x_n) = p_N(n)\, p_{\mathcal{X}|N}(\{x_1, \ldots, x_n\}\,|\,n) \quad \text{for } n = 0, 1, 2, \ldots. \tag{D.20}$$

Janossy densities were encountered (but left unnamed) early in Chapter 2, (2.10). Using the ordered argument list as in (2.13) gives

$$j_n(x_1, \ldots, x_n) = n!\, p_N(n)\, p_{\mathcal{X}|N}(x_1, \ldots, x_n\,|\,n) \quad \text{for } n = 0, 1, 2, \ldots. \tag{D.21}$$

Intuitively, from [17, p. 125],

$$j_n(x_1, \ldots, x_n) = \Pr\left[\begin{array}{c} \text{Exactly } n \text{ points in a realization} \\ \text{with one point in each infinitesimal} \\ [x_i + dx_i),\ \ i = 1, \ldots, n \end{array}\right]. \tag{D.22}$$

Now, for the finite point process $\Xi_{k|k}^{D}$, $p_N(m) = 1$ and $p_N(n) = 0$ if $n \neq m$, so only one of the Janossy functions is nonzero. The Janossy densities are

$$j_n(x_1, \ldots, x_n) = \begin{cases} m!\, p_{k|k}(x_1, \ldots, x_m), & \text{if } n = m, \\ 0, & \text{if } n \neq m, \end{cases} \tag{D.23}$$

where $p_{k|k}(x_1, \ldots, x_m)$ is the posterior pdf given by (D.11). The first moment intensity is denoted in [17] by $m_1(x)$. From [17, Lemma 5.4.III], it is given in terms of the Janossy density functions by

$$m_1(x) = \sum_{n=0}^{\infty} \frac{1}{n!} \int_{\mathcal{S}^+} \cdots \int_{\mathcal{S}^+} j_{n+1}(x, x_1, \ldots, x_n)\, dx_1 \cdots dx_n. \tag{D.24}$$

From (D.23), only the term $n = m - 1$ is nonzero, so that

$$m_1(x) = \frac{1}{(m-1)!} \int_{\mathcal{S}^+} \cdots \int_{\mathcal{S}^+} m!\, p_{k|k}(x, x_1, \ldots, x_{m-1})\, dx_1 \cdots dx_{m-1}. \tag{D.25}$$

The integral (D.25) is exactly m times the integral in (D.12), so the first moment approximation to $\Xi_{k|k}^{D}$ is identical to the intensity (D.14).

Appendix E
MMIF: Marked Multitarget Intensity Filter

This appendix derives the marked multitarget intensity filter (MMIF) recursion for linear Gaussian target and measurement models via the EM method. Targets are modeled as PPPs that are "marked" with measurements.

As seen from Section 8.1, measurement-marked target PPPs are equivalent to ordinary PPPs on the Cartesian product of the measurement and target spaces. These joint PPPs are superposed, and the target states estimated via the EM method. As mentioned in Section 6.2.2, the MMIF satisfies the "at most one measurement per target rule" in the mean.

E.1 Target Modeling

The multiple target state of L targets at time t_k is the vector

$$x_k \equiv (x_k(1), \ldots, x_k(L)) , \tag{E.1}$$

where $x_k(\ell) \in \mathbb{R}^{n_x}$ for $1 \leq \ell \leq L$. All targets move according to a linear Gauss-Markov model. Target states are estimated at the discrete times $t_0 < t_1 < t_2 < \ldots$, where t_0 corresponds to a starting time at which the a priori target pdfs are specified. The pdf of a target in state $x_{k-1}(\ell)$ at time t_{k-1} transitioning to state $x_k(\ell)$ at time t_k is

$$\Psi_{k-1}(x_k(\ell) \mid x_{k-1}(\ell)) = \mathcal{N}(x_k(\ell); F_{k-1}(\ell) x_{k-1}(\ell), Q_{k-1}(\ell)) , \tag{E.2}$$

where the system matrix $F_{k-1}(\ell) \in \mathbb{R}^{n_x \times n_x}$ and the process noise covariance matrix $Q_{k-1}(\ell) \in \mathbb{R}^{n_x \times n_x}$ are specified. The target motion model is equivalent to $x_k(\ell) = F_{k-1}(\ell) x_{k-1}(\ell) + u_{k-1}(\ell)$, where the process noise $u_{k-1}(\ell) \in \mathbb{R}^{n_x}$ is zero mean Gaussian distributed with covariance matrix $Q_{k-1}(\ell)$. The process noises are assumed independent from target to target.

Each target is modeled as a PPP. It is assumed, recursively, that the intensity function of target ℓ at time t_{k-1} is

$$f^{\ell}_{k-1|k-1}(x) \;=\; \widehat{I}_{k-1|k-1}(\ell)\,\mathcal{N}\left(x\,;\,\widehat{x}_{k-1|k-1}(\ell),\,P_{k-1|k-1}(\ell)\right), \qquad (E.3)$$

where the MAP estimate $\hat{x}_{k-1|k-1}(\ell)$, its covariance matrix $P_{k-1|k-1}(\ell)$, and intensity $\widehat{I}_{k-1|k-1}(\ell)$ are known. Under the target motion model (E.2), the predicted detected target intensity function at time t_k is

$$f^{\ell}_{k|k-1}(x) \;=\; P^{D}_{k}(\ell)\,I_k(\ell)\,\mathcal{N}\left(x\,;\,\widehat{x}_{k|k-1}(\ell),\,P_{k|k-1}(\ell)\right), \qquad (E.4)$$

where $P^{D}_{k}(\ell)$ is the probability of detecting target ℓ at time t_k and is assumed independent of target state x. Also, the predicted state and covariance matrix of target ℓ are

$$\widehat{x}_{k|k-1}(\ell) \;=\; F_{k-1}(\ell)\,\widehat{x}_{k-1|k-1}(\ell) \qquad (E.5)$$

$$P_{k|k-1}(\ell) \;=\; F_{k-1}(\ell)\,P_{k-1|k-1}(\ell)\,F^{T}_{k-1}(\ell)\,+\,Q_{k-1}(\ell). \qquad (E.6)$$

The coefficient $I_k(\ell)$ is estimated from data at time t_k as part of the MMIF recursion.

E.2 Joint Measurement-Target Intensity Function

An arbitrary measurement $z \in \mathbb{R}^{n_z}$ at time t_k originates either from one of the L targets or from the background clutter. If it originates from the ℓ-th target with state x, then the pdf of z conditioned on x is

$$p_{Z|X_k(\ell)}(z\,|\,x) \;=\; \mathcal{N}\left(z\,;\,H_k(\ell)\,x,\,R_k(\ell)\right), \qquad (E.7)$$

where the measurement matrix $H_k(\ell) \in \mathbb{R}^{n_z \times n_x}$ and the measurement noise covariance matrix $R_k(\ell) \in \mathbb{R}^{n_z \times n_z}$ are both specified. The measurement model is equivalent to $z = H_k(\ell)\,x + v_k(\ell)$, where the measurement noise $v_k(\ell) \in \mathbb{R}^{n_z}$ is zero mean Gaussian distributed with covariance matrix $R_k(\ell)$. The measurement and target process noises are assumed independent.

Measurements are modeled as marks that are associated with targets that are realizations of a target PPP. Marked processes are described in a general setting in Chapter 8. As seen from the Marking Theorem of Section 8.1, a measurement-marked target PPP is equivalent to a PPP on the Cartesian product of the measurement and target spaces, that is, on $\mathbb{R}^{n_z} \times \mathbb{R}^{n_x}$. The measurement process is not assumed to be a PPP.

The intensity function of the joint measurement-target PPP of target ℓ in state $x \in \mathbb{R}^{n_x}$ at time t_k is, from the expression (8.2),

$$\lambda^{\ell}_{k|k}(z,\,x) \;=\; f^{\ell}_{k|k-1}(x)\,p_{Z|X_k(\ell)}(z\,|\,x). \qquad (E.8)$$

From the basic property of PPPs, the expected number of marked detected targets, that is, the number of targets with a measurement, is the multiple integral over $\mathbb{R}^{n_z} \times \mathbb{R}^{n_x}$:

$$\int_{\mathbb{R}^{n_z} \times \mathbb{R}^{n_x}} \lambda_{k|k}^{\ell}(z, x)\, dz\, dx \;=\; \int_{\mathbb{R}^{n_x}} f_{k|k-1}^{\ell}(x)\, dx \;=\; P_k^D(\ell)\, I_k(\ell)\,. \tag{E.9}$$

This statement is equivalent to the "at most one measurement per target rule," but only in the mean.

Substituting (E.4) and (E.7) gives the joint measurement-target intensity function

$$\begin{aligned}
\lambda_{k|k}^{\ell}(z, x) &= P_k^D(\ell)\, I_k(\ell)\, \mathcal{N}\left(x\,;\, \widehat{x}_{k|k-1}(\ell),\, P_{k|k-1}(\ell)\right) \mathcal{N}\left(z\,;\, H_k(\ell)\, x_k(\ell),\, R_k(\ell)\right) \\
&= P_k^D(\ell)\, I_k(\ell)\, \mathcal{N}\left(x\,;\, \widehat{x}_{k|k}(z\,;\, \ell),\, P_{k|k}(\ell)\right) \mathcal{N}\left(z\,;\, \widehat{z}_{k|k-1}(\ell),\, S_{k|k}(\ell)\right),
\end{aligned} \tag{E.10}$$

where, using $\widehat{x}_{k|k-1}(\ell)$ and $P_{k|k-1}(\ell)$ above, the usual Kalman filter equations give

$$\begin{aligned}
S_{k|k}(\ell) &= R_k(\ell) + H_k(\ell)\, P_{k|k-1}(\ell)\, H_{k-1}^T \\
\widehat{z}_{k|k-1}(\ell) &= H_k(\ell)\widehat{x}_{k|k-1}(\ell) \\
W_k(\ell) &= P_{k|k-1}(\ell)\, H_k^T(\ell)\left\{H_k(\ell)\, P_{k|k-1}(\ell)\, H_{k-1}^T(\ell) + R_k(\ell)\right\}^{-1} \\
P_{k|k}(\ell) &= P_{k|k-1}(\ell) - W_k(\ell)\, H_k(\ell)\, P_{k|k-1}(\ell) \\
\widehat{x}_{k|k}(z\,;\, \ell) &= F_{k-1}(\ell)\, \widehat{x}_{k-1|k-1}(\ell) + W_k(\ell)\left\{z - \widehat{z}_{k|k-1}(\ell)\right\}.
\end{aligned} \tag{E.11}$$

The joint measurement-target PPPs are independent because measurements are independent when conditioned on target state, and because targets are assumed independent. The measurement clutter intensity function is

$$\lambda_{k|k}^0(z) \;=\; I_k(0)\, q_k(z)\,, \tag{E.12}$$

where $q_k(z)$ is a specified clutter pdf, i.e., $\int_{\mathbb{R}^{n_z}} q_k(z)\, dz \;=\; 1$. In the language of Section 8.1, the clutter model is a compound PPP. To ease the notational burden later in the EM method, let $\lambda_{k|k}^0(z) \;\equiv\; \lambda_{k|k}^0(z, \varnothing)$.

The joint measurement-multitarget PPP at time t_k is the superposition of target and clutter intensity functions:

$$\begin{aligned}
\lambda_{k|k}(z, x) &= \lambda_{k|k}^0(z) + \sum_{\ell=1}^{L} \lambda_{k|k}^{\ell}(z, x) \\
&= I_k(0)\, q_k(z) \\
&\quad + \sum_{\ell=1}^{L} P_k^D(\ell)\, I_k(\ell)\, \mathcal{N}\left(x\,;\, \widehat{x}_{k|k}(z\,;\, \ell),\, P_{k|k}(\ell)\right) \mathcal{N}\left(z\,;\, \widehat{z}_{k|k-1}(\ell),\, S_{k|k}(\ell)\right).
\end{aligned} \tag{E.13}$$

This sum parameterizes the likelihood function of the MMIF filter. The EM method uses it in the next section to derive a recursion for estimating target states and the intensity coefficients.

E.2.1 Likelihood Function

Denote the number of measurements at time t_k by $m_k \geq 1$ and the measurements themselves by

$$z_k(1 : m_k) = \{z_k(1), \ldots, z_k(m_k)\},$$

where $z_k(j) \in \mathbb{R}^{n_z}$, $j = 1, \ldots, m_k$. (Details for the special case $m_k = 0$ are omitted.) In a joint measurement-target PPP, every measurement z is always paired with a point x in target state space, but whether or not x corresponds to a target or to clutter is unknown. Denote the target states associated with the measurements (marks) by

$$x_k(1 : m_k) = \{x_k(1), \ldots, x_k(m_k)\},$$

where $x_k(j) \in \mathbb{R}^{n_x}$, $j = 1, \ldots, m_k$. The paired data are

$$\mathcal{Z}_k = \{(z_k(j), x_k(j)) : j = 1, \ldots, m_k\}.$$

Because the target model is a PPP, the data $\mathcal{Z}_k$ are a realization of the measurement-target PPP with intensity (E.13). Its likelihood function is

$$p(\mathcal{Z}_k) = e^{-\int_{\mathbb{R}^{n_z} \times \mathbb{R}^{n_x}} \lambda_{k|k}(z,x)\,dz\,dx} \prod_{j=1}^{m_k} \lambda_{k|k}(z_k(j), x_k(j))$$

$$= e^{-I_k(0) - \sum_{\ell=1}^{L} P_k^D(\ell)\, I_k(\ell)} \prod_{j=1}^{m_k} \left\{ \lambda_{k|k}^0(z_k(j)) + \sum_{\ell=1}^{L} \lambda_{k|k}^\ell(z_k(j), x_k(j)) \right\},$$

$$\tag{E.14}$$

where (E.9) is used in the last equation.

The difficulty is that there are as many unknown target states as there are data, while there are L target modes and a clutter mode. Denote the states of the L target modes by

$$\chi_k(1 : L) = \{\chi_k(1), \ldots, \chi_k(L)\}, \tag{E.15}$$

where $\chi_k(j) \in \mathbb{R}^{n_x}$, $j = 1, \ldots, m_k$. The clutter mode is mode zero, and its state is $\chi_k(0) = \varnothing$. The unobserved target state $x_k(j)$ of the measurement $z_k(j)$ corresponds to one of the L target modes or to clutter. Let σ_j denote the index of this mode, so that $\sigma_j \in \{0, 1, \ldots, L\}$. It is now assumed that

$$x_k(j) = \chi_k(\sigma_j), \quad j = 1, \ldots, m_k. \tag{E.16}$$

In other words, measurements that arise from the same mode have exactly the same target state. The constraints (E.16) violates the exact form of the "at most one measurement per target rule", but it is not violated *in the mean*. The target states to be estimated are $\chi_k(1:L)$.

The superposition in (E.14) is a clear indication of the utility of the EM method for computing MAP estimates. In EM parlance, (E.14) is the incomplete data pdf. It is natural (indeed, other choices seem contrived here) to let the indices $\sigma \equiv \{\sigma_1, \ldots, \sigma_{m_k}\}$ denote the missing data. The complete data pdf is defined by

$$p\left(\mathcal{Z}_k, \sigma\right) = e^{-I_k(0) - \sum_{\ell=1}^{L} P_k^D(\ell) I_k(\ell)} \prod_{j=1}^{m_k} \lambda_{k|k}^{\sigma_j}\left(z_k(j), \chi_k(\sigma_j)\right). \tag{E.17}$$

Let $I_k(0:L) \equiv (I_k(0), I_k(1), \ldots, I_k(L))$. The posterior pdf of σ is, by the definition of conditioning,

$$\begin{aligned} p\left(\sigma \mid \chi_k(1:L), I_k(0:L)\right) &= \frac{p\left(\mathcal{Z}_k, \sigma\right)}{p\left(\mathcal{Z}_k\right)} \\ &= \prod_{j=1}^{m_k} w_{\sigma_j}(z_k(j); \chi_k(1:L), I_k(0:L)), \end{aligned} \tag{E.18}$$

where, for $1 \le \ell \le L$, the weights for an arbitrary measurement z are given by

$$\begin{aligned} &w_\ell(z; \chi_k(1:L), I_k(0:L)) \\ &= \frac{P_k^D(\ell)\, I_k(\ell)\, \mathcal{N}\left(\chi_k(\ell); \widehat{x}_{k|k}(z; \ell), P_{k|k}(\ell)\right) \mathcal{N}\left(z; \widehat{z}_{k|k-1}(\ell), S_{k|k}(\ell)\right)}{I_k(0)q_k(z) + \sum_{\ell=1}^{L} P_k^D(\ell)I_k(\ell)\mathcal{N}\left(\chi_k(\ell); \widehat{x}_{k|k}(z; \ell), P_{k|k}(\ell)\right) \mathcal{N}\left(z; \widehat{z}_{k|k-1}(\ell), S_{k|k}(\ell)\right)}. \end{aligned} \tag{E.19}$$

The weight for $\ell = 0$ is

$$\begin{aligned} &w_0(z; \chi_k(1:L), I_k(0:L)) \\ &= \frac{I_k(0)\, q_k(z)}{I_k(0)q_k(z) + \sum_{\ell=1}^{L} P_k^D(\ell)I_k(\ell)\mathcal{N}\left(\chi_k(\ell); \widehat{x}_{k|k}(z; \ell), P_{k|k}(\ell)\right) \mathcal{N}\left(z; \widehat{z}_{k|k-1}(\ell), S_{k|k}(\ell)\right)}. \end{aligned} \tag{E.20}$$

The coefficient $e^{-I_k(0) - \sum_{\ell=0}^{L} P_k^D(\ell) I_k(\ell)}$ cancels out in the weight calculation. The weights are ratios of intensities. They are the probabilities that the measurement z is generated by target ℓ, or by clutter if $\ell = 0$.

Let $r = 0, 1, \ldots$ be the EM iteration index, and let $\chi_k^{(0)}(1:L)$ and $I_k^{(0)}(0:L)$ be specified initial values of the target states and their intensity coefficients. The EM auxiliary function is the conditional expectation

$$Q\left(\chi_k(1:L),\, I_k(0:L) \mid \chi_k^{(r)}(1:L),\, I_k^{(r)}(0:L)\right)$$

$$= \sum_{j=1}^{m_k} \{\log\, p(\mathcal{Z}_k,\sigma)\}\; w_\ell\left(z;\, \chi_k^{(r)}(1:L),\, I_k^{(r)}(0:L)\right). \qquad \text{(E.21)}$$

Proceeding algebraically in the manner used frequently in Chapter 3 and dropping terms that do not depend on $\chi_k(1:L)$ and $I_k(0:L)$ gives the simplified expression

$$Q\left(\chi_k(1:L),\, I_k(0:L) \mid \chi_k^{(r)}(1:L),\, I_k^{(r)}(0:L)\right) = -I_k(0) - \sum_{\ell=1}^{L} P_k^D(\ell)\, I_k(\ell)$$

$$+ \sum_{\ell=0}^{L} \sum_{j=1}^{m_k} w_\ell\left(z;\, \chi_k^{(r)}(1:L),\, I_k^{(r)}(0:L)\right)\, \log\, \lambda_{k|k}^{\ell}\,(z_k(j),\, \chi_k(\ell)).$$

$$\text{(E.22)}$$

Maximizing the auxiliary function with respect to $\chi_k(1:L)$ and $I_k(0:L)$ gives the EM recursion. Details are straightforward and are omitted.

E.3 MMIF Recursion

Using (3.38) gives the EM update for the intensity coefficient of the ℓ-th target as

$$I_k^{(r+1)}(\ell) = \frac{1}{P_k^D(\ell)} \sum_{j=1}^{m_k} w_\ell\left(z_k(j);\, \chi_k^{(r)}(1:L),\, I_k^{(r)}(0:L)\right). \qquad \text{(E.23)}$$

The factor $P_k^D(\ell)$ cancels the same factor in the weights (E.19). For clutter, $\ell = 0$, the updated intensity coefficient is

$$I_k^{(r+1)}(0) = \sum_{j=1}^{m_k} w_0\left(z_k(j);\, \chi_k^{(r)}(1:L),\, I_k^{(r)}(0:L)\right). \qquad \text{(E.24)}$$

These updates accord well with the interpretation of the weights.

Finding the updated state for target ℓ is little different. Setting the gradient with respect to $\chi_k(\ell)$ equal to zero and solving gives the update

$$\chi_k^{(r+1)}(\ell) = \frac{\sum_{j=1}^{m_k} w_\ell\left(z_k(\ell);\, \chi_k^{(r)}(1:L),\, I_k^{(r)}(0:L)\right)\, \widehat{x}_{k|k}(z_k(j);\, \ell)}{\sum_{j=1}^{m_k} w_\ell\left(z_k(\ell);\, \chi_k^{(r)}(1:L),\, I_k^{(r)}(0:L)\right)}.$$

$$\text{(E.25)}$$

A more intuitive way to write the result is to substitute for $\widehat{x}_{k|k}(z_k(j);\ell)$ using (E.11). By linearity, the updated state is given by the Kalman filter

$$\chi_k^{(r+1)}(\ell) = F_{k-1}(\ell)\widehat{x}_{k-1|k-1}(\ell) + W_k(\ell)\left\{\widetilde{z}_{k|k}^{(r+1)}(\ell) - \widehat{z}_{k|k-1}(\ell)\right\}, \quad (E.26)$$

where the "synthetic" measurement for target ℓ is defined by

$$\widetilde{z}_{k|k}^{(r+1)}(\ell) = \frac{\sum_{j=1}^{m_k} w_\ell\left(z_k(\ell);\ \chi_k^{(r)}(1:L),\ I_k^{(r)}(0:L)\right) z_k(j)}{\sum_{j=1}^{m_k} w_\ell\left(z_k(\ell);\ \chi_k^{(r)}(1:L),\ I_k^{(r)}(0:L)\right)}. \quad (E.27)$$

This concludes one iteration of the EM algorithm.

On convergence, at say iteration r_{last}, the MAP estimates of the state of target ℓ and its intensity are

$$\widehat{I}_{k|k}(\ell) = I_{k|k}^{(r_{\mathrm{last}})}(\ell) \qquad 0 \le \ell \le L, \quad (E.28)$$

$$\widehat{x}_{k|k}(\ell) = \chi_{k|k}^{(r_{\mathrm{last}})}(\ell) \qquad 1 \le \ell \le L. \quad (E.29)$$

EM iteration stopping criteria are discussed elsewhere.

The M-step of the EM update recursions are explicitly solvable because the surveillance region is the entire measurement space $\mathbb{R}^{n_z}$. Bounded regions are necessary in practice. As mentioned in Section 3.2.4, the update equations are replaced by appropriately modified versions of (3.34) and (3.37), respectively, for bounded regions. If all targets are well inside the surveillance region, the equations here are excellent approximations.

Appendix F
Linear Filter Model

The output of a linear filter with stationary input signal is exponentially distributed when the passband of the filter is a subset of the input signal band. This appendix derives this result as the limit of a PPP model. The points of the PPP realizations are in the spectral (frequency) domain.

The frequency domain output of a linear filter with a classical stationary Gaussian input signal is equivalent to a compound PPP, assuming the input signal bandwidth is larger than the filter bandwidth. In this case the pdf of the instantaneous filter output power, a^2, in the filter passband is

$$p(a^2 \,;\, S^2) \;=\; \frac{1}{S^2} \, \exp\left(-\frac{a^2}{S^2} \right). \tag{F.1}$$

The parameter of this exponential distribution is equal to the signal power, $S^2 < \infty$, in the filter passband.

F.1 PPP Signal Model

The input signal is modeled in the frequency domain, *not* the time domain. (This contrasts sharply with the time domain model mentioned in Section 8.1.2.) The frequency domain PPP "emits" points, or shots, across the entire signal band. The only shots that pass through the filter are those in the filter passband. The shots that pass through the filter comprise a noncausal PPP process in a frequency domain equal to that of the filter passband. In essence, the filter acts as a thinning process on the input PPP. The filtered, or thinned, points are not observed directly in the filter output. Instead, each point carries a complex-valued mark, or phase, and the measurement is the coherent sum u of the marks of the points that pass through the filter. Thus, the measurement u is known, but the number n of shots in the filter output and the individual marks comprising the coherent sum are all unknown.

Let $\lambda(\omega)$ be the filtered PPP intensity—it is defined in the frequency domain over the filter passband, B. It is shown in this appendix that the function $\lambda(\omega)$ is the signal power spectrum.

Let $v \equiv \int_B \lambda(x)\,dx$ be the mean number of points in B. Now, suppose there are n points with marks $u_k = (c_k, s_k)^T$, where c_k and s_k are the in-phase and quadrature components of the k-th mark, respectively. The marks are assumed i.i.d. with pdf

$$p(u_k) \equiv \mathcal{N}\left(\begin{bmatrix} c_k \\ s_k \end{bmatrix}; \begin{bmatrix} 0 \\ 0 \end{bmatrix}, \frac{\hbar^2}{2}\begin{bmatrix} 1 & 0 \\ 0 & 1 \end{bmatrix}\right)$$

$$= \frac{1}{\pi\,\hbar^2}\exp\left(-\frac{c_k^2 + s_k^2}{\hbar^2}\right). \tag{F.2}$$

The mark power is the variance of its squared magnitude, or $\hbar^2$. (The choice of the symbol $\hbar^2$ to represent mark power is intended to suggest that the limit $\hbar \to 0$ will eventually be taken.) The pdf of the coherent mark sum $u = \sum_{k=1}^{n} u_k \equiv [c, s]^T$ is

$$p(u\,|\,n) = \frac{1}{\pi\,n\,\hbar^2}\exp\left(-\frac{u^T u}{n\,\hbar^2}\right). \tag{F.3}$$

The pdf of the joint event (u, n) is, from (2.4) and (F.3),

$$p(u, n) = p(n)\,p(u\,|\,n)$$

$$= \begin{cases} e^{-v}, & \text{if } n = 0 \\ e^{-v}\,\dfrac{v^n}{n!}\,\dfrac{1}{\pi\,n\,\hbar^2}\,\exp\left(-\dfrac{u^T u}{n\,\hbar^2}\right), & \text{if } n \geq 1. \end{cases}$$

Marginalizing over n gives

$$p(u) = \sum_{n=0}^{\infty} p(u, n)$$

$$= \sum_{n=0}^{\infty} p(n)\,p(u\,|\,n)$$

$$= e^{-v}\left\{1 + \frac{1}{\pi\,\hbar^2}\sum_{n=1}^{\infty}\frac{v^n}{n!\,n}\exp\left(-\frac{u^T u}{n\,\hbar^2}\right)\right\}. \tag{F.4}$$

F.2 Poisson Limit

Substituting the constraint

$$v\,\hbar^2 = S^2 \quad \Leftrightarrow \quad \hbar^2 = v^{-1}\,S^2 \tag{F.5}$$

into (F.4) gives

$$p(u) = e^{-v} \left\{ 1 + \frac{v}{\pi\,S^2} \sum_{n=1}^{\infty} \frac{v^n}{n!\,n} \exp\left(-\frac{(u^T u)\,v}{n\,S^2}\right) \right\}. \tag{F.6}$$

Taking the limit as $v \to \infty$ yields the result (see below)

$$p(u; S^2) = \lim_{v\to\infty} p(u) = \frac{1}{\pi\,S^2} \exp\left(-\frac{u^T u}{S^2}\right). \tag{F.7}$$

Changing to power $a^2 = u^T u$ gives the pdf (F.1).

The limit (F.7) shows that as mark power $\hbar^2 \to 0$ and mean shot intensity $v \to \infty$ in such a way that the total power S^2 is constant, the coherent sum of marks is identically distributed to the output of the linear filter with stationary Gaussian input.

The limit (F.7) is obtained by taking the Fourier transform of $p(u) \equiv p(c, s)$ in (F.4). Let $\delta(\cdot)$ denote the Dirac delta function. Then,

$$\begin{aligned}
\Phi(\omega_1, \omega_2) &\equiv \int_{-\infty}^{\infty}\int_{-\infty}^{\infty} p(c, s)\, e^{i\,(c\,\omega_1 + s\,\omega_2)}\, dc\, ds \\
&= e^{-v}\left\{ \delta(\omega_1)\,\delta(\omega_2) + \sum_{n=1}^{\infty} \frac{v^n}{n!} \int_{-\infty}^{\infty}\int_{-\infty}^{\infty} \right. \\
&\qquad\qquad \left. \frac{1}{\pi\,n\,\hbar^2} \exp\left(-\frac{c^2 + s^2}{n\,\hbar^2}\right) e^{i\,(c\,\omega_1 + s\,\omega_2)} \right\} dc\, ds \\
&= e^{-v}\left\{ \delta(\omega_1)\,\delta(\omega_2) + \sum_{n=1}^{\infty} \frac{v^n}{n!} \exp\left(-\frac{n\,\hbar^2}{4}\left(\omega_1^2 + \omega_2^2\right)\right) \right\} \\
&= e^{-v}\left\{ \delta(\omega_1)\,\delta(\omega_2) - 1 + \exp\left[v \exp\left(-\frac{\hbar^2}{4}\left(\omega_1^2 + \omega_2^2\right)\right)\right] \right\}.
\end{aligned}$$

The interchange of summation and double integral in the first step is justified because the series is absolutely convergent. Substituting (F.5) and taking the limit gives

$$\begin{aligned}
\lim_{v\to\infty} \Phi(\omega_1, \omega_2) &= \lim_{v\to\infty} e^{-v}\{\delta(\omega_1)\,\delta(\omega_2) - 1\} \\
&\qquad + \lim_{v\to\infty} \exp\left[-v + v\exp\left(-\frac{S^2}{4\,v}\left(\omega_1^2 + \omega_2^2\right)\right)\right] \\
&= \lim_{v\to\infty} \exp\left[-v + v\left(1 - \frac{S^2}{4\,v}\left(\omega_1^2 + \omega_2^2\right) + O\left(v^{-2}\right)\right)\right] \\
&= \exp\left[-\frac{S^2}{4}\left(\omega_1^2 + \omega_2^2\right)\right].
\end{aligned}$$

The limit (F.7) follows immediately from the Fourier inversion formula.

F.2.1 Utility

Frequency domain PPPs that model the outputs of filters with nonoverlapped pass-bands are independent. This follows from the interpretation of the PPPs as thinned versions of the same PPP. (See the end of Section 2.8.) In contrast, for the usual time domain model, the complex-valued outputs of the cells of a discrete Fourier transform are *asymptotically* independent if the windows are not overlapped, and if the input is a wideband stationary Gaussian signal.

The marked PPP model of signal power spectrum supports the development of algorithms based on the method of EM.

Glossary

Affine Sum Used in parameter estimation problems in which the intensity function is of the form $f_0(x) + \Sigma_i f_i(x ; \theta_i)$, where the estimated parameters are $\{\theta_i\}$. The word *affine* refers to the fact that no parameters are estimated for the term $f_0(x)$.

Augmented Target State Space Typically, a target state space $S^+ = S \cup \phi$ comprising a continuous component S, such as $S \subset \mathbb{R}^n$, and a discrete component ϕ not in S. The points in S^+ represent mutually exclusive and exhaustive statistical hypotheses about the target. The state ϕ is interpreted as the hypothesis that a target generates measurements statistically indistinguishable from clutter, i.e., the target is a "clutter target". More generally, a finite or countable number of discrete hypotheses can be used.

Bayes-Markov Filter A sequential estimation method that recursively determines the posterior density, or pdf, of the available data. Point estimates and some measure of the area of uncertainty (AOU) are extracted from the posterior density in different ways, depending on the application; however, point estimates and their AOUs characterize the posterior density only for the linear-Gaussian Kalman filter.

Binomial Point Process An i.i.d. point process in which the number of points in a specified set $\mathcal{R} \subset \mathcal{S}$ is binomially distributed with parameter equal to the integral over $\mathcal{R}$ of a specified pdf on $\mathcal{S}$. It provides an interesting contrast to the Poisson point process.

Campbell's Theorem A classic theorem (1909) that gives an explicit form for the expected value of the random sum $\Sigma_i f(x_i)$, where $\{x_i\}$ are the points of a realization of a PPP. It is a special case of Slivnyak's Theorem.

Clutter Point measurements that do not originate from any physical target under track. Clutter can be persistent, as in ground clutter, and it can be statistical in nature, that is, arise from the locations of threshold crossings of fluctuations in some ambient noise background.

Cramér-Rao Bound (CRB) A lower bound on the variance of any unbiased parameter estimate. It is derived directly from the likelihood function of the data.

Data to Target Assignment Problem A problem that arises with tracking single targets in clutter, and multiple targets either with or without clutter.

Dirac Delta Function Not really a function at all, but an operator that performs a point evaluation of the integrand of an integral. Often defined as a limit of a sequence of test functions.

Expectation-Maximization A method used to obtain ML and MAP parameter estimation algorithms. It is especially well suited to likelihood functions of processes that are sums or superpositions of other, simpler processes. Under broad conditions, it is guaranteed convergent to a local maximum of the likelihood function.

Finite Point Process A geometrical distribution of random occurrences of finitely many points, the number and locations of which are typically random. In general, finite point processes are *not* Poisson point processes.

Fisher Information Matrix (FIM) The inverse of the Cramér-Rao Bound matrix.

Gaussian Mixture A Gaussian sum whose integral is one. Equivalently, a Gaussian sum whose weights sum to one. Every Gaussian mixture is a pdf.

Gaussian Sum A weighted sum of multivariate Gaussian probability density functions, where the weights are non-negative. Gaussian sums are not, in general, pdfs.

Generalized Functions These are not truly functions at all, but operators. The concept of point masses leads to the need to define integrals over discrete sets in a nontrivial way. One way to do this is measure theory, another is generalized functions. The classic example is the Dirac delta function $\delta(x)$; see *Test Functions* below. The classic book [71] by Lighthill is charming, short, and very readable.

Homogeneous Poisson Point Process A PPP whose intensity is a (non-negative) constant.

Independent Increments A concept defined for stochastic processes $X(t)$ which states that the random variables $X(b) - X(a)$ and $X(d) - X(c)$ are independent if the intervals (a, b) and (c, d) are disjoint.

Independent Scattering A concept defined for point processes which states that the numbers and locations of points of a point process in two different sets are independent if the sets are disjoint. (Sometimes called independent increments, despite the confusion in terminology with stochastic processes.)

Intensity The defining parameter of a PPP. In general, intensity is the sum of a nonnegative ordinary function and at most countably many Dirac delta functions.

Intensity Filter A multi-target tracking filter that recursively updates the intensity function of a PPP approximation to the target state point process.

Intensity Function The intensity for orderly PPPs is an ordinary function, typically denoted by $\lambda(x)$. Intuitively, $\lambda(x)\, dx$ specifies the expected number of points of the

PPP that occur in an infinitesimal volume dx located at x. The expectation is over the ensemble of all possible PPP realizations.

Iterative Majorization The numerical analysis interpretation of the method of Expectation-Maximization.

Likelihood Function Any positive scalar multiple of the pdf of a realization of a random variable when thought of as a function of a parameter to be estimated. In applications, the realization is typically called the measured data.

Likelihood Ratio The ratio of two different likelihood functions evaluated afor the same data. The numerator is the likelihood function of a random variable character-ized by an assumption called the alternative hypothesis, or H_1, and the denominator is the likelihood function of a random variable characterized by the null hypothesis, or H_0.

Markov Chain Monte Carlo (MCMC) A statistical technique in which a sample is drawn from a pdf $p(x)$ by finding a realization of a specially-devised Markov chain that is run for "many" steps. This realization is a sample from the long term probability vector $q(x)$ of the Markov chain. The procedure works because the chain is devised so that (1) $q(x)$ exists, and (2) $q(x)$ is identical to $p(x)$. Typically, MCMC is used only when $p(x)$ cannot be sampled from directly.

Measure Theory A beautiful elaboration of the concept of length, volume, etc. that underpins the highest level of axiomatic mathematical rigor for point processes. Readers unacquainted with it will undoubtedly find the language difficult at first. This is the price paid for rigor, and the reason for books like this one.

Microtarget A concept related to the interpretation of PPPs as models for real targets. The intensity function of the multitarget PPP is that of the microtargets, and the peaks in the intensity function correspond to the locations of real targets. (The word *microtarget* is analogous to the concept of *microstate* in statistical mechanics.)

Observed Information Matrix (OIM) A surrogate for the FIM that is potentially useful when the FIM is not easily evaluated. It is the negative Hessian matrix of the likelihood function of the data, evaluated not at the true value of the parameter but at the ML parameter estimate. The expectation of the OIM over all data is *not* the FIM in general.

Orderly Point Process A finite point process in which the points in a realization are distinct with probability one. An orderly PPP is a PPP in which the intensity is an intensity function.

Point Process A geometrical distribution of random occurrences of points. Both the number and locations of the points are typically random. In general, point pro-cesses are *not* Poisson point processes.

Poisson Distribution A discrete pdf with a single parameter μ defined on the non-negative integers 0, 1, 2, ... by $\Pr[n] = e^{-\mu} \mu^n / n!$. (Readers new to point processes need to be vigilant when reading to avoid confusing the Poisson distribution with the Poisson point process.)

Poisson's Gambit   A term that describes the act of modeling the number of trials in a sequence of Bernoulli trials as Poisson distributed. The name is pertains to situations wherein the Poisson assumption is imposed by the modeler rather than arising naturally in the application. The advantage of the assumption is, e.g., that the numbers of occurrences of heads and tails in a series of coin flips are independent under Poisson's gambit.

Poisson Point Process (PPP)   A special kind of point process characterized (parameterized) by an intensity function $\lambda(x)$ that specifies both the number and probability density of the locations of points. The number of points in a bounded set $\mathcal{R}$ is Poisson distributed with parameter $\mu = \int_{\mathcal{R}} \lambda(x)\,dx$. Conditioned on n, the points are i.i.d. in $\mathcal{R}$ with pdf $\lambda(x)/\mu$. Compare to Binomial point processes (BPP).

Positron Emission Tomography (PET)   A widely used medical imaging methodology that is based estimating the spatial intensity of positron decay. The well known Shepp-Vardi algorithm (1982) is the basis of most, if not all, intensity estimation algorithms.

Posterior Cramér-Rao Bound (PCRB)   A lower bound on the variance of an unbiased nonlinear tracking filter.

Probability Density Function (pdf)   A function that represents the probability that a realization x of a random variable X falls in the infinitesimal region dx.

Probability Hypothesis Density (PHD)   An additive theory of Bayesian evidence accrual first proposed by Stein and Winter [121].

PHD Filter   A multitarget sequential tracking filter which avoids the data to target assignment problem by using a Poisson point process (PPP) to approximate the multitarget state. The correspondence of data to individual target tracks is not maintained. Target birth and clutter processes are assumed known a priori.

Radon-Nikodym Derivative   A measure theoretic term that, with appropriate qualifications and restricted to $\mathbb{R}^n$, is another name for the likelihood ratio of the data under two different hypotheses.

Random Sum   See *Campbell's Theorem* and *Slivnyak's Theorem*.

Sequential Monte Carlo   A generic name for particle methods.

Slivnyak's Theorem   An important theorem (1962) about the expected value of a random sum that depends on how a point in a PPP realization relates to other points in the same realization. The random sum is $\Sigma_i f(x_i, \{x_1, \ldots, x_n\} \setminus x_i)$. Campbell's Theorem is the special case $f(x_i, \cdot) \equiv f(x_i)$.

Single Photon Emission Computed Tomography (SPECT)   A widely used medical imaging technique for estimating the spatial intensity of radioisotope decay based on multiple gamma (Anger) camera snapshots. The physics differs from PET in that only one gamma photon arises from each decay.

Target An entity characterized by a point in a target state space. Typically, the target state evolves sequentially, that is, its state changes over time (or other time-like variable).

Target State Space A set whose elements completely characterizes the properties of a target that are of interest in an application. Typically, this set is the vector space $\mathbb{R}^n$, or some subset thereof, that represents target kinematic properties or other properties, e.g., radar cross section. This space is sometimes augmented by discrete attributes that represent specific categorical properties, e.g., target identity.

Test Functions A sequence of infinitely differentiable functions used in proofs involving generalized functions. For example, for the Dirac delta function $\delta(x)$, the test sequence is often taken to be the sequence of Gaussian pdfs $\mathcal{N}\left(x\,;\,0,\,\sigma_n^2\right)$, where $\sigma_n \to 0$, so that for any continuous function $f(x)$, $\int_{\mathbb{R}} f(x)\,\delta(x)\,\mathrm{d}x \equiv \lim_{\sigma_n \to 0} \int_{\mathbb{R}} f(x)\,\mathcal{N}\left(x\,;\,0,\,\sigma_n^2\right)\,\mathrm{d}x = f(0)$.

Test Sets A collection of "simple" sets, e.g., intervals, that generate the Borel sets, which in turn are used to define measurable sets. A common style of proof is to demonstrate a result on a sufficiently rich class of test sets, and then invoke appropriate general limit theorems to extend the result to measurable sets.

Transmission Tomography A method of imaging the variation in spatial density of an object known only from measurements of the line integrals of the density function for a collection of straight lines.

List of Acronyms

AOU	Area Of Uncertainty
CFTP	Coupling From The Past
CRB	Cramér-Rao Bound
CT	Computed Tomography (Emission Tomography)
DFT	Discrete Fourier Transform
EM	Expectation-Maximization
FIM	Fisher Information Matrix
FOV	Field of View
GEM	Generalized EM
HCR	Hammersley-Chapman-Robbins
HMM	Hidden Markov Model
i.i.d.	Independent and Identically Distributed
MAP	Maximum *A Posteriori*
MCMC	Markov Chain Monte Carlo
MHT	Multiple Hypothesis Tracking
ML	Maximum Likelihood
MMIF	Marked Multitarget Intensity Filter
MMPP	Markov Modulated Poisson Process
MRI	Magnetic Resonance Imaging
MTT	Multiple Target Tracking
OIM	Observed Information Matrix
PCRB	Posterior Cramér-Rao Bound
pdf	Probability Density Function
PET	Positron Emission Tomography
PHD	Probability Hypothesis Density
PMHT	Probabilistic Multiple Hypothesis Tracking
PMT	Photo-Multiplier Tubes
PPP	Poisson Point Process
RACS	RAndom Closed Set
SAGE	Space Alternating Generalized EM
SDE	Stochastic Differential Equation
SMC	Sequential Monte Carlo

SNR	Signal to Noise Ratio
SPECT	Single Photon Emission Computed Tomography
SPRT	Sequential Probability Ratio Test
SVD	Singular Value Decomposition
TBD	Track Before Detect
TOF	Time Of Flight

References

1. D. A. Abraham and A. P. Lyons. Novel physical interpretations of K-distributed reverberation. *IEEE Journal of Oceanic Engineering*, JOE-27(4):800–813, 2002.
2. A. Baddeley. Stochastic geometry: An introduction and reading-list. *International Statistical Review*, 50(2):179–193, 1982.
3. A. Baddeley and E. B. V. Jensen. *Stereology for Statisticians*. Chapman & Hall/CRC, Boca Raton, FL, 2004.
4. Y. Bar-Shalom and T. E. Fortmann. *Tracking and Data Association*. Academic, Boston, MA, 1988.
5. E. W. Barankin. Locally best unbiased estimates. *Annals of Mathematical Statistics*, 20: 477–501, 1949.
6. M. S. Bartlett. Smoothing periodograms from time series with continuous spectra. *Nature*, 161:686–687, May 1948.
7. M. S. Bartlett. The spectral analysis of two-dimensional point processes. *Biometrica*, 51: 299–311, 1964.
8. M. S. Bartlett. *The Statistical Analysis of Spatial Pattern*. Chapman and Hall, New York, 1975.
9. R. E. Bellman. *Adaptive Controls Processes: A Guided Tour*. Princeton University Press, Princeton, NJ, 1961.
10. B. Bollobás. *Random Graphs*. Cambridge University Press, New York, 2001.
11. N. R. Campbell. The study of discontinuous phenomena. *Proceedings of the Cambridge Philosophical Society*, 15:117–136, 1909.
12. M. Á. Carreira-Perpiñán. Gaussian mean shift is an EM algorithm. *IEEE Trans. Pattern Analysis and Machine Intelligence*, PAMI-29(5):767–776, May 2007.
13. V. Cevher and I. Kaplan. Pareto frontiers of sensor networks for localization. In *Proceedings of the 2008 International Conference on Information Processing in Sensor Networks*, pages 27–38, St. Louis, Missouri, 2008.
14. D. G. Chapman and H. Robbins. Minimum variance estimation without regularity assumptions. *Annals of Mathematical Statistics*, 22(4):581–586, 1951.
15. E. Çinlar. On the superposition of m-dimensional point processes. *Journal of Applied Probability*, 5:169–176, 1968.
16. N. A. C. Cressie. *Statistics for Spatial Data*. Wiley, New York, Revised edition, 1993.
17. D. J. Daley and D. Vere-Jones. *An Introduction to the Theory of Point Processes*, volume I: Elementary Theory and Methods. Springer, New York, Second edition, 2003.
18. W. B. Davenport, Jr., and W. L. Root. *An Introduction to the Theory of Random Signals and Noise*. McGraw-Hill, New York, 1958.
19. J. de Leeuw. Applications of convex analysis to multidimensional scaling. In J. R. Barra, F. Brodeau, G. Romier, and B. van Cutsem, editors, *Recent Developments in Statistics*, pages 133–145. North-Holland, Amsterdam, 1977.

20. J. de Leeuw and W. J. Heiser. Convergence of correction matrix algorithms for multidimensional scaling. In J. C. Lingoes, editor, *Geometric Representations of Relational Data*, pages 735–752. Mathesis Press, Ann Arbor, MI, 1977.

21. J. de Leeuw and W. J. Heiser. Multidimensional scaling with restrictions on the configuration. In P. R. Krishnaiah, editor, *Multivariate Analysis*, volume 5, pages 501–522. North-Holland, Amsterdam, 1980.

22. A. H. Delaney and Y. Bresler. A fast and accurate iterative reconstruction algorithm for parallel-beam tomography. *IEEE Transactions on Image Processing*, IP-5(5):740–753, 1996.

23. A. P. Dempster, N. M. Laird, and D. B. Rubin. Maximum likelihood from incomplete data via the EM algorithm. *Journal of the Royal Statistical Society, B*, 39:1–39, 1977.

24. P. Diaconis. The Markov chain Monte Carlo revolution. *Bulletin (New Series) of the American Mathematical Society*, 46:179–205, 2009.

25. S. W. Dufour. *Intersections of Random Convex Regions*. PhD thesis, Department of Statistics, Stanford University, Stanford, CA, 1972.

26. B. Efron and D. V. Hinkley. Assessing the accuracy of the maximum likelihood estimator: Observed versus expected fisher information (with discussion). *Biometrika*, 65(3):457–487, 1978.

27. A. Einstein. On the method of theoretical physics. *Philosophy of Science*, 1(2):163–169, 1934.

28. C. L. Epstein. *Introduction to the Mathematics of Medical Imaging*. SIAM Press, Philadelphia, PA, second edition, 2007.

29. A. Erdélyi, editor. *Higher Transcendental Functions*, volume 2. Bateman Manuscript Project, New York, 1953.

30. O. Erdinc, P. Willett, and Y. Bar-Shalom. A physical space approach for the probability hypothesis density and cardinalized probability hypothesis density filters. In *Proceedings of the SPIE Conference on Signal Processing of Small Targets*, Orlando, FL, volume 6236, April 2006.

31. O. Erdinc, P. Willett, and Y. Bar-Shalom. The bin-occupancy filter and its connection to the PHD filters. IEEE Transactions on Signal Processing, 57: 4232–4246, 2009.

32. K. J. Falconer. Applications of a result on spherical integration to the theory of convex sets. *The American Mathematical Monthly*, 90(10):690–693, 1983.

33. P. Faure. Theoretical model of reverberation noise. *Journal of the Acoustical Society of America*, 36(2):259–266, 1964.

34. J. A. Fessler. Statistical image reconstruction methods for transmission tomography. In M. Sonka and J. M. Fitzpatrick, editors, *Handbook of Medical Imaging*, SPIE, Bellingham, Washington, volume 2, pages 1–70, 2000.

35. J. A. Fessler and A. O. Hero. Space-alternating generalized expectation- maximization algorithm. *IEEE Transactions on Signal Processing*, SP-42(10):2664–2677, 1994.

36. P. M. Fishman and D. L. Snyder. The statistical analysis of space-time point processes. *IEEE Transactions on Information Theory*, IT-22:257–274, 1976.

37. D. Fränken, M. Schmidt, and M. Ulmke. "Spooky action at a distance" in the cardinalized probability hypothesis density filter. *IEEE Transactions on Aerospace and Electronic Systems*, AES-45(4):1657–1664, October 2009.

38. K. Fukunaga and L. D. Hostetler. The estimation of the gradient of a density function, with application in pattern recognition. *IEEE Transactions on Information Theory*, IT-21(1): 32–40, January 1975.

39. I. I. Gikhman and A. V. Skorokhod. *Introduction to the Theory of Random Processes*. Dover, Mineola, NY, Unabridged republication of 1969 edition, 1994.

40. J. R. Goldman. Stochastic point processes: Limit theorems. *The Annals of Mathematical Statistics*, 38(3):771–779, 1967.

41. I. R. Goodman, R. P. S. Mahler, and H. T. Nguyen. *Mathematics of Data Fusion*. Kluwer, Dordrecht, 1997.

42. G. R. Grimmett and D. D. Stirzaker. *Probability and Random Processes*. Oxford University Press, Oxford, Third edition, 2001.

43. M. Haenggi. On distances in uniformly random networks. *IEEE Transactions on Information Theory*, IT-51:3584–3586, 2005.

44. P. Hall. *Introduction to the Theory of Coverage Processes*. Wiley, New York, 1988.

45. J. M. Hammersley. On estimating restricted parameters. *Journal of the Royal Statistical Society, Series B*, 12(2):192–240, 1950.

46. W. Härdle and L. Simar. *Applied Multivariate Statistical Analysis*. Springer, Berlin, 2003.

47. S. I. Hernandez. *State Estimation and Smoothing for the Probability Hypothesis Density Filter*. Tech. Report ECSTR10-13, Victoria University of Wellington, School of Engineering and Computer Science, April 14, 2010. (Available from `http://ecs.victoria.ac.nz/Main/TechnicalReportSeries`) (Also submitted as a PhD thesis, Victoria University of Wellington, New Zealand, 2010.)

48. A. O. Hero. Poisson models and mean-squared error for correlator estimators of time delay. *IEEE Transactions on Information Theory*, IT-34:843–858, 1988.

49. A. O. Hero. Lower bounds on estimator performance for energy-invariant parameters of multidimensional Poisson processes. *IEEE Transactions on Information Theory*, IT-35: 287–303, 1989.

50. A. O. Hero and J. A. Fessler. A fast recursive algorithm for computing CR-type bounds for image reconstruction problems. In *Proceedings of the IEEE Nuclear Science Symposium and Medical Imaging Conference, Orlando, FL*, pages 1188–1190, 1992.

51. A. O. Hero and J. A. Fessler. A recursive algorithm for computing Cramér-rao-type bounds on estimator variance. *IEEE Transactions on Information Theory*, IT-40:843–848, 1994.

52. P. G. Hoel, S. C. Port, and C. J. Stone. *Introduction to Probability Theory*. Houghton Mifflin, Boston, MA, Fourth edition, 1971.

53. E. Jakeman and R. J. A. Tough. Non-Gaussian models for the statistics of scattered waves. *Advances in Physics*, 37(5):471–529, 1988.

54. A. Jazwinski. *Stochastic Processes and Filtering Theory*. Dover, Mineola, NY, Unabridged republication of 1970 edition, 2007.

55. T. Jebara. *Machine Learning: Discriminative and Generative*. Kluwer, Boston, MA, 2004.

56. M. E. Johnson. *Multivariate Statistical Simulation*. Wiley, New York, 1987.

57. A. F. Karr. *Point Processes and Their Statistical Inference*. Marcel Dekker, New York, Second edition, 1991.

58. K. Kastella. *A maximum likelihood estimator for report-to-track association*. In Proceedings of the SPIE, Signal and Data Processing or Small Targets, Orlando, Florida, volume 1954, 386–393, 1993.

59. K. Kastella. *Event averaged maximum likelihood estimation and mean-field theory in multi-target tracking*. IEEE Transactions on Automatic Control, 50:1070–1073, 1995.

60. K. Kastella. *Discrimination Gain for Sensor Management in Multitarget Detection and Tracking*. IMACS Conference on Computational Engineering in Systems Applications, Symposium on Control, Optimization and supervision, volume 1, Lille, France, 9–12 July pages 167–172, 1996.

61. M. G. Kendall and P. A. P. Moran. *Geometrical Probability*. Griffin, London, 1963.

62. A. I. Khinchine. *Mathematical Methods in the Theory of Queueing*. Griffon, London, 1955. Translated from Russian, 1960.

63. J. F. C. Kingman. *Poisson Processes*. Clarendon Press, Oxford, 1993.

64. G. E. Kopec. Formant tracking using hidden Markov models and vector quantization. *IEEE Transactions on Acoustics, Speech, and Signal Processing*, ASSP-34:709–729, 1986.

65. K. Lange and R. Carson. EM reconstruction algorithms for emission and transmission tomography. *Journal of Computer Assisted Tomography*, 8(2):306–316, 1984.

66. I. Langner. *Development of a parallel computing optimized head movement correction method in positron emission tomography.* Master's thesis, University of Applied Sciences Dresden and Research Center, Dresden-Rossendorf, 2003. `http://www.jens-langner.de/ftp/MScThesis.pdf`.

67. L. Lazos and R. Poovendran. Stochastic coverage in heterogeneous sensor networks. *ACM Transactions on Sensor Networks*, 2(3):325–358, 2006.

68. K. Lewin. The research center for group dynamics at Massachusetts Institute of Technology. *Sociometry*, 8: 126–136, 1945.

69. T. A. Lewis. Finding the observed information matrix when using the EM algorithm. *Journal of the Royal Statistical Society, Series B (Methodological)*, 44(2):226–233, 1982.

70. R. M. Lewitt and S. Matej. Overview of methods for image reconstruction from projections in emission computed tomography. *Proceedings of the IEEE*, 91(10):1588–1611, 2003.

71. M. J. Lighthill. *Introduction to Fourier Analysis and Generalized Functions.* Cambridge University Press, London, 1958.

72. L. B. Lucy. An iterative technique for the rectification of observed distributions. *The Astronomical Journal*, 79:745–754, 1974.

73. T. E. Luginbuhl. *Estimation of General, Discrete-Time FM Signals.* PhD thesis, Department of Electrical Engineering, University of Connecticut, Storrs, CT, 1999.

74. R. P. S. Mahler. Multitarget Bayes filtering via first-order multitarget moments. *IEEE Transactions on Aerospace and Electronic Systems*, AES-39:1152–1178, 2003.

75. R. P. S. Mahler. PHD filters of higher order in target number. *IEEE Transactions on Aerospace and Electronic Systems*, AES-43:1523–1543, 2007.

76. R. P. S. Mahler. *Statistical Multisource-Multitarget Information Fusion.* Artech House, Boston, MA, 2007.

77. B. Matérn. Spatial variation. *Meddelanden fran Statens Skogsforskningsinstitut (Communications of the State Forest Research Institute)*, 49(5):163–169, 1960.

78. B. Matérn. *Spatial Variation.* Number 36 in Lecture Notes in Statistics. Springer, New York, second edition, 1986.

79. G. Matheron. *Random Sets and Integral Geometry.* John Wiley & Sons, New York, 1975.

80. G. J. McLachlan and T. Krishnan. *The EM Algorithm and Extensions.* Wiley, New York, 1997.

81. G. J. McLachlan and D. Peel. *Finite Mixture Models.* Wiley, New York, 2000.

82. M. I. Miller, D. L. Snyder, and T. R. Miller. Maximum-likelihood reconstruction for single-photon emission computed-tomography. *IEEE Transactions on Nuclear Science*, NS-32(1):769–778, 1985.

83. P. Mitra. Spectral analysis: Point processes, August 16, 2006. http://wiki.neufo.org/neufo/jsp/Wiki?ParthaMitra.

84. J. Møller and R. P. Waagepetersen. *Statistical Inference and Simulation for Spatial Point Processes.* Chapman & Hall/CRC, Boca Raton, FL, 2004.

85. M. R. Morelande, C. M. Kreucher, and K. Kastella. *A bayesian approach to multiple target detection and tracking.* IEEE Transactions on Signal Processing, SP-55:1589–1604, 2007.

86. N. Nandakumaran, T. Kirubarajan, T. Lang, and M. McDonald. Gaussian mixture probability hypothesis density smoothing with multiple sensors. *IEEE Transactions on Aerospace and Electronic Systems*. Accepted for publication, 2010.

87. N. Nandakumaran, T. Kirubarajan, T. Lang, M. McDonald, and K. Punithakumar. Multitarget tracking using probability hypothesis density smoothing. *IEEE Transactions on Aerospace and Electronic Systems*. submitted, March 2008.

88. J. K. Nelson, E. G. Rowe, and G. C. Carter. Detection capabilities of randomly-deployed sensor fields. *International Journal of Distributed Sensor Networks*, 5(6):708 – 728, 2009.

89. R. Niu, P. Willett, and Y. Bar-Shalom. Matrix CRLB scaling due to measurements of uncertain origin. *IEEE Transactions on Signal Processing*, SP-49:1325–1335, 2001.

90. B. Øksendal. *Stochastic Differential Equations, An Introduction with Applications.* Springer, Berlin, Fourth edition, 1995.

91. J. A. O'Sullivan and J. Benac. Alternating minimization algorithms for transmission tomography. *IEEE Transactions on Medical Imaging*, MI-26:283–297, 2007.

92. C. Palm. Intensitätsschwankungen im fernsprechverkehr. *Ericsson Techniks*, 44:1–189, 1943.

93. A. Papoulis. *Probability, Random Variables, and Stochastic Processes.* McGraw-Hill, New York, 1965.

94. M. D. Penrose. On k-connectivity for a geometric random graph. *Random Structures and Algorithms*, 15(2):145–164, 1999.

95. A. Popescu. Markov-Modulated Poisson Processes. http://www.itm.bth.se/~adrian/courses/modern_techniques_networking/assignments/Tools.html.

96. J. G. Propp and D. B. Wilson. Exact sampling with coupled markov chains and applications to statistical mechanics. *Random Structures and Algorithms*, 9(1/2):223–252, 1996.

97. S. L. Rathbun and N. Cressie. Asymptotic properties of estimators for the parameters of a spatial inhomogeneous poisson point process. *Advances in Applied Probability*, 26:122–154, 1994.

98. H. E. Rauch, F. Tung, and C. T. Striebel. Maximum likelihood estimates of linear dynamic systems. *AIAA Journal*, 3:1445–1450, 1965.

99. R. A. Redner and Homer F. Walker. Mixture densities, maximum likelihood, and the EM algorithm. *SIAM Review*, 26(2):195–239, 1984.

100. S. Resnick. *Adventures in Stochastic Processes, with Illustrations*. Birkhäuser, Boston, MA, 1992.

101. S. O. Rice. Mathematical analysis of random noise. *Bell System Technical Journal*, 23–24:1–162, 1944.

102. W. H. Richardson. Bayesian-based iterative method of image restoration. *Journal of the Optical Society of America*, 62:55–59, 1972.

103. B. D. Ripley. *Spatial Statistics*. John Wiley & Sons, New York, 1981.

104. B. Ristic, S. Arulampalam, and N. Gordon. *Beyond the Kalman Filter*. Artech House, Boston, 2004.

105. H. E. Robbins. On the measure of a random set. *The Annals of Mathematical Statistics*, 15:70–74, 1944.

106. H. E. Robbins. On the measure of a random set. II. *The Annals of Mathematical Statistics*, 16:342–347, 1945.

107. E. G. Rowe and T. A. Wettergren. Coverage and reliability of randomly distributed sensor systems with heterogeneous detection range. *International Journal of Distributed Sensor Networks*, 5(4):303–320, 2009.

108. L. A. Santaló. *Integral Geometry and Geometric Probability*. Cambridge University Press, Cambridge, Second edition, 2004.

109. R. T. Seeley. Spherical harmonics. *The American Mathematical Monthly*, 73:115–121, 1966. Slaught Memorial Supplement.

110. L. A. Shepp and Y. Vardi. Maximum likelihood reconstruction for emission tomography. *IEEE Transactions on Medical Imaging*, MI-1(2):113–122, 1982.

111. H. Sidenbladh. Multi-target particle filtering for probability hypothesis density. In *Proceedings of the International Conference on Information Fusion, Cairns, Australia*, pages 800–806. ISIF, 2003.

112. M. Šimandl, J. Královec, and P. Tichavský. Filtering, predictive and smoothing Cramér-Rao bounds for discrete-time nonlinear dynamic systems. *Automatica*, 37:1703–1716, 2001.

113. I. M. Slivnyak. Some properties of stationary flows of homogeneous random events. *Theory of Probability and Its Applications*, 7:336–341, 1962. Russian original in *Teoriya Veroyatnostei i ee Primeneniya*, 347–352, 1962.

114. B. J. Slocumb and D. L. Snyder. Maximum likelihood estimation applied to quantum-limited optical position keeping. In *Proceedings of the SPIE Technical Symposium on Optical Engineering and Photonics in Aerospace Sensing Orlando, FL*, pages 165–176, 1990.

115. B. J. Slocumb. Position-sensing algorithms for optical communications. Master's thesis, Department of Electrical Engineering, Washington University, St. Louis, MO, 1988.

116. W. B. Smith and R. R. Hocking. A simple method for obtaining the information matrix for a multivariate normal distribution. *The American Statistician*, 22:18–20, 1968.

117. D. L. Snyder and P. M. Fishman. How to track a swarm of fireflies by observing their flashes. *IEEE Transactions on Information Theory*, IT-21:692–695, 1975.

118. D. L. Snyder. *Random Point Processes*, Wiley, New York, 1975.

119. D. L. Snyder and M. I. Miller. *Random Point Processes in Time and Space*. Springer, New York, Second edition, 1991.

120. H. Solomon. *Geometric Probability*. Society for Industrial and Applied Mathematics, Philadelphia, PA, 1978.

121. M. C. Stein and C. L. Winter. An additive theory of probabilistic evidence accrual, 1993. Report LA-UR-93-3336, Los Alamos National Laboratories.

122. L. D. Stone, T. L. Corwin, and C. A. Barlow. *Bayesian Multiple Target Tracking*. Artech House, Inc., Norwood, MA, 1999.

123. D. Stoyan, W. S. Kendall, and Joseph Mecke. *Stochastic Geometry and its Applications*. Wiley, Chichester, second edition, 1995.

124. R. L. Streit. Multisensor multitarget intensity filter. In *Proceedings of the International Conference on Information Fusion, Cologne, Germany*. ISIF, pp 1694–1701 30 June–3 July 2008.

125. R. L. Streit. PHD intensity filtering is one step of a MAP estimation algorithm for positron emission tomography. In *Proceedings of the International Conference on Information Fusion, Seattle*. ISIF, pp 308–315 6 July – 9 July 2009.

126. R. L. Streit and T. E. Luginbuhl. Maximum likelihood method for probabilistic multi-hypothesis tracking. In *Proceedings of the SPIE Conference on Signal and Data Processing of Small Targets*, volume 2235, pages 394–405, Orlando, FL, 1991.

127. R. L. Streit and T. E. Luginbuhl. A probabilistic multi-hypothesis tracking algorithm without enumeration and pruning. In *Proceedings of the Sixth Joint Service Data Fusion Symposium*, pages 1015–1024, Laurel, Maryland, 1993.

128. R. L. Streit and T. E. Luginbuhl. Probabilistic multi-hypothesis tracking, 1995. Technical Report 10,428, Naval Undersea Warfare Center, Newport, RI.

129. R. L. Streit and Tod E. Luginbuhl. Estimation of Gaussian mixtures with rotationally invariant covariance matrices. *Communications in Statistics: Theory and Methods*, 26:2927–2944, 1997.

130. R. L. Streit and L. D. Stone. Bayes derivation of multitarget intensity filters. In *Proceedings of the International Conference on Information Fusion, Cologne, Germany*. ISIF, pp.1686–1693 30 June–3 July 2008.

131. M. Ter-Pogossian. *The Physical Aspects of Diagnostic Radiology*. Hoeber Medical Division, Harper and Rowe, New York, 1967.

132. H. R. Thompson. Distribution of distance to Nth neighbour in a population of randomly distributed individuals. *Ecology*, 37:391–394, 1956.

133. P. Tichavský, C. H. Muravchic, and A. Nehorai. Posterior Cramér-Rao bounds for discrete-time nonlinear filtering. *IEEE Transactions on Signal Processing*, SP-46:1386–1396, 1998.

134. H. L. Van Trees. *Detection, Estimation, and Modulation Theory—Part I*. Wiley, New York, 1968.

135. H. L. Van Trees and Kristine L. Bell, editors. *Bayesian Bounds for Parameter Estimation and Nonlinear Filtering and Tracking*. Wiley, 2007.

136. M. N. M. van Lieshout. *Markov Point Processes and Their Applications*. Imperial College Press, London, 2000.

137. B.-N. Vo, S. Singh, and A. Doucet. Sequential Monte Carlo methods for multi-target filtering with random finite sets. *IEEE Transactions on Aerospace and Electronic Systems*, AES-41:1224–1245, 2005.

138. B.-T. Vo, B.-N. Vo, and A. Cantoni. The cardinalized probability hypothesis density filter for linear Gaussian multi-target models. In *Proceedings of the 40th Annual Conference on Information Sciences and Systems, Princeton, NJ*, pp.681–686 March 22–24 2006.

139. B.-T. Vo, B.-N. Vo, and A. Cantoni. Analytic implementations of the cardinalized probability hypothesis density filter. *IEEE Transactions on Signal Processing*, SP-55:3553–3567, 2007.

140. W. G. Warren. The center-satellite concept as a basis for ecological sampling (with discussion). In G. P. Patil, E. C. Pielou, and W. E. Waters, editors, *Statistical Ecology*, volume 2, pages 87–118. Pennsylvania State University Press, University Park, IL, 1971. (ISBN 0-271-00112-7).

141. Y. Watanabe. Derivation of linear attenuation coefficients from CT numbers for low-energy photons. *Physics in Medicine Biology*, 44:2201–2211, 1999.

142. T. A. Wettergren and M. J. Walsh. Localization accuracy of track-before-detect search strategies for distributed sensor networks. *EURASIP Journal on Advances in Signal Processing*, Article ID 264638:15, 2008.

Index

Breinigsville, PA USA
03 November 2010
248464BV00011B/112/P